Attention all "Enzyme Handbook" users:

A file with the complete volume indexes Vols. 1 through 10 in delimited ASCII format is available for downloading at no charge from the Springer EARN mailbox. Delimited ASCII format can be imported into most databanks.

The file has been compressed using the popular shareware program "PKZIP" (Trademark of PKware Inc., PKZIP is available from most BBS and shareware distributors).

This file is distributed without any expressed or implied warranty.

To receive this file send an e-mail message to:
SVSERV@DHDSPRI6.BITNET

The message must be:
GET /CHEMISTRY/ENZ_HB.ZIP

SVSERV is an automatic data distribution system. It responds to your message. The following commands are available:

HELP	returns a detailed instruction set for the use of SVSERV,
DIR *(name)*	returns a list of files available in the directory "name",
INDEX *(name)*	same as "DIR"
CD *<name>*	changes to directory "name",
SEND *<filename>*	invokes a message with the file "filename",
GET *<filename>*	same as "SEND".

D. Schomburg · D. Stephan (Eds.)
GBF – Gesellschaft für Biotechnologische Forschung

Enzyme Handbook 10

Class 1.1: Oxidoreductases

EC 1.1.1.150 – EC 1.1.99.26

for EC 1.1.1.1 – EC 1.1.1.149
see Vol. 9

Springer-Verlag Berlin Heidelberg GmbH

Professor Dr. Dietmar Schomburg
Dr. Dörte Stephan
GBF – Gesellschaft für Biotechnologische Forschung mbH
Mascheroder Weg 1
38124 Braunschweig
FRG

This collection of datasheets was generated from the database „BRENDA"

ISBN 978-3-642-47755-3   ISBN 978-3-642-57756-7 (eBook)
DOI 10.1007/978-3-642-57756-7

CIP data applied for

Media conversion, printing and bookbinding: Brühlsche Universitätsdruckerei, Giessen
Production of the plasticfiles: Lux-Plastik oHG, Murnau
SPIN: 10505109   51/3020 - 5 4 3 2 1 0 - Printed on acid-free paper

Preface

Recent progress on enzyme immobilisation, enzyme production, coenzyme regeneration and enzyme engineering has opened up fascinating new fields for the potential application of enzymes in a large range of different areas. As more progress in research and application of enzymes has been made the lack of an up-to-date overview of enzyme molecular properties has become more apparent. Therefore, we started the development of an enzyme data information system as part of protein-design activities at GBF. The present book "Enzyme Handbook" represents the printed version of this data bank. In future a computer searchable version will be also available.

The enzymes in this Handbook are arranged according to the Enzyme Commission list of enzymes. Some 3000 "different" enzymes will be covered. Frequently enzymes with very different properties are included under the same EC number. Although we intend to give a representative overview on the characteristics and variability of each enzyme the Handbook is not a compendium. The reader will have to go to the primary literature for more detailed information. Naturally it is not possible to cover all the numerous literature references for each enzyme (for special enzymes up to 40000) if the data representation is to be concise as is intended.

It should be mentioned here that the literature data are extracted from literature and critically evaluated by qualified scientists. On the other hand the original authors' nomenclature for enzyme forms and subunits is retained as is their nomenclature for organisms and strains even if the organism is reclassified in the meantime. The cross references to the protein sequence data bank and to the Brookhaven protein 3D structure data bank are taken directly from their data files without further verification by the authors. In order to keep the tables concise redundant information is avoided as far as possible (e.g. if K_m values are measured in the presence of an obvious cosubstrate, only the name of the cosubstrate is given in parentheses as a commentary without reference to its specific role).

The authors are grateful to the following biologists and chemists for invaluable help in the compilation of data: Margit Salzmann, Cornelia Munaretto, Dr. Ida Schomburg, Dr. Astrid Beermann and Astrid Haberz. In addition we would like to thank Mrs. C. Munaretto and Dr. I. Schomburg for the correction of the final manuscript.

Braunschweig, Spring 1995   Dörte Stephan
                                             Dietmar Schomburg

V

BRENDA – Compilation of Enzyme Data

To collect basic characteristics of enzymes – is that not a kind of archaic activity in the times of molecular biology and computer-aided data banks providing sequences of nucleic acids and proteins with little more delay than a few days as well as their three-dimensional structures? What should be the purpose of compiling turnover numbers, Michaelis constants, substrate specificities, sources, synonyms etc. of enzymes from sometimes remote publications? The answer sounds as simple as surprising: The aim of the compilation of data is to make use of the overwhelming abundance of structural knoweldge we owe to the new techniques of molecular biology.

Admittedly, it was not primarily enzymology which caused the explosion of knowledge in biology during the last decade. This was due to the advance of molecular biology which enabled us to isolate genes, to amplify them *ad libidum* and to elucidate their primary structure within days only. Also, the optimization and automatization of techniques for the analysis of macromolecules has provided detailed insights into a large variety of complex biomolecules nobody would have anticipated in the early seventies. Due to powerful computers it has now become feasible to propose fairly realistic models of macromolecules based solely on primary structures and homology considerations.

Nevertheless – or therefore – it appears as mandatory as rewarding to know the brave world of enzymology in which one had and often still has to come along without any detailed structural knowledge. We should not ignore that nature has not generated the multiplicity of structures, because it simply felt obliged to the principle of diversification or because it wanted to test our computing capacity to handle sequence data. It had to create new structures to cope with the steadily changing demands of a variable environment. Thus, amino acid sequences, folding of peptide chains and conformational details are only the technical tools of nature to catalyse specific biological functions. In consequence, *it is the functional profile of an enzyme which enables a biologist or physician to analyze a metabolic pathway and its disturbance; it is the substrate specificity of an enzyme which tells an analytical biochemist how to design an assay; it is the stability, specificity and efficiency of an enzyme which determines its usefulness in the biotechnical transformation of a molecule.* And the sum of all these functional data will have to be considered when the designer of artificial biocatalysts has to choose the optimum prototype to start with.

Unfortunately, it is by no means as simple to design (organize) a meaningful and systematic compilation of functional enzymological data as to enter sequences of amino acids or nucleotides into a data base. Functional data are less well defined, are never devoid of a trace of ambiguity, their selection remains inevitably subjective, and their complexity requires simplification. The present compilation of enzymological data, therefore, can and will not be a substitute for original publications but rather offer a key to the literature. But I do think that the Enzyme Handbook is indeed an excellent key to open or reopen the mysterious world of enzyme to all those who there have to find the solutions of their problems: to biologists, physicians, structural biochemists, biochemical analysts, biotechnologists and also to the molecular biologists.

Braunschweig, Spring 1993   Leopold Flohé
                                             GBF, Scientific Director

List of Abbreviations

A — adenosine
Ac — acetyl
ACP — acyl-carrier-protein
ADP — adenosine 5'-diphosphate
Ala — alanine
All — allose
Alt — altrose
AMP — adenosine 5'-monophosphate
Ara — arabinose
Arg — arginine
Asn — asparagine
Asp — aspartic acid
ATP — adenosine 5'-triphosphate
Bicine — N,N'-bis(2-hydroxyethyl) glycine
C — cytidine
cal — calorie
CDP — cytidine 5'-diphosphate
CDTA — trans-1,2-diaminocyclo-hexane-N,N,N,N-tetra-aceticacid
CHAPS — 3-[(3-cholamidopropyl)-dimethylammonio]-1-propanesulfonate
CHAPSO — 3-[(3-cholamidopropyl)-dimethylammonio]-2-hydroxy-1-propane-sulfonate
CMP — cytidine 5'-monophosphate
CoA — coenzyme A
CTP — cytidine 5'-triphosphate
Cys — cysteine
d — deoxy-
D- and L- — prefixes indicating configuration
DFP — diisopropylfluorophosphate
DNA — deoxyribonucleic acid
DPN — diphosphopyridinium nucleotide (now NAD)
DTNB — 5,5'-dithiobis(2-nitrobenzoate)
DTT — dithiothreitol (i.e. Cleland's reagent)
e — electron
EC — number of enzyme in Enzyme Commission's system
E. coli — Escherichia coli
EDTA — ethylene diaminetetraacetate

EGTA — ethylene glycol bis (β-amino-ethylether) tetraacetate
EPR — electron paramagnetic resonance
ER — endoplasmic reticulum
Et — ethyl
EXAFS — extended X-ray absorption fine structure
FAD — flavin-adenine dinucleotide
FMN — flavin mononucleotide (ribo-flavin 5'-monophosphate)
FPLC — fast protein liquid chroma-tography
Fru — fructose
Fuc — fucose
G — guanosine
Gal — galactose
GDP — guanosine 5'-diphosphate
Glc — glucose
GlcN — glucosamine
GlcNAc — N-acetylglucosamine
Gln — glutamine
Glu — glutamic acid
Gly — glycine
Glygly — glycylglycine
GMP — guanosine 5'-monophosphate
GSH — glutathione
GSSG — oxidized glutathione
GTP — guanosine 5'-triphosphate
Gul — gulose
h — hour
H_4 — tetrahydro
HEPES — 4-(2-hydroxyethyl)-1-piper-azineethane sulfonic acid
His — histidine
HPLC — high performance liquid chromatography
Hyl — hydroxylysine
Hyp — hydroxyproline
IAA — iodoacetamide
Ig — immunoglobulin
Ile — isoleucine
Ido — idose
IDP — inosine 5'-diphosphate
IMP — inosine 5'-monophosphate

List of Abbreviations

ir	irreversible	r	reversible
ITP	inosine 5'-triphosphate	Rha	rhamnose
K_m	Michaelis constant	Rib	ribose
L-	see D-	RNA	ribonucleic acid
Leu	leucine	mRNA	messenger RNA
Lys	lysine	rRNA	ribosomal RNA
Lyx	lyxose	tRNA	transfer RNA
M	mol/l	Sar	N-methylglycine
m-	meta-		(sarcosine)
Man	mannose	SDS-PAGE	sodium dodecyl sulphate
MES	2-(N-morpholino)ethane		polyacrylamide gel
	sulfonate		electrophoresis
Met	methionine	Ser	serine
min	minute	SFK-525A	2-diethylaminoethyl-2,2-
MOPS	3-(N-morpholino)		diphenylvalerate
	propane sulfonate	sp.	species
Mur	muramic acid	T	ribosylthymine
MW	molecular weight	$t_{1/2}$	time for half-completion
NAD	nicotinamide-adenine		of reaction
	dinucleotide	Tal	talose
NADH	reduced NAD	TDP	ribosylthymine
NADP	NAD phosphate		5'-diphosphate
NADPH	reduced NADP	TEA	triethanolamine
NAD(P)H	indicates either NADH	THF	tetrahydrofolate
	or NADPH	Thr	threonine
NDP	nucleoside 5'-diphosphate	TMP	ribosylthymine
NEM	N-ethylmaleimide		5'-monophosphate
Neu	neuraminic acid	Tos-	tosyl-(p-toluenesulfonyl-)
NMN	nicotinamide	TPN	triphosphopyridinium
	mononucleotide		nucleotide (now NADP)
NMP	nucleoside	Tris	tris(hydroxymethyl)-
	5'-monophosphate		aminomethane
NTP	nucleoside 5'-triphosphate	Trp	tryptophan
o-	ortho-	TTP	ribosylthymine
Orn	ornithine		5'-triphosphate
p-	para-	Tyr	tyrosine
PCMB	p-chloro-mercuribenzoate	U	uridine
PEG	polyethylene glycol	U/mg	µmol/(mg·min)
PEP	phosphoenolpyruvate	UDP	uridine 5'-diphosphate
pH	$-\log_{10} [H^+]$	UMP	uridine 5'-monophosphate
Ph	phenyl	UTP	uridine 5'-triphosphate
Phe	phenylalanine	UV	ultraviolet
PIXE	proton-induced	Val	valine
	X-ray emission	Xaa	symbol for an amino
PMSF	phenylmethane-		acid of unknown consti-
	sulfonylfluoride		tution in peptide formula
Pro	proline	XAS	X-ray absorption
Q_{10}	factor for the change in		spectroscopy
	reaction rate for a 10°	XTP	xanthosine 5'-triphosphate
	temperature increase	Xyl	xylose

X

Index

(Alphabetical order of Enzyme names)

EC-No.	Name	EC-No.	Name

1.1.1.72 Glycerol dehydrogenase (NADP+)
1.1.1.156 Glycerol 2-dehydrogenase (NADP+)
1.1.99.5 Glycerol-3-phosphate dehydrogenase
1.1.1.8 Glycerol-3-phosphate dehydrogenase (NAD+)
1.1.1.94 Glycerol-3-phosphate dehydrogenase (NAD(P)+)
1.1.1.177 Glycerol-3-phosphate 1-dehydrogenase (NADP+)
1.1.3.21 Glycerol-3-phosphate oxidase
1.1.99.14 Glycolate dehydrogenase
1.1.1.185 L-Glycol dehydrogenase
1.1.1.26 Glyoxylate reductase
1.1.1.79 Glyoxylate reductase(NADP+)
1.1.1.45 L-Gulonate 3-dehydrogenase
1.1.3.8 L-Gulonolactone oxidase
1.1.1.164 Hexadecanol dehydrogenase
1.1.3.5 Hexose oxidase
1.1.1.23 Histidinol dehydrogenase
1.1.1.155 Homoisocitrate dehydrogenase
1.1.1.3 Homoserine dehydrogenase
1.1.99.6 D-2-Hydroxy-acid dehydrogenase
1.1.3.15 (S)-2-Hydroxy-acid oxidase
1.1.99.24 Hydroxyacid-oxoacid transhydrogenase
1.1.1.35 3-Hydroxyacyl-CoA dehydrogenase
1.1.1.152 3alpha-Hydroxy-5beta-androstan-17-one 3alpha-dehydrogenase
1.1.1.97 3-Hydroxybenzyl-alcohol dehydrogenase
1.1.1.30 3-Hydroxybutyrate dehydrogenase
1.1.1.61 4-Hydroxybutyrate dehydrogenase
1.1.1.157 3-Hydroxybutyryl-CoA dehydrogenase

1.1.1.52 3alpha-Hydroxycholanate dehydrogenase
1.1.1.241 6-endo-Hydroxycineole dehydrogenase
1.1.1.166 Hydroxycyclohexanecarb-oxylate dehydrogenase
1.1.1.226 4-Hydroxycyclohexanecarb-oxylate dehydrogenase
1.1.99.26 3-Hydroxycyclohexanone dehydrogenase
1.1.1.66 omega-Hydroxydecanoate 1-dehydrogenase
1.1.1.232 5-Hydroxyeicosatetra-enoate dehydrogenase
1.1.1.98 (R)-2-Hydroxy-fatty-acid dehydrogenase
1.1.1.99 (S)-2-Hydroxy-fatty-acid dehydrogenase
1.1.99.2 2-Hydroxyglutarate dehydrogenase
1.1.1.230 3alpha-Hydroxyglycyr-rhetinate dehydrogenase
1.1.1.31 3-Hydroxyisobutyrate dehydrogenase
1.1.1.167 Hydroxymalonate dehydrogenase
1.1.3.19 4-Hydroxymandelate oxidase
1.1.1.178 3-Hydroxy-2-methylbutyryl-CoA dehydrogenase
1.1.1.170 3beta-Hydroxy-4beta-methylcholestenecarboxy-late 3-dehydrogenase (decarboxylating)
1.1.1.88 Hydroxymethylglutaryl-CoA reductase
1.1.1.34 Hydroxymethylglutaryl-CoA reductase (NADPH)
1.1.1.60 2-Hydroxy-3-oxopropionate reductase
1.1.1.222 (R)-4-Hydroxyphenyl-lactate dehydrogenase
1.1.1.237 Hydroxyphenylpyruvate reductase
1.1.3.27 Hydroxyphytanate oxidase
1.1.1.59 3-Hydroxypropionate dehydrogenase

EC-No.	Name	EC-No.	Name

1 NOMENCLATURE

EC number
1.1.1.150

Systematic name
21-Hydroxysteroid:NAD⁺ 21-oxidoreductase

Recommended name
21-Hydroxysteroid dehydrogenase (NAD⁺)

Synonymes
Dehydrogenase, 21-hydroxy steroid

CAS Reg. No.
37250-75-2

---

2 REACTION AND SPECIFICITY

Catalysed reaction
Pregnan-21-al + NADH →
→ pregnan-21-ol + NAD⁺

Reaction type
Redox reaction

Natural substrates

Substrate spectrum
1 21-Dehydrocortisol + NADH (stereospecificity [2, 3], transfer of hydrogen from 4-pro-S-position of NADH to 21-pro-S-position of cortisol [3]) [1–3]
2 More (acts on a number of 21-hydroxycorticosteroids, e.g.: 21-dehydro-cortisone, 21-dehydro-11-deoxycortisol, 21-dehydro-DELTA¹-cortisol, 21-dehydro-11-deoxycorticosterone, 21-dehydrocorticosterone, not: ace-taldehyde, pyruvic acid, benzaldehyde, salicylaldehyde, methylglyoxal, glyoxal) [1]

Product spectrum
1 Cortisol + NAD⁺
2 ?

Inhibitor(s)
Ag⁺ [1]; p-Substituted mercuribenzoate [1]; Hg²⁺ [1]; Androsterone (inhibiti-on of enzyme I, no inhibition of enzyme II) [1]; Testosterone (inhibition of en-zyme I, no inhibition of enzyme II) [1]; 17beta-Estradiol (inhibition of enzyme I, no inhibition of enzyme II) [1]; NAD⁺ [1]; NADP⁺ [1]

Cofactor(s)/prosthetic group(s)/activating agents
NADH [1–3]

Metal compounds/salts

Turnover number (min⁻¹)

Specific activity (U/mg)
More [1]

K_m-value (mM)
0.105 (21-dehydrocorticosterone) [1]; 0.111 (21-dehydro-11-deoxycortisol) [1]; 0.172 (21-dehydro-DELTA¹-cortisol) [1]; 0.174 (21-dehydro-11-deoxycorticosterone) [1]; 0.182 (21-dehydrocortisol) [1]; 0.221 (21-dehydrocortisone) [1]

pH-optimum
6.2 (enzyme II) [1]; 6.9 (enzyme I) [1]

pH-range
5.5–7.2 (about 65% of activity maximum at pH 5.5 and 7.2, enzyme II) [1]; 5.5–8.2 (about 75% of activity maximum at pH 5.5 and 8.2, enzyme I) [1]

Temperature optimum (°C)
30 (assay at) [1]

Temperature range (°C)

---

3 ENZYME STRUCTURE

Molecular weight

Subunits

Glycoprotein/Lipoprotein
–

---

4 ISOLATION/PREPARATION

Source organism
Sheep [1–3]

Source tissue
Liver [1, 3]

Localisation in source

Purification
Sheep (partial, 2 enzyme forms: I and II) [1]

Crystallization

–

Cloned

–

Renaturated

–

---

5 STABILITY

pH

Temperature (°C)
56 (5 min, 30% loss of activity) [1]

Oxidation

Organic solvent

General stability information

Storage

---

6 CROSSREFERENCES TO STRUCTURE DATABANKS

PIR/MIPS code

Brookhaven code

---

7 LITERATURE REFERENCES

[1] Monder, C., White, A.: J. Biol. Chem.,240,71–77 (1965)
[2] Orr, J.C., Monder, C.: J. Biol. Chem.,250,7547–7553 (1975)
[3] Orr, J.C., Monder, C.: J. Steroid Biochem.,6,297–299 (1975)

1 NOMENCLATURE

EC number
1.1.1.151

Systematic name
21-Hydroxysteroid:NADP⁺ 21-oxidoreductase

Recommended name
21-Hydroxysteroid dehydrogenase (NADP⁺)

Synonymes
Dehydrogenase, 21-hydroxy steroid (nicotinamide adenine dinucleotide phosphate)
21-Hydroxy steroid dehydrogenase (nicotinamide adenine dinucleotide phosphate)
NADP-21-hydroxysteroid dehydrogenase [1]

CAS Reg. No.
37250-76-3

2 REACTION AND SPECIFICITY

Catalysed reaction
Pregnan-21-al + NADPH →
→ pregnan-21-ol + NADP⁺

Reaction type
Redox reaction

Natural substrates

Substrate spectrum
1 21-Dehydrocortisol + NADPH (r) [1]
2 More (acts on a number of 21-hydroxycorticosteroids, e.g. 21-dehydro-cortisone, 21-dehydro-11-deoxycortisol, 21-dehydro-DELTA¹-cortisol, 21-dehydro-11-deoxycorticosterone, 21-dehydrocortisone, reduction of benzaldehyde, salicylaldehyde, glyoxal and methylglyoxal is also catalyzed, not: acetaldehyde, pyruvic acid) [1]

Product spectrum
1 Cortisol + NADP⁺ (r) [1]
2 ?

Inhibitor(s)
 Ag⁺ [1]; Hg²⁺ [1]; NAD⁺ [1]; NADP⁺ [1]; p-Substituted mercuribenzoate [1];
 More (not: androsterone, testosterone, 17beta-estradiol) [1]

Cofactor(s)/prosthetic group(s)/activating agents
 NADPH [1]; NADP⁺ [1]

Metal compounds/salts

Turnover number (min⁻¹)

Specific activity (U/mg)
 More [1]

K_m-value (mM)
 0.118 (21-dehydro-11-deoxycorticosterone) [1]; 0.137
 (21-dehydro-11-deoxycortisol, 21-dehydro-DELTA¹-cortisol) [1]; 0.177 (21-de-
 hydrocortisol) [1]; 0.233 (21-dehydrocorticosterone) [1]; 0.461 (21-dehydro-
 cortisone) [1]

pH-optimum
 5.9–6.1 [1]

pH-range
 5.2–7.4 (about 50% of activity maximum at pH 5.2 and 7.4) [1]

Temperature optimum (°C)
 30 (assay at) [1]

Temperature range (°C)

---

3 ENZYME STRUCTURE

Molecular weight

Subunits

Glycoprotein/Lipoprotein
 –

---

4 ISOLATION/PREPARATION

Source organism
 Sheep [1]

Source tissue
 Liver [1]

Localisation in source
 Sheep (partial) [1]

Purification

Crystallization

–

Cloned

–

Renaturated

–

---

5 STABILITY

pH

Temperature (°C)
56 (5 min, 80% loss of activity) [1]

Oxidation

Organic solvent

General stability information

Storage

---

6 CROSSREFERENCES TO STRUCTURE DATABANKS

PIR/MIPS code

Brookhaven code

---

7 LITERATURE REFERENCES

[1] Monder, C., White, A.: J. Biol. Chem.,240,71–77 (1965)

1 NOMENCLATURE

EC number
1.1.1.152

Systematic name
3alpha-Hydroxy-5beta-steroid:NAD+ 3-oxidoreductase

Recommended name
3alpha-Hydroxy-5beta-androstan-17-one 3alpha-dehydrogenase

Synonymes
Etiocholanolone 3alpha-dehydrogenase
Dehydrogenase, etiocholanolone 3alpha-
3alpha-Hydroxy-5beta-steroid dehydrogenase

CAS Reg. No.
37250-77-4

2 REACTION AND SPECIFICITY

Catalysed reaction
3alpha-Hydroxy-5beta-androstan-17-one + NAD+ →
→ 5beta-androstan-3,17-dione + NADH

Reaction type
Redox reaction

Natural substrates

Substrate spectrum
1 3alpha-Hydroxy-5beta-androstan-17-one + NAD+ (i.e. etiocholanolone) [1]

Product spectrum
1 5beta-Androstan-3,17-dione + NADH [1]

Inhibitor(s)

Cofactor(s)/prosthetic group(s)/activating agents
NAD+ [1]

Metal compounds/salts

Turnover number (min^{-1})

Specific activity (U/mg)

K_m-value (mM)

pH-optimum

pH-range

Temperature optimum (°C)

Temperature range (°C)

---

3 ENZYME STRUCTURE

Molecular weight
47000 (Pseudomonas testosteroni, meniscus depletion method, gel filtration) [1]

Subunits

Glycoprotein/Lipoprotein
–

---

4 ISOLATION/PREPARATION

Source organism
Pseudomonas testosteroni (strain STDH-m) [1]

Source tissue

Localisation in source

Purification

Crystallization
–

Cloned
–

Renaturated
–

5 STABILITY

pH

Temperature (°C)

Oxidation

Organic solvent

General stability information

Storage

---

6 CROSSREFERENCES TO STRUCTURE DATABANKS

PIR/MIPS code

Brookhaven code

---

7 LITERATURE REFERENCES

[1] Roe, C.R., Kaplan, N.O.: Biochemistry,8,5093–5103 (1969)

---

1 NOMENCLATURE

EC number
1.1.1.153

Systematic name
7,8-Dihydrobiopterin:NADP+ oxidoreductase

Recommended name
Sepiapterin reductase

Synonymes
Reductase, sepiapterin

CAS Reg. No.
9059-48-7

2 REACTION AND SPECIFICITY

Catalysed reaction
Sepiapterin + NADPH →
→ 7,8-dihydrobiopterin + NADP+ (ordered bi-bi mechanism [6])

Reaction type
Redox reaction

Natural substrates
Sepiapterin + NADPH (terminal step in the biosynthetic pathway for tetrahydrobiopterin [1, 6–9, 13–15], catalyzes the complete reduction of pyrovoylte-
trahydropterin to tetrahydrobiopterin [15], responsible for conversion of lac-
toyltetrahydropterin to tetrahydrobiopterin [1, 6–9, 13–15])

Substrate spectrum
 1 Sepiapterin + NADPH (r [4, 5], equilibrium lies much in favor of dihydrobiopterin formation [4, 6], in the reverse reaction, L-erythro and L-threo-dihy-
   droneopterin are also oxidized, the corresponding L-isomers are inactive
   [4]) [1–15]
 2 Isosepiapterin + NADPH [4]
 3 Xanthopterin B$_2$ + NADPH [4]
 4 NADPH + carbonyl compounds of non-pteridine derivatives (including
   some vicinal dicarbonyl compounds, e.g. benzil, phenylpropanedione
   [11], p-quinone, menadione, methylglyoxal, phenylglyoxal, 4-nitrobenzal-
   dehyde, acetophenone, acetoin, propiophenone, benzylacetone [12],
   non-specific NADPH-dependent carbonyl reductase activity [11, 12]) [11,
   12]
 5 More (enzyme also catalyzes the isomerization of 6-1'-oxo-2'-hydroxypro-
   pyl(6-lactoyltetrahydropterin) to 6-1'-hydroxy-2'-oxopropyltetrahydropterin
   [10], enzyme is quite similar to general aldo-keto reductases, especially
   to carbonyl reductase [12], not: 6-acetyl-7-methyldihydropterin [4]) [4, 10,
   12]

Product spectrum
 1 7,8-Dihydrobiopterin + NADP$^+$ (r [4, 5]) [1–15]
 2 6-(1-Hydroxypropyl)-dihydropterin + NADP$^+$ [4]
 3 6-(1',2'-Dihydropropyl)-dihydrolumazine + NADP$^+$ [4]
 4 ?
 5 ?

Inhibitor(s)
Acetic acid (0.1 M) [4]; Sodium acetate [4]; Sepiapterin (above 0.05 mM)
[4]; NADPH (above 0.1 mM) [4]; p-Chloromercuribenzoate [4]; Monoiodoacetate [4]; N-Ethylmaleimide [4]; KCN (slight) [4]; NaN$_3$ (slight) [4]; 2,4-Di-
nitrophenol (slight) [4]; Pterin (slight) [4]; Leucopterin (slight) [4]; 6-Methyl-
pterin-H$_2$ [4]; 7-Methylpterin (slight) [4]; 6,7-Dimethylpterin [4]; 6,7-Diethyl-
pterin-H$_2$ [4]; 6-Hydroxymethylpterin [4]; 6-Carboxypterin [4, 12]; Biopterin
(slight) [4]; Isosepiapterin [4]; Ethacrynic acid [12]; Unconjugated pteridines (e.g. isosepiapterin, 6,7-dimethylpteridin-H$_2$) [4]; Biopterin-H$_2$ [4];
L-Erythro-neopterin (slight) [4]; L-Erythro-neopterin-H$_2$ [4]; Propionic acid [4,
12]; Benzoic acid [4]; Propionamide (slight) [4]; D,L-Lactic acid (slight) [4];
Pyruvic acid (slight) [4]; N-Acetylserotonin [7, 12]; Rutin [12]; Dicoumarol
[12]; Indomethacin [12]; Catecholamine [14]; Indoleamine [14]

Cofactor(s)/prosthetic group(s)/activating agents
NAD$^+$ (activity one-fifth of that with NADP$^+$) [4]; NADP$^+$ [4]; NADPH [1–15];
NADH (activity one-fourth of that with NADPH (horse), activity one-half of
that with NADPH (rat)) [4]; More (the enzyme does not contain flavin) [12]

Metal compounds/salts

Turnover number (min⁻¹)

Specific activity (U/mg)
18 [6]; More [4]

K_m-value (mM)
0.021 (sepiapterin) [4]; 0.015 (sepiapterin) [6]; 0.020 (sepiapterin) [7]; 0.014 (NADPH) [4]; 0.0017 (NADPH) [6]; 0.005–0.006 (NADPH) [7]; 0.14 (proprio-phenone) [12]; 0.043 (benzylacetone) [12]; 0.057 (dihydrobiopterin) [4]; 0.077 (NADP⁺) [4]; 0.0218 (phenylpropanedione) [11]; 0.020 (benzil) [11]; 1.0 (diacetyl) [11]; 0.021 (1-phenyl-1,2-propanedione (+ NADPH)) [12]; 0.118 (1-phenyl-1,2-propanedione (+ NADH)) [12]; 1.02 (diacetyl (+NADH)) [12]; 0.0183 (benzil) [12]; 4.4 (methylglyoxal) [12]; 0.59 (phenylglyoxal) [12]; 5.19 (2,4-pentanedione) [12]; 0.75 (p-quinone) [12]; 0.18 (menadione) [12]; 0.00082 (9,10-phenanthrene quinone) [12]; 0.43 (p-nitrobenzaldehyde) [12]; 1.1 (acetophenone) [12]; 29 (acetoin) [12]; More [11]

pH-optimum
5.5 [4, 6]; 6.4 (reduction of sepiapterin) [4]; 8.6 (optimum of isomerization of 6–1'-oxo-2'-hydroxypropyl(6-lactoyltetrahydropterin) to 6–1'-hydroxy-2'-oxo-propyltetrahydropterin) [10]; 10.4 (oxidation of 7,8-dihydrobiopterin) [4]

pH-range
5.5–7.4 (5.5: optimum, 7.4: 80% of activity maximum) [6]

Temperature optimum (°C)

Temperature range (°C)

---

3 ENZYME STRUCTURE

Molecular weight
55000 (rat, gel filtration) [6]

Subunits
Dimer (2 × 27500, identical, rat, SDS-PAGE) [6]

Glycoprotein/Lipoprotein
–

---

4 ISOLATION/PREPARATION

Source organism
Dog [5]; Human (low activity [5]) [5, 7, 8, 13]; Rat (expressed in E. coli [1]) [1–6, 10–12, 14, 15]; Horse [4]; Guinea pig (low activity) [5]; Mouse [5, 9]; Rabbit (low activity) [5]; Chicken [5]; Pig (low activity) [5]

Source tissue
    Cultured human amniotic fibroblasts [7]; Reticulocytes [9]; Cultured human mononuclear blood cells [7]; Macrophages [8]; Fibroblasts [8]; Cultured cells (T24 or THP-1 [8]) [7, 8]; Liver [2–5, 13]; Blood [4, 5]; Spleen [9]; Brain [14]; Erythrocytes [4–6, 9–12, 15]; Bone marrow [9]; More (not in leukocytes and ascites tumor cells) [5]

Localisation in source
    Cytoplasm (horse liver) [4]

Purification
    Rat [3, 6]; Horse [4]

Crystallization
    –

Cloned
    (expressed in E. coli) [1]

Renaturated
    –

---

5 STABILITY

pH
    5.5 (unstable) [4]

Temperature (°C)

Oxidation

Organic solvent

General stability information

Storage

---

6 CROSSREFERENCES TO STRUCTURE DATABANKS

PIR/MIPS code
    PIR2:JQ1176 (human); PIR2:A36024 (rat (fragment)); PIR2:A36400 (rat (fragments))

Brookhaven code

7 LITERATURE REFERENCES

[1] Citron, B.A., Milstien, S., Gutierrez, J.C., Levine, R.A., Yanak, B.L., Kaufman, S.: Proc. Natl. Acad. Sci. USA,87,6436–6440 (1990)
[2] Nagai, M.: Arch. Biochem. Biophys.,126,426–435 (1968)
[3] Matsubara, M., Katoh, S., Akino, M., Kaufman, S.: Biochim. Biophys. Acta,122,202–212 (1966)
[4] Katoh, S.: Arch. Biochem. Biophys.,146,202–214 (1971)
[5] Katoh, S., Arai, Y., Taketani, T., Yamada, S.: Biochim. Biophys. Acta,370,378–388 (1974)
[6] Sueoka, T., Katoh, S.: Biochim. Biophys. Acta,717,265–271 (1982)
[7] Fere, J., Naylor, E.W.: Biochem. Biophys. Res. Commun.,148,1475–1481 (1987)
[8] Werner, E.R., Werner-Felmayer, G., Fuchs, D., Hausen, A., Reibnegger, G., Yim, J.J., Pfleiderer, W., Wachter, H. : J. Biol. Chem.,265,3189–3192 (1990)
[9] Kerler, F., Hültner, L., Ziegler, I., Katzenmaier, G., Bacher, A.: J. Cell. Physiol.,142, 268–271 (1990)
[10] Katoh, S., Sueoka, T.: J. Biochem.,101,275–278 (1987)
[11] Katoh, S., Sueoka, T.: Biochem. Biophys. Res. Commun.,118,859–866 (1984)
[12] Sueoka, T., Katoh, S.: Biochim. Biophys. Acta,843,193–198 (1985)
[13] Curtius, H.-C., Heintel, D., Ghisla, S., Kuster, T., Leimbacher, W., Niederwieser, A.: Eur. J. Biochem.,148,413–419 (1985)
[14] Katoh, S., Sueoka, T., Yamada, S.: Biochem. Biophys. Res. Commun.,105,75–81 (1982)
[15] Smith, G.K.: Arch. Biochem. Biophys.,255,254–266 (1987)

1 NOMENCLATURE

EC number
1.1.1.154

Systematic name
(S)-Ureidoglycolate:NAD(P)$^+$ oxidoreductase

Recommended name
Ureidoglycolate dehydrogenase

Synonymes
Dehydrogenase, ureidoglycolate

CAS Reg. No.
62213-62-1

2 REACTION AND SPECIFICITY

Catalysed reaction
(S)-Ureidoglycolate + NAD(P)$^+$ →
→ oxalureate + NAD(P)H

Reaction type
Redox reaction

Natural substrates
(S)-Ureidoglycolate + NAD(P)H (enzyme in degradation of allantoin under anaerobic conditions) [1]

Substrate spectrum
1 (S)-Ureidoglycolate + NAD(P)$^+$ (specific for (S)-ureidoglycolate, reverse reaction not detected) [1]

Product spectrum
1 Oxalureate + NAD(P)H [1]

Inhibitor(s)
L-Lactate [1]; Glycolate [1]; Glyoxylate [1]; Zn^{2+} [1]

Cofactor(s)/prosthetic group(s)/activating agents
NAD$^+$ [1]; NADP$^+$ [1]

Metal compounds/salts

Turnover number (min⁻¹)

Specific activity (U/mg)
10.6 [1]

Kₘ-value (mM)
2.16 (NAD⁺) [1]; 4.5 ((S)-ureidoglycolate) [1]; 0.1 (NADP⁺) [1]

pH-optimum
8.0–8.4 [1]

pH-range
7.2–8.8 (7.2: about 45% of activity maximum, 8.8: about 65% of activity maximum) [1]

Temperature optimum (°C)
22 (assay at room temperature) [1]

Temperature range (°C)

---

3 ENZYME STRUCTURE

Molecular weight

Subunits

Glycoprotein/Lipoprotein
–

---

4 ISOLATION/PREPARATION

Source organism
Arthrobacter allantoicus [1]; Streptococcus allantoicus [1]; E. coli [1]

Source tissue

Localisation in source

Purification
Arthrobacter allantoicus [1]

Crystallization
–

Cloned
–

Renaturated
–

5 STABILITY

pH

6–8.5 (30°C, 20 min, no loss of activity) [1]; 6 (below, 30°C, 20 min, decrease in activity) [1]

Temperature (°C)

4 (30% glycerol, 3 h, stable) [1]; 30 (unstable above) [1]

Oxidation

Organic solvent

General stability information

Storage

---

6 CROSSREFERENCES TO STRUCTURE DATABANKS

PIR/MIPS code

Brookhaven code

---

7 LITERATURE REFERENCES

[1] Van der Drift, C., Van Helvoort, P.E.M., Vogels, G.D.: Arch. Biochem. Biophys.,145, 465–469 (1971)

---

1 NOMENCLATURE

EC number
1.1.1.155

Systematic name
(-)-1-Hydroxy-1,2,4-butanetricarboxylate:NAD$^+$ oxidoreductase (decarboxylating)

Recommended name
Homoisocitrate dehydrogenase

Synonymes
Dehydrogenase, homoisocitrate
Homoisocitric dehydrogenase [2]

CAS Reg. No.
9067-90-7

2 REACTION AND SPECIFICITY

Catalysed reaction
(-)-1-Hydroxy-1,2,4-butanetricarboxylate + NAD$^+$ →
→ 2-oxoadipate + CO$_2$ + NADH

Reaction type
Redox reaction
Oxidative decarboxylation
Reductive carboxylation

Natural substrates
1-Hydroxy-1,2,4-butanetricarboxylate + NAD$^+$ (production of 2-oxoadipic acid (a precursor of lysine biosynthesis) [2], alpha-aminoadipate pathway for biosynthesis of lysine [4]) [2, 4]

Substrate spectrum
1 1-Hydroxy-1,2,4-butanetricarboxylate + NAD$^+$ (i.e. homoisocitrate [1], r [1]) [1–4]
2 More (catalyzes keto-enol tautomerization of tritiated alpha-ketoadipate, not: ethanol, isocitrate, malate, glutamate) [1]

Product spectrum
1 2-Oxoadipate + CO$_2$ + NADH [1]
2 ?

Inhibitor(s)
Homoisocitrate (1.2 mM [1]) [1, 2]; 2-Oxoadipate [2]; Oxaloglutarate [2]

Cofactor(s)/prosthetic group(s)/activating agents
NAD^+ ($NADP^+$ can replace NAD^+ [3]) [1–3]; NADH [1, 2]; alpha-Ketoadipate (required for reductive carboxylation) [1]; $NaHCO_3$-CO_2 (required for reductive carboxylation) [1]

Metal compounds/salts
K^+ (required for optimal activity) [1, 2]; Mg^{2+} (stimulation of oxidative decarboxylation, required for reductive carboxylation [1], required [2, 3]) [1–3]; Mn^{2+} (stimulation of oxidative decarboxylation) [1, 2]

Turnover number (min^{-1})

Specific activity (U/mg)
3.8 [1]; 6.304 [2]

K_m-value (mM)
0.065 (NADH) [1]; 1.4 (homoisocitric acid) [3]; 0.01 (below, homoisocitrate) [1]; 1.5 (alpha-ketoadipate) [1]; 0.33 (NAD^+) [1]

pH-optimum
7.0 (reductive carboxylation) [1]; 7.8 (homocitrate + NAD^+) [3]; 8.3–8.8 (oxidative decarboxylation) [1]

pH-range
6.1–7.7 (6.1: about 60% of activity maximum, 7.7: about 50% of activity maximum, reductive carboxylation) [1]; 7.4–9.3 (7.4: about 50% of activity maximum, 9.3: about 90% of activity maximum, oxidative decarboxylation) [1]

Temperature optimum (°C)
20 (assay at) [2]; 30 (assay at) [1]

Temperature range (°C)

---

3 ENZYME STRUCTURE

Molecular weight
48000 (Saccharomyces cerevisiae, gel filtration [1], Sacharomycopsis lipolytica, native or denaturing acrylamide gel electrophoresis [2]) [1, 2]

Subunits
Monomer (1 × 48000, Saccharomycopsis lipolytica, native or denaturing acrylamide gel electrophoresis) [2]
Dimer (2 × 48000, Saccharomycopsis lipolytica, at high protein concentration or in presence of substrate the enzyme exists as dimer) [2]

Glycoprotein/Lipoprotein
–

4 ISOLATION/PREPARATION

Source organism
Saccharomyces cerevisiae [1, 3]; Saccharomycopsis lipolytica [2]; Schizo-saccharomyces pombe [4]

Source tissue
Cell [1]

Localisation in source

Purification
Saccharomyces cerevisiae (partial [3]) [1, 3]; Saccharomycopsis lipolytica [2]

Crystallization
−

Cloned
−

Renaturated
−

5 STABILITY

pH

Temperature (°C)

Oxidation

Organic solvent

General stability information
Mn^{2+} stabilizes [2]; Mg^{2+} stabilizes [2]

Storage
−20°C, stable as ammonium sulfate precipitate [2]; 0°C, pH 7.2, rapid loss of activity [2]

6 CROSSREFERENCES TO STRUCTURE DATABANKS

PIR/MIPS code

Brookhaven code

7 LITERATURE REFERENCES

[1] Rowley, B., Tucci, A.F.: Arch. Biochem. Biophys.,141,499–510 (1970)
[2] Gaillardin, C.M., Ribet, A.-M., Heslot, H.: Eur. J. Biochem.,128,489–494 (1982)
[3] Broquist, H.P.: Methods Enzymol.,17B,118–119 (1971) (Review)
[4] Ye, Z.-H., Bhattacharjee, J. K.: J. Bacteriol.,170,5968–5970 (1988)

1 NOMENCLATURE

EC number
1.1.1.156

Systematic name
Glycerol:NADP⁺ 2-oxidoreductase (glycerone-forming)

Recommended name
Glycerol 2-dehydrogenase (NADP⁺)

Synonymes
Dihydroxyacetone reductase
Reductase, dihydroxyacetone (reduced nicotinamide adenine dinucleotide phosphate)
Dihydroxyacetone reductase (NADPH)
DHA oxidoreductase [5]

CAS Reg. No.
39342-20-6

---

2 REACTION AND SPECIFICITY

Catalysed reaction
Glycerol + NADP⁺ →
→ glycerone + NADPH

Reaction type
Redox reaction

Natural substrates
Dihydroxyacetone + NADPH (enzyme involved in glycerol synthesis [3, 4], possible role in xerotolerance [4], enzyme of glycerol cycle [5], involved in synthesis of glycerol during early spore germination [6]) [3–6]

Substrate spectrum
1 Glycerol + NADP⁺ (r [1, 3]) [1–6]
2 More (low activity with: 1,2-propanediol [1, 3], 2,3-butanediol [1, 3], 1,2-butanediol [1], 1,2,3-butanetriol [1], iso-erythritol [3], meso-erythritol [6], L-arabitol [6]) [1, 3, 6]

Product spectrum
1 Dihydroxyacetone + NADPH
2 ?

Inhibitor(s)
EDTA (slight) [1]; KCl (high ionic strength, above 100 mM) [6]; $MgCl_2$ [6]; $MnCl_2$ [6]; L-Cysteine [1]; Dithiothreitol [1]; 2-Mercaptoethanol [1]; p-Chloro-mercuribenzoate [1]; 8-Hydroxyquinoline [1]

Cofactor(s)/prosthetic group(s)/activating agents
NADP⁺ (pro-S stereospecific [3], completely specific for [1, 3, 6]) [1–4, 6]; NADPH (completely specific for [1, 3, 6]) [1–4, 6]

Metal compounds/salts
More (neither monovalent nor divalent cations activate the enzyme) [1]

Turnover number (min⁻¹)

Specific activity (U/mg)
15.7 [1]; 31.1 [6]

K_m-value (mM)
0.34 (NADH (+ dihydroxyacetone)) [1]; 51 (glycerol + (NADP⁺)) [1]; 0.1 (NADP⁺ (+ glycerol)) [1]; 0.67 (dihydroxyacetone + (NADPH)) [1]; 1400 (glycerol, pH 9.0) [3]; 0.015 (NADP⁺, pH 9.0) [3]; 0.8 (dihydroxyacetone) [3]; 0.004 (NADPH) [3]; 3.9 (dihydroxyacetone) [6]; 0.01 (NADPH) [6]; 0.17 (NADP⁺) [6]; 2000 (glycerol) [6]

pH-optimum
6.0 (reduction of dihydroxyacetone) [1]; 6–6.5 (reduction of dihydroxyaceto-ne) [6]; 7.5 (reduction of dihydroxyacetone) [3]; 9.0 (oxidation of glycerol) [3]; 9.4 (oxidation of glycerol) [6]; 9.5 (oxidation of glycerol) [1]

pH-range
5.0–8.0 (5.0: about 70% of activity maximum, 8.0: about 50% of activity ma-ximum, reduction of dihydroxyacetone) [6]; 6.0–9.5 (at pH 6.0 and 9.5: about 55% of activity maximum, reduction of dihydroxyacetone) [3]; 8.0–10.5 (at pH 8.0 and 10.5: about 50% of activity maximum, oxidation of glycerol) [3]

Temperature optimum (°C)
20 (assay at) [5]; 25 (assay at) [6]

Temperature range (°C)

---

3 ENZYME STRUCTURE

Molecular weight
34000–39000 (Phycomyces blakesleeanus, gel filtration, SDS-PAGE) [6]
57000 (Schizosaccharomyces pombe, gel filtration) [1]
100000 (Mucor javanicus, gel filtration) [6]

Subunits
  Monomer (1 × 39000, Phycomyces blakesleeanus, SDS-PAGE) [6]
  Dimer (1 × 25000 + 1 × 30000, Schizosaccharomyces pombe, SDS-PAGE)
  [1]

Glycoprotein/Lipoprotein
  –

4 ISOLATION/PREPARATION

Source organism
  Schizosaccharomyces pombe [1, 2, 4]; Schizosaccharomyces malidevor-
  ans [2]; Schizosaccharomyces octosporus [2]; Phycomyces blakesleeanus
  [6]; Dunaliella parva (unicellular alga) [3, 5]; Mucor javanicus [6]; More (not:
  Euglena gracilis) [3]

Source tissue
  Cell [1]; Mycelium [6]; Sporangiospores [6]

Localisation in source
  Cytoplasm [5]

Purification
  Schizosaccharomyces pombe [1]; Phycomyces blakesleeanus [6]

Crystallization
  –

Cloned
  –

Renaturated
  –

5 STABILITY

pH
  4 (unstable) [1]; 7–11 (stable) [1]

Temperature (°C)
  50 (5 min, pH 7, 50% loss of activity [1], 10 min, 10% loss of activity, enzyme in unheated cell extract [4]) [1, 4]; 60 (10 min, 30% loss of activity, enzyme in unheated cell extract) [4]; 65 (5 min, pH 7 complete loss of activity)
  [1]; 70 (10 min, complete loss of activity, enzyme in crude cell extract) [4]

Oxidation

Organic solvent

General stability information
Glycerol protects against loss of activity both at high temperature and at low pH [1]

Storage
–70°C, several months [1]; 4°C, 2 weeks, purified enzyme: stable, crude extract: 90% loss of activity [1]; –15°C, in diluted form the enzyme gradually loses activity [6]

---

6 CROSSREFERENCES TO STRUCTURE DATABANKS

PIR/MIPS code

Brookhaven code

---

7 LITERATURE REFERENCES

[1] Marshall, J.H., Kong, Y.-C., Sloan, J., May, J.W.: J. Gen. Microbiol.,135,697–701 (1989)
[2] Vasiliadis, G.E., Sloan, J., Marshall, J.H., May, J.W. : Arch. Microbiol.,147,263–267 (1987)
[3] Ben-Amotz, A., Avron, M.: FEBS Lett.,29,153–155 (1973)
[4] Kong, Y.-C., May, J.W., Marshall, J.H.: J. Gen. Microbiol.,131,1571–1579 (1985)
[5] Gimmler, H., Lotter, G.: Z. Naturforsch.,37c,1107–1114 (1982)
[6] Van Laere, A.: FEMS Microbiol. Lett.,30,377–381 (1985)

1 NOMENCLATURE

EC number
1.1.1.157

Systematic name
(S)-3-Hydroxybutanoyl-CoA:NADP+ oxidoreductase

Recommended name
3-Hydroxybutyryl-CoA dehydrogenase

Synonymes
beta-Hydroxybutyryl coenzyme A dehydrogenase
L(+)-3-Hydroxybutyryl-CoA dehydrogenase [8]
BHBD [11]
Dehydrogenase, L-3-hydroxybutyryl coenzyme A (nicotinamide adenine
dinucleotide phosphate)
L-(+)-3-Hydroxybutyryl-CoA dehydrogenase
beta-Hydroxybutyryl-CoA dehydrogenase

CAS Reg. No.
39319-78-3

2 REACTION AND SPECIFICITY

Catalysed reaction
3-Acetoacetyl-CoA + NADPH →
→ (S)-3-hydroxybutanoyl-CoA + NADP+

Reaction type
Redox reaction

Natural substrates
3-Acetoacetyl-CoA + NADH (metabolic pathway of butyrate formation [3, 4,
6], enzyme involved in fatty acyl-CoA beta-oxidizing system [4, 10]) [3–6, 8,
10]

Enzyme Handbook © Springer-Verlag Berlin Heidelberg 1995
Duplication, reproduction and storage in data banks are only
allowed with the prior permission of the publishers

Substrate spectrum

1 3-Acetoacetyl-CoA + NAD(P)H (r [8], coenzyme: NADPH [1, 2, 6], NADH [3–6], N-acetyl-S-acetoacetyl-cysteamine reduced with 15% the rate of acetoacetyl-CoA [8], oxidation of 3-hydroxybutyryl-CoA (at pH 9.5) proceeds with 7% the rate of acetoacetyl-CoA reduction (at pH 6.5) [8], Clostridium kluyveri: NADPH-dependent enzyme specific for 3-hydroxybutyryl-CoA, NAD-linked enzyme acts also on 3-hydroxycaproyl-CoA [2], not: 3-hydroxyvaleryl-CoA, 3-hydroxycaproyl-CoA [8]) [1–11]

Product spectrum

1 (S)-3-Hydroxybutanoyl-CoA + NADP+ [8]

Inhibitor(s)

Cofactor(s)/prosthetic group(s)/activating agents
NADPH (specific for [2, 8]) [1, 2, 6–8, 10]; NADH [3–7, 9]; NADP+ [8]

Metal compounds/salts
Selenium (incorporation of selenium into the enzyme occurs randomly and is not required for any specific function [1, 2], deposited throughout the methionine rich protein as selenomethionine [1]) [1, 2]

Turnover number (min⁻¹)

Specific activity (U/mg)
292 [1, 2]; 450 [8]

K$_m$-value (mM)
0.050 (acetoacetyl-CoA) [8]; 0.070 (NADPH) [8]

pH-optimum
6.5 (acetoacetyl-CoA reduction) [8]; 9.5 (3-hydroxybutyryl-CoA oxidation) [8]

pH-range
5–8 (5: about 35% of activity maximum, 8: about 30% of activity maximum, acetoacetyl-CoA reduction) [8]

Temperature optimum (°C)
25 (assay at) [2, 8]

Temperature range (°C)

---

3 ENZYME STRUCTURE

Molecular weight
215000–220000 (Clostridium kluyveri, sedimentation equilibrium, gel filtration) [8]

Subunits
Octamer (8 × 26000, Clostridium kluyveri, SDS-PAGE) [2]

Glycoprotein/Lipoprotein
–

---

4 ISOLATION/PREPARATION

Source organism
Clostridium pasteurianum [9]; Clostridium butylicum [9]; Clostridium butyri-
cum [9]; Clostridium tetanomorphum [9]; Clostridium roseum [9]; Clostridi-
um rubrum [9]; Clostridium kluyveri (NAD+ dependent activity and NADP+
dependent activity [7]) [1, 2, 7–9]; Rat [4, 10]; Clostridium acetobutylicum
[3, 11]; Dasytricha ruminantium [5]; Butyriovibrio fibrisolvens [6]

Source tissue
Liver [4, 10]

Localisation in source
Peroxisomes (core) [4]; Mitochondria (distribution in) [10]

Purification
Clostridium kluyveri (simultaneous single-step purification [2]) [1, 2, 8]

Crystallization
–

Cloned
[11]

Renaturated
–

---

5 STABILITY

pH

Temperature (°C)
45 (50% loss of activity after 50 min (NADP+-dependent activity), 80% loss
of activity after 20 min (NAD+-dependent activity)) [7]

Oxidation

Organic solvent

General stability information

Storage
-20°C [8]; 0°C, NADP+-dependent enzyme 6 days stable, NAD+-dependent
enzyme 2 days, 40 % loss of activity [7]

6 CROSSREFERENCES TO STRUCTURE DATABANKS

PIR/MIPS code
PIR2:A43723 (Clostridium acetobutylicum)

Brookhaven code

7 LITERATURE REFERENCES

[1] Hartmanis, M.G.N., Sliwkowski, M.X.: Curr. Top. Cell. Regul.,27,479–486 (1985)
[2] Sliwkowski, M.X., Hartmanis, M.G.N.: Anal. Biochem.,141,344–347 (1984)
[3] Hartmanis, M.G.N., Gatenbeck, S.: Appl. Environ. Microbiol.,47,1277–1283 (1984)
[4] Hayashi, H., Hino, S., Yamasaki, F.: Eur. J. Biochem.,120,47–51 (1981)
[5] Yarlett, N., Lloyd, D., Williams, A.G.: Biochem. J.,228,187–192 (1985)
[6] Miller, T.L., Jenesel, S.E.: J. Bacteriol.,138,99–104 (1979)
[7] Hillmer, P., Gottschalk, G.: Biochim. Biophys. Acta,334,12–23 (1974)
[8] Madan, V.K., Hillmer, P., Gottschalk, G.: Eur. J. Biochem.,32,51–56 (1973)
[9] von Hugo, H., Schoberth, S., Madan, V.K., Gottschalk, G.: Arch. Mikrobiol.,87,
    189–202 (1972)
[10] Haddock, B.A., Yates, D.W., Garland, P.B.: Biochem. J.,119,565–573 (1970)
[11] Youngleson, J.S., Jones, D.T., Woods, D.R.: J. Bacteriol.,171,6800–6807 (1989)

1 NOMENCLATURE

EC number
1.1.1.158

Systematic name
UDP-N-acetylmuramate:NADP+ oxidoreductase

Recommended name
UDP-N-acetylmuramate dehydrogenase

Synonymes
UDP-N-acetylenol-pyruvoylglucosamine reductase
Reductase, uridine diphosphoacetylpyruvoylglucosamine
Uridine diphospho-N-acetylglucosamine-enolpyruvate reductase
UDP-N-acetylenolpyruvoylglucosamine reductase
UDP-N-acetylglucosamine-enoylpyruvate reductase [4]
UDP-GlcNAc-enoylpyruvate reductase [3]
Uridine-5'-diphospho-N-acetyl-2-amino-2-deoxy-3-O-lactylglucose:NADP-oxi-doreductase [5, 6]

CAS Reg. No.
39307-28-3

2 REACTION AND SPECIFICITY

Catalysed reaction
UDP-N-acetyl-3-O-(1-carboxyvinyl)-D-glucosamine + NADPH →
→ UDP-N-acetylmuramate + NADP+

Reaction type
Redox reaction

Natural substrates
More (enzyme catalyzes one of the cytoplasmic steps of E. coli peptidogly-can synthesis) [3, 4]

Substrate spectrum
1 UDP-N-acetyl-3-O-(1-carboxyvinyl)-D-glucosamine + NADPH [5]

Product spectrum
1 UDP-N-acetyl-3-D-muramate + NADP+ [5]

Inhibitor(s)

Cs^+ [6]; Mg^{2+} [6]; Mn^{2+} [6]; p-Chloromercuribenzoate [5]; N-Ethylmaleimide [5]; Iodoacetamide [5]; Li^+ (slight, in presence of K^+) [6]; Na^+ (slight, in presence of K^+) [6]

Cofactor(s)/prosthetic group(s)/activating agents

NADH (reductase activity much higher in presence of NADPH than in presence of NADH (5%)) [5]; NADPH [5]; FAD (not a flavoprotein [1], greatly stimulated by FAD in presence of K^+ [2], prosthetic group [5], flavoprotein, sodium dithionite, sodium borohydride and to a lesser extent NADH can replace NADPH) [2, 5]; More (no effect: FMN) [2]

Metal compounds/salts

Tl^+ (enzyme greatly stimulated by monovalent cations, $Tl^+ > K^+ > NH_4^+ > Rb^+ >> Li^+, Na^+$) [6]; K^+ (enzyme greatly stimulated by monovalent cations, $Tl^+ > K^+ > NH_4^+ > Rb^+ >> Li^+, Na^+$, effect on pH optimum and K_m [6], stimulation by FAD in presence of K^+ [2]) [2, 6]; NH_4^+ (enzyme greatly stimulated by monovalent cations, $Tl^+ > K^+ > NH_4^+ > Rb^+ >> Li^+, Na^+$) [6]; Rb^+ (enzyme greatly stimulated by monovalent cations, $Tl^+ > K^+ > NH_4^+ > Rb^+ >> Li^+, Na^+$) [6]; Li^+ (enzyme greatly stimulated by monovalent cations, $Tl^+ > K^+ > NH_4^+ > Rb^+ >> Li^+, Na^+$) [6]; Na^+ (enzyme greatly stimulated by monovalent cations, $Tl^+ > K^+ > NH_4^+ > Rb^+ >> Li^+, Na^+$) [6]

Turnover number (min⁻¹)

Specific activity (U/mg)

199.1 [1]

K_m-value (mM)

0.04 (UDP-N-acetyl-3-O-(1-carboxyvinyl)-D-glucosamine) [5]; 0.049 (NADPH) [5]

pH-optimum

8.1–8.5 [5, 6]

pH-range

7.1–9.9 (at pH 7.1 and 9.9: about 60% of activity maximum) [5]

Temperature optimum (°C)

40 (increase of activity up to 40°C) [5]

Temperature range (°C)

40–50 (40°C: activity maximum, 50°C: about 30% of activity maximum) [5]

---

3 ENZYME STRUCTURE

Molecular weight

35000 (E. coli, gel filtration) [1]
35500 (Enterobacter cloacae NRC 492, gel filtration) [6]

Subunits
Dimer (1 × 21500 + 1 × 13500, E. coli, SDS-PAGE) [1]

Glycoprotein/Lipoprotein
–

---

4 ISOLATION/PREPARATION

Source organism
E. coli [1, 3, 4]; Staphylococcus epidermidis [2]; Enterobacter cloacae NRC 492 [5, 6]

Source tissue

Localisation in source

Purification
E. coli [1]; Staphylococcus epidermidis [2]; Enterobacter cloacae NRC 492 [5]

Crystallization
–

Cloned
–

Renaturated
–

---

5 STABILITY

pH

Temperature (°C)

Oxidation
Very sensitive to oxidation, rapid loss of activity in absence of added dithiothreitol [5]

Organic solvent

General stability information

Storage
–20°C [1]; –70°C, several months [2]; –20°C, pH 8.0, presence of dithiothreitol, up to 2 months without loss of activity [5]

6 CROSSREFERENCES TO STRUCTURE DATABANKS

PIR/MIPS code

Brookhaven code

7 LITERATURE REFERENCES

[1] Anwar, R.A., Vlaovic, M.: Can. J. Biochem.,57,188–196 (1979)
[2] Wickus, G.G., Rubenstein, P.A., Warth, A.D., Strominger, J.L.: J. Bacteriol.,113, 291–294 (1973)
[3] Mengin-Lecreulx, D., van Heijenoort, J.: J. Bacteriol.,163,208–212 (1985)
[4] Mengin-Lecreulx, D., Flouret, B., van Heijenoort, J.: J. Bacteriol.,154,1284–1290 (1983)
[5] Taku, A., Gunetileke, K.G., Anwar, R.A.: J. Biol. Chem.,245,5012–5016 (1970)
[6] Taku, A., Anwar, R.A.: J. Biol. Chem.,248,4971–4976 (1973)

1 NOMENCLATURE

EC number
1.1.1.159

Systematic name
7alpha-Hydroxysteroid:NAD+ 7-oxidoreductase

Recommended name
7alpha-Hydroxysteroid dehydrogenase

Synonymes
7alpha-Hydroxy steroid dehydrogenase
Dehydrogenase, 7alpha-hydroxy steroid
7alpha-HSDH [1]

CAS Reg. No.
39361-64-3

2 REACTION AND SPECIFICITY

Catalysed reaction
3alpha,7alpha,12alpha-Trihydroxy-5beta-cholanate + NAD+ →
→ 3alpha,12alpha-dihydroxy-7-oxo-5beta-cholanate + NADH

Reaction type
Redox reaction

Natural substrates
Bile acids + NAD+ (serve as energy sources in absence of glucose) [7]

Substrate spectrum
1 Cholic acid + NAD+ (i.e. 3alpha,7alpha,12alpha-trihydroxy-5beta-cholana-
  te, r [4]) [2, 4, 6, 7, 9]
2 Glycocholic acid + NAD+ (i.e.
  N-[3alpha,7alpha,12alpha-trihydroxy-24-oxocholan-24-yl]glycine) [2, 4, 6,
  7, 9]
3 Taurocholic acid + NAD+ (i.e.
  2[[3alpha,7alpha,12alpha-trihydroxy-24-oxo-5beta-cholan-24-yl]amino]et-
  hane sulfonic acid) [2, 4, 6, 8, 9]
4 Chenodeoxycholic acid + NAD+ (i.e. 3alpha,7alpha-dihydroxy-5beta-cho-
  lan-24-oic acid) [2, 4, 6–9]
5 Glycochenodeoxycholic acid + NAD+ (i.e.
  N-[3alpha-7alpha-dihydroxy-24-oxocholan-24-yl]glycine, no reaction [6])
  [2, 4, 7–9]

6 Taurochenodeoxycholic acid + NAD⁺ [2, 8, 9]
7 Dehydrocholic acid + NADH [4]
8 12-Ketochenodeoxycholic acid + NAD⁺ [4, 6]
9 7,12-Diketolithocholic acid + NADH [4]
10 More (wide variety of 5alpha- or 5beta-cholanoic acids, overview) [6]

Product spectrum
1 3alpha,12alpha-Dihydroxy-7-oxo-5beta-cholan-24-oic acid + NADH [4]
2 N-[3alpha,12alpha-Dihydroxy-7,24-dioxo-5beta-cholan-24-yl]glycine + NADH [7]
3 2[[3alpha,12alpha-Dihydroxy-7,24-dioxo-5beta-cholan-24-yl]amino]ethane sulfonic acid + NADH
4 3alpha-Hydroxy-7-oxo-5beta-cholan-24-oic acid + NADH
5 N-[3alpha-Hydroxy-7,24-dioxocholan-24-yl]glycine + NADH
6 2[[3alpha-Hydroxy-7,24-dioxo-5beta-cholan-24-yl]amino]ethane sulfonic acid + NADH
7 3,12-Diketolithocholic acid + NAD⁺ [4]
8 7,12-Diketochenodeoxycholic acid + NADH [4]
9 12-Ketochenodeoxycholic acid + NAD⁺ [4]
10 ?

Inhibitor(s)
$CoCl_2$ [2]; $HgCl_2$ [2]; $FeCl_3$ [2]; $MgCl_2$ [2]; $ZnCl_2$ [2]; $BaCl_2$ [2]; $CuCl_2$ [2]; NH_4Cl [2]; Sodium citrate [2]; Potassium oxalate [2]; EDTA [2, 7]; Sodium perchlorate [2]; Sodium periodate [2]; Sodium persulfate [2]; Ascorbic acid [2]; Sodium lauryl sulfate [2]; Triton X-100 [2]; Tween [2]; Mercaptoethanol [7]

Cofactor(s)/prosthetic group(s)/activating agents
NAD⁺ (one enzyme form specific for NAD⁺, other form specific for NADP⁺ [7]) [4, 7–9]; NADH (B-stereospecific) [3]; NADP⁺ (one enzyme form specific for NADP⁺, other form specific for NAD⁺ [7]) [7, 8]; 2-Mercaptoethanol (activation) [2]; Dithiothreitol (activation) [2]

Metal compounds/salts
NaCl (activation) [2]; KCl (activation) [2]; $CaCl_2$ (activation) [2]; $MnCl_2$ (activation) [2]

Turnover number (min⁻¹)

Specific activity (U/mg)
3750 [1]; 450 [2]

K_m-value (mM)
0.0065 (chenodeoxycholic acid) [5]; 0.06 (chenodeoxycholic acid) [9];
0.085 (glycochenodeoxycholic acid) [9]; 0.09 (cholic acid) [5]; 0.1
(3alpha,7alpha-dihydroxy-5beta-cholanoic acid, 3alpha,7alpha-dihydroxy-
5beta-cholanoyl glycine, 3alpha,7alpha-dihydroxy-5beta-cholanoyl taurine)
[8]; 0.22 (cholic acid) [4]; 0.24 (taurochenodeoxycholic acid) [9]; 0.25
(NADH (+ dehydrocholic acid)) [4]; 0.27 (NAD$^+$ (+ cholic acid)) [4];
0.32–0.34 (3alpha,7alpha,12alpha-trihydroxy-5beta-cholanoic acid,
3alpha,7alpha,12alpha-trihydroxy-5beta-cholanoyl glycine, 3alpha,7alpha,
12alpha-trihydroxycholanoyl taurine) [8]; 0.38 (12-ketolithocholic acid) [4];
0.4 (NAD$^+$ (+ 12-ketolithocholic acid)) [4]; 0.42 (NADP$^+$) [5]; 0.45 (NAD$^+$ (+
chenodeoxycholic acid)) [8]; 0.7 (dehydrocholic acid) [4]; 0.8 (cholic acid)
[9]; 0.99 (7,12-diketolithocholic acid) [4]; 1.0 (glycocholic acid, taurocholic
acid) [9]; 1.3 (NADH (+ 7,12-diketolithocholic acid)) [4]

pH-optimum
7.0–9.0 (NADP$^+$-dependent activity) [8]; 8.5–9.0 (oxidation of 7alpha-hy-
droxy group) [7]; 9.4–9.6 [1]; 9.5–10.0 (NAD$^+$-dependent activity) [8];
9.5–11.5 [5]; 10.5 (oxidation of cholic acid) [4]; More (value for reduction of
dehydrocholic acid below 5.2, not measurable due to insolubility of substrate) [4]

pH-range

Temperature optimum (°C)
25–30 [4]

Temperature range (°C)
25–45 [4]

3 ENZYME STRUCTURE

Molecular weight
127000 (Bacteroides fragilis, NADP$^+$-dependent form, gel filtration) [7]
104000 (Brevibacterium fuscum, gel filtration) [4]
80000 (Bacteroides fragilis, NAD$^+$-specific form, gel filtration) [7]
54000 (E. coli, gel filtration) [1]

Subunits
Tetramer (4 × 27000, Brevibacterium fuscum, SDS-PAGE) [4]

Glycoprotein/Lipoprotein
–

4 ISOLATION/PREPARATION

Source organism
E. coli [1–3, 6, 9]; Brevibacterium fuscum [4]; Clostridium absonum [5];
Bacteroides fragilis [7, 8]

Source tissue

Localisation in source
Soluble part of cell [1]

Purification
E. coli [1]; Brevibacterium fuscum [4]; Bacteroides fragilis (partial) [7]

Crystallization
–

Cloned
–

Renaturated
–

5 STABILITY

pH
6–11 [4]

Temperature (°C)
30 (stable below) [4]; 40 (inactivation at) [4]; 60 (10 min stable) [1]; 65
(inactivation of NADP$^+$-dependent form, NAD$^+$-dependent form stable) [7];
75 (30 min, presence of substrates, NAD$^+$-dependent form stable) [7];
80 (10 min, complete inactivation) [1]

Oxidation

Organic solvent

General stability information
NAD$^+$-dependent form stable to freezing/thawing, NADP$^+$-dependent form
unstable [8]; Stabilization by glutathione, EDTA [9]; Inactivation by freezing
[9]

Storage

6 CROSSREFERENCES TO STRUCTURE DATABANKS

PIR/MIPS code

PIR2:JT0951 (Escherichia coli); PIR2:A38527 (Escherichia coli (strain HB101)); PIR2:A42468 (Eubacterium sp. (strain VPI 12708)); PIR2:A36439 (Eubacterium sp. (strain VPI12708) (fragment))

Brookhaven code

7 LITERATURE REFERENCES

[1] Prabha, V., Gupta, M., Seiffge, D., Gupta, K.G.: Can. J. Microbiol.,36,131–135 (1990)
[2] Prabha, V., Gupta, M., Gupta, K.G.: Can. J. Microbiol.,35,1076–1080 (1989)
[3] Ottolina, G., Riva, S., Carrea, G., Danieli, B., Buckmann, A.F.: Biochim. Biophys. Acta,998,173–178 (1989)
[4] Kinoshita, S., Kadota, K., Inoue, T., Sawada, H., Taguchi, H.: J. Ferment. Technol.,66, 145–152 (1988)
[5] MacDonald, I.A., Roach, P.D.: Biochim. Biophys. Acta,665,262–269 (1981)
[6] Haslewood, E.S., Haslewood, G.A.D.: Biochem. J.,157,207–210 (1976)
[7] Hylemon, P.B., Sherrod, J.A.: J. Bacteriol.,122,418–424 (1975)
[8] MacDonald, I.A., Williams, C.N., Mahoney, D.E., Christie, W.M.: Biochim. Biophys. Acta,384,12–24 (1975)
[9] MacDonald, I.A., Williams, C.N., Mahony, D.E.: Biochim. Biophys. Acta,309,243–253 (1973)

1 NOMENCLATURE

EC number
1.1.1.160

Systematic name
(+,-)-5-[(tert-Butylamino)-2'-hydroxypropoxy]-1,2,3,4-tetrahydro-1-naphthol: NADP$^+$ oxidoreductase

Recommended name
Dihydrobunolol dehydrogenase

Synonymes
Dehydrogenase, dihydrobunolol
Bunolol reductase

CAS Reg. No.
62213-61-0

---

2 REACTION AND SPECIFICITY

Catalysed reaction
(+,-)-5-[(tert-Butylamino)-2'-hydroxypropoxy]-3,4-dihydro-1(2H)-naphthalenone + NADPH →
→ (+,-)-5-[(tert-butyl-amino)-2'-hydroxypropoxy]-1,2,3,4-tetrahydro-1-naphthol + NADP$^+$

Reaction type
Redox reaction

Natural substrates
(+,-)-5-[(tert-Butylamino)-2'-hydroxypropoxy]-3,4-dihydro-1(2H)-naphthalenone + NADPH (i.e. bunolol, metabolism of bunolol, a potent beta-adrenoceptor blocking agent in human liver) [1]

Substrate spectrum
1 (+,-)-5-[(tert-Butylamino)-2'-hydroxypropoxy]-3,4-dihydro-1(2H)-naphthalenone + NADPH (i.e. bunolol, r, also acts more slowly with NAD$^+$) [1]

Product spectrum
1 (+,-)-5-[(tert-Butylamino)-2'-hydroxypropoxy]-1,2,3,4-tetrahydro-1-naphthol + NADP$^+$ (i.e. dihydrobunolol) [1]

Inhibitor(s)

Cofactor(s)/prosthetic group(s)/activating agents
   NADPH [1]; NADP+ [1]; NAD+ (slow) [1]

Metal compounds/salts

Turnover number (min^{-1})

Specific activity (U/mg)

K_m-value (mM)

pH-optimum
   6.7 (assay at) [1]

pH-range

Temperature optimum (°C)
   37 (assay at) [1]

Temperature range (°C)

3 ENZYME STRUCTURE

Molecular weight

Subunits

Glycoprotein/Lipoprotein
   –

4 ISOLATION/PREPARATION

Source organism
   Human [1]

Source tissue
   Liver [1]

Localisation in source
   Cytoplasm [1]

Purification

Crystallization
   –

Cloned
   –

Renaturated
   –

5 STABILITY

pH

Temperature (°C)

Oxidation

Organic solvent

General stability information

Storage

6 CROSSREFERENCES TO STRUCTURE DATABANKS

PIR/MIPS code

Brookhaven code

7 LITERATURE REFERENCES

[1] Leinweber, F.-J., Greenough, R.C., Schwender, C.F., Kaplan, H.R., di Carlo, F.J.: Xenobiotica,2,191–202 (1972)

1 NOMENCLATURE

EC number
1.1.1.161

Systematic name
5beta-Cholestane-3alpha,7alpha,12alpha,26-tetraol:NAD⁺ 26-oxidoreductase

Recommended name
Cholestanetetraol 26-dehydrogenase

Synonymes
Dehydrogenase, cholestanetetrol 26-
5beta-Cholestane-3 alpha,7alpha,12alpha,26-tetrol dehydrogenase [2]
TEHC-NAD oxidoreductase [3]

CAS Reg. No.
62213-60-9

2 REACTION AND SPECIFICITY

Catalysed reaction
5beta-Cholestane-3alpha,7alpha,12alpha,26-tetraol + NAD⁺ →
→ 3alpha,7alpha,12alpha-trihydroxy-5beta-cholestan-26-al + NADH

Reaction type
Redox reaction

Natural substrates

Substrate spectrum
1 5beta-Cholestane-3alpha,7alpha,12alpha,26-tetraol + NAD⁺ (r [1]) [1–3]

Product spectrum
1 3alpha,7alpha,12alpha-Trihydroxy-5beta-cholestan-26-al + NADH [1–3]

Inhibitor(s)
1,10-Phenanthroline [2]; Isobutyramide [2]; More (product inhibition) [2]

Cofactor(s)/prosthetic group(s)/activating agents
NAD⁺ [1, 2]; NADH [1]

Metal compounds/salts

Enzyme Handbook © Springer-Verlag Berlin Heidelberg 1995
Duplication, reproduction and storage in data banks are only
allowed with the prior permission of the publishers

Turnover number (min⁻¹)

Specific activity (U/mg)
  0.672 [2]

K_m-value (mM)

pH-optimum
  10.6 [2]

pH-range
  8.3–11.2 (8.3: about 20% of activity maximum, 11.2: about 60% of activity
  maximum) [2]

Temperature optimum (°C)
  37 (assay at) [1]

Temperature range (°C)

---

3 ENZYME STRUCTURE

Molecular weight

Subunits

Glycoprotein/Lipoprotein
  –

---

4 ISOLATION/PREPARATION

Source organism
  Rat (both alcohol:NAD⁺ oxidoreductase (EC 1.1.1.1) activity and 5beta-cho-
  lestane-3alpha,7alpha,12alpha,26-tetraol:NAD⁺ 26-oxidoreductase activity
  (EC 1.1.1.161) are catalyzed by the same active side of the same enzyme
  protein [3]) [1, 3]; Human (both alcohol:NAD⁺ oxidoreductase (EC 1.1.1.1)
  activity and 5beta-cholestane-3alpha,7alpha,12alpha,26-tetraol:NAD⁺
  26-oxidoreductase activity (EC 1.1.1.161) are catalyzed by the same active
  side of the same enzyme protein) [2]

Source tissue
  Liver [1–3]

Localisation in source

Purification
  Rat (partial) [1]; Human [2]

Crystallization
  –

Cloned

–

Renaturated

–

5 STABILITY

pH

Temperature (°C)

Oxidation

Organic solvent

General stability information

Storage

6 CROSSREFERENCES TO STRUCTURE DATABANKS

PIR/MIPS code

Brookhaven code

7 LITERATURE REFERENCES

[1] Masui, T., Herman, R., Staple, E.: Biochim. Biophys. Acta,117,266–268 (1966)
[2] Okuda, A., Okuda, K.: J. Biol. Chem.,258,2899–2905 (1983)
[3] Okuda, K., Takigawa, N.: Biochem. Biophys. Res. Commun.,33,788–793 (1968)

1 NOMENCLATURE

EC number
1.1.1.162

Systematic name
Erythritol:NADP+ oxidoreductase

Recommended name
Erythrulose reductase

Synonymes
Reductase, D-erythrulose
D-Erythrulose reductase
D-Threitol:NADP+ oxidoreductase (in reference 2 the enzymatic reaction product of D-erythrulose is incorrectly identified as erythritol, gas-liquid and thin-layer chromatographic data have confirmed the product to be D-threitol, not erythritol, for this reason the systematic name D-threitol:NADP+ oxidoreductase should be used instead of erythritol:NADP+ oxidoreductase which has been designated by the IUB Enzyme commision) [1]

CAS Reg. No.
52064-49-0

2 REACTION AND SPECIFICITY

Catalysed reaction
D-Erythrulose + NAD(P)H →
→ D-threitol + NAD(P)+ (in reference 2 the enzymatic reaction product of D-erythrulose is incorrectly identified as erythritol, gas-liquid and thin-layer chromatographic data have confirmed the product to be D-threitol, not erythritol, for this reason the systematic name D-threitol:NADP+ oxidoreductase should be used instead of erythritol:NADP+ oxidoreductase which has been designated by the IUB Enzyme commission) [1]

Reaction type
Redox reaction

Natural substrates
D-Erythrulose + NAD(P)+ (one route of formation of tetritols in mammalia) [3]

Substrate spectrum
  1 D-Erythrulose + NAD(P)H (the equilibrium of the reaction strongly favors
    the reduction of D-erythrulose, the reverse reaction is slightly detectable
    when carried out at pH 7.5–9.0 using NADP+, erythritol does not serve as
    substrate [1, 3], reversibility could not be demonstrated [2], highly speci-
    fic for D-erythrulose [1–3]) [1–3]

Product spectrum
  1 D-Threitol + NAD(P)+ (in reference 2 the enzymatic reaction product of
    D-erythrulose is incorrectly identified as erythritol, gas-liquid and thin-layer
    chromatographic data have confirmed the product to be D-threitol, not
    erythritol, for this reason the systematic name D-threitol:NADP+ oxidore-
    ductase should be used instead of erythritol:NADP+ oxidoreductase
    which has been designated by the IUB Enzyme commission [1]) [1, 3]

Inhibitor(s)
  NADP+ [1, 5]; 2',5'-ADP [1, 5]; 2'-AMP [1, 5]

Cofactor(s)/prosthetic group(s)/activating agents
  NADPH [1, 2]; NADH (also utilized more slowly) [1, 2]; NADP+ (enzyme contains 2–3 mol of bound NADP+) [1, 3]

Metal compounds/salts

Turnover number (min⁻¹)

Specific activity (U/mg)
  71.3 [1]; 205 [3]; 123 [2]

K_m-value (mM)
  0.22 (NADH) [1, 2]; 0.0068 (NADPH) [1, 2]; 0.36 (D-erythrulose (+ NADH))
  [1]; 0.38 (D-erythrulose) [3]; 0.067 (NADH) [3]; 0.0079 (NADPH) [3]

pH-optimum
  5.8 (D-erythrulose + NADPH) [3]; 5.85 (D-erythrulose + NADH) [1, 2]; 6.25
  (D-erythrulose + NADH) [1, 2]; 6.3 (D-erythrulose + NADH) [3]

pH-range

Temperature optimum (°C)
  25 (assay at) [3]

Temperature range (°C)

3 ENZYME STRUCTURE

Molecular weight
  90000 (bovine, sedimentation equilibrium analysis, gel filtration, sucrose
  density gradient centrifugation) [1, 2]
  96000 (chicken, sedimentation equilibrium analysis) [3]

Subunits
  Tetramer (4 × 22000, bovine, SDS-PAGE [1, 2], 4 × 22400, chicken,
  SDS-PAGE [3]) [1–3]

Glycoprotein/Lipoprotein
  –

---

4 ISOLATION/PREPARATION

Source organism
  Bovine [1, 2, 4, 5]; Human [1]; Chicken [1, 3]

Source tissue
  Liver [1–5]; Pancreas [1]; Heart [1]; Kidney [1]

Localisation in source

Purification
  Bovine [1, 2]; Chicken [3]

Crystallization
  [1, 3, 4]

Cloned
  –

Renaturated
  –

---

5 STABILITY

pH
  8.5 (maximal stability) [4]

Temperature (°C)
  0 (23.5 h, 90% loss of activity, regains about 55–65% of its original activity
  after 60 min) [2]; 12 (inactivated below) [4]; 21 (23.5 h, 20–25% loss of activity) [2]; More (cold inactivation [1, 2, 4], cold inactivation accelerated by
  increasing salt concentration and decreasing enzyme concentration [1],
  chicken: quite stable, even when a dilute solution at high ionic strength is in-
  cubated at low temperatures [3]) [1–4]

Oxidation
  Photooxidation (in presence of rose-bengal [1, 5], protection by NADP+ [1,
  5], and to a lesser extent by 2',5'-ADP) [1, 5]

Organic solvent

General stability information
2,5'-ADP protects against rose bengal-sensitized photoinactivation, to a lesser extent than NADP+ [5]; NADP+ protects against heat inactivation or rose bengal-sensitized photoinactivation and cold inactivation [1, 2, 5]; NADP+ protects against pH inactivation [4]

Storage
4°C, as crystalline suspension in 1.43 mM ammonium sulfate, pH 8.2, containing 15 mM dithiothreitol and 0.2 mM NADP+, several months [1]

6 CROSSREFERENCES TO STRUCTURE DATABANKS

PIR/MIPS code

Brookhaven code

7 LITERATURE REFERENCES

[1] Uehara, K., Hosomi, S.: Methods Enzymol.,89,232–237 (1982) (Review)
[2] Uehara, K., Tanimoto, T., Sato, H.: J. Biochem.,75,333–345 (1974)
[3] Uehara, K., Mannen, S., Hosomi, S., Miyashita, T.: J. Biochem.,87,47–55 (1980)
[4] Uehara, K., Tanimoto, T.: J. Biochem.,78,519–526 (1975)
[5] Uehara, K., Mannen, S., Hosomi, S.: J. Biochem.,85,1003–1008 (1979)

1 NOMENCLATURE

EC number
1.1.1.163

Systematic name
Cyclopentanol:NAD+ oxidoreductase

Recommended name
Cyclopentanol dehydrogenase

Synonymes
Dehydrogenase, cyclopentanol

CAS Reg. No.
37364-12-8

2 REACTION AND SPECIFICITY

Catalysed reaction
Cyclopentanol + NAD+ →
→ cyclopentanone + NADH

Reaction type
Redox reaction

Natural substrates
Cyclopentanol + NAD+ (growth on cyclopentanol as sole carbon source) [1]

Substrate spectrum
1 Cyclopentanol + NAD+ [1]

Product spectrum
1 Cyclopentanone + NADH [1]

Inhibitor(s)

Cofactor(s)/prosthetic group(s)/activating agents
NAD+ [1]

Metal compounds/salts

Turnover number (min^{-1})

Specific activity (U/mg)

K$_m$-value (mM)

pH-optimum
   9.5 [1]

pH-range
   7.5–10.5 (7.5: about 20% of activity maximum, 10.5: about 65% of activity maximum) [1]

Temperature optimum (°C)

Temperature range (°C)

---

3 ENZYME STRUCTURE

Molecular weight

Subunits

Glycoprotein/Lipoprotein
   –

---

4 ISOLATION/PREPARATION

Source organism
   Pseudomonas sp. NCIB 9872 [1]

Source tissue
   Cells [1]

Localisation in source
   Soluble [1]

Purification

Crystallization

Cloned
   –

Renaturated
   –

5 STABILITY

pH

Temperature (°C)

Oxidation

Organic solvent

General stability information

Storage

6 CROSSREFERENCES TO STRUCTURE DATABANKS

PIR/MIPS code

Brookhaven code

7 LITERATURE REFERENCES

[1] Griffin, M., Trudgill, P.W.: Biochem. J.,129,595–603 (1972)

1 NOMENCLATURE

EC number
1.1.1.164

Systematic name
Hexadecanol:NAD⁺ oxidoreductase

Recommended name
Hexadecanol dehydrogenase

Synonymes
Dehydrogenase, hexadecanol

CAS Reg. No.
62213-59-6

2 REACTION AND SPECIFICITY

Catalysed reaction
Hexadecanal + NADH →
→ hexadecanol + NAD⁺

Reaction type
Redox reaction

Natural substrates

Substrate spectrum
1 Hexadecanal + NADH [1]
2 More (the liver enzyme acts on long-chain alcohols from C_8 to C_{16}, the Euglena enzyme also oxidizes the aldehydes to fatty acids) [1, 2]

Product spectrum
1 Hexadecanol + NAD⁺
2 ?

Inhibitor(s)

Cofactor(s)/prosthetic group(s)/activating agents
NADH [1]; NADPH (to a lesser extent) [1]

Metal compounds/salts

Turnover number (min^{-1})

Specific activity (U/mg)

K_m-value (mM)

pH-optimum

pH-range

Temperature optimum (°C)

Temperature range (°C)

---

3 ENZYME STRUCTURE

Molecular weight

Subunits

Glycoprotein/Lipoprotein

–

---

4 ISOLATION/PREPARATION

Source organism
  Euglena gracilis Z [1]; Rat [2]

Source tissue
  Cell [1]; Liver [2]

Localisation in source

Purification

Crystallization

–

Cloned

–

Renaturated

–

5 STABILITY

pH

Temperature (°C)

Oxidation

Organic solvent

General stability information

Storage

6 CROSSREFERENCES TO STRUCTURE DATABANKS

PIR/MIPS code

Brookhaven code

7 LITERATURE REFERENCES

[1] Kolattukudy, P.E.: Biochemistry,9,1095–1102 (1970)
[2] Stoffel, W., LeKim, D., Heyn, G.: Hoppe-Seyler's Z. Physiol. Chem.,351,875–883
     (1970)

1 NOMENCLATURE

EC number
1.1.1.165

Systematic name
2-Butyne-1,4-diol:NAD+ 1-oxidoreductase

Recommended name
2-Alkyn-1-ol dehydrogenase

Synonymes
Dehydrogenase, 2-alkyn-1-ol

CAS Reg. No.
54576-94-2

2 REACTION AND SPECIFICITY

Catalysed reaction
2-Butyne-1,4-diol + NAD+ →
→ 4-hydroxy-2-butynal + NADH

Reaction type
Redox reaction

Natural substrates

Substrate spectrum
1 2-Butyne-1,4-diol + NAD+ [1]
2 1,4-Butanediol + NAD+ (at 94% of the reaction rate with 2-butyne-1,4-diol) [1]
3 n-Propanol + NAD+ (at 18% of the reaction rate with 2-butyne-1,4-diol) [1]
4 1,2-Propanediol + NAD+ (at 29% of the reaction rate with 2-butyne-1,4-diol) [1]
5 1,2-Butanediol + NAD+ (at 12% of the reaction rate with 2-butyne-1,4-diol) [1]
6 1,2,4-Butanetriol + NAD+ (at 24% of the reaction rate with 2-butyne-1,4-diol) [1]
7 1,6-Hexanediol + NAD+ (at 12% of the reaction rate with 2-butyne-1,4-diol) [1]
8 2-Propene-1-ol + NAD+ (at 18% of the reaction rate with 2-butyne-1,4-diol) [1]
9 2-Butene-1-ol + NAD+ (at 8% of the reaction rate with 2-butyne-1,4-diol, mixture of cis- and trans-) [1]

10 trans-2-Butene-1,4-diol + NAD$^+$ (at 33% of the reaction rate with
    2-butyne-1,4-diol) [1]
11 1-Butene-3-ol + NAD$^+$ (at 12% of the reaction rate with 2-butyne-1,4-diol)
    [1]
12 2-Propyne-1-ol + NAD$^+$ (at 76% of the reaction rate with
    2-butyne-1,4-diol) [1]
13 3-Butyne-1-ol + NAD$^+$ (at 6% of the reaction rate with 2-butyne-1,4-diol)
    [1]
14 1-Butyne-1-ol + NAD$^+$ (at 6% of the reaction rate with 2-butyne-1,4-diol)
    [1]
15 2-Butyne-1-ol + NAD$^+$ (at 118% of the reaction rate with
    2-butyne-1,4-diol) [1]
16 2,4-Hexadiyne-1,6-diol + NAD$^+$ (at 58% of the reaction rate with
    2-butyne-1,4-diol) [1]

Product spectrum
    1 4-Hydroxy-2-butynal + NADH [1]
    2 ?
    3 ?
    4 ?
    5 ?
    6 ?
    7 ?
    8 ?
    9 ?
   10 ?
   11 ?
   12 ?
   13 ?
   14 ?
   15 ?
   16 ?

Inhibitor(s)
    Iodoacetic acid [1]; p-Chloromercuribenzoate [1]; KCN (slight) [1]; NaN$_3$
    (slight) [1]; CuCl$_2$ [1]; HgCl$_2$ [1]; ZnCl$_2$ [1]; NiCl$_2$ [1]

Cofactor(s)/prosthetic group(s)/activating agents
    NAD$^+$ [1]; NADP$^+$ (more slowly) [1]

Metal compounds/salts

Turnover number (min^{-1})

Specific activity (U/mg)
  More [1]

K_m-value (mM)
  9.1 (2-butyne-1,4-diol) [1]

pH-optimum
  8.2 (2-butyne-1,4-diol + NAD$^+$) [1]

pH-range

Temperature optimum (°C)

Temperature range (°C)

---

3 ENZYME STRUCTURE

Molecular weight

Subunits

Glycoprotein/Lipoprotein
  –

---

4 ISOLATION/PREPARATION

Source organism
  Fusarium merismoides (B11) [1]

Source tissue
  Mycelium [1]

Localisation in source

Purification
  Fusarium merismoides (B11) [1]

Crystallization
  –

Cloned
  –

Renaturated
  –

3

5 STABILITY

pH
  5.8–8.5 (30°C, 2.5 h, stable) [1]

Temperature (°C)
  40 (1 h, stable below) [1]; 60 (1 h, 80% loss of activity) [1]

Oxidation

Organic solvent

General stability information
  Purified enzyme is highly labile in presence of dithiothreitol [1]

Storage
  5°C, 2 days [1]

---

6 CROSSREFERENCES TO STRUCTURE DATABANKS

PIR/MIPS code

Brookhaven code

---

7 LITERATURE REFERENCES

[1] Miyoshi, T., Sato, H., Harada, T.: Biochim. Biophys. Acta,358,231–239 (1974)

1 NOMENCLATURE

EC number
1.1.1.166

Systematic name
(1S,3R,4S)-3,4-Dihydroxycyclohexane-1-carboxylate:NAD⁺ 3-oxidoreductase

Recommended name
Hydroxycyclohexanecarboxylate dehydrogenase

Synonymes
Dehydrogenase, dihydroxycyclohexanecarboxylate
Dihydroxycylohexanecarboxylate dehydrogenase
(-)-t-3,t-4-Dihydroxycyclohexane-c-1-carboxylate-NAD oxidoreductase [1]

CAS Reg. No.
55467-53-3

2 REACTION AND SPECIFICITY

Catalysed reaction
(1S,3R,4S)-3,4-Dihydroxycyclohexane-1-carboxylate + NAD⁺ →
→ (1S,4S)-4-hydroxy-3-oxocyclohexane-1-carboxylate + NADH [1])

Reaction type
Redox reaction

Natural substrates
(-)-Quinate + NAD⁺ [1]

Substrate spectrum
1 (-)-Quinate + NAD⁺ (r, (acts on hydroxycyclohexanecarboxylates having
  an equatorial carboxyl group at C-1, an axial hydroxyl group at C-3 and
  an equatorial hydroxyl or carbonyl group at C-4, including (-)-quinate and
  (-)-shikimate) [1]
2 (-)-Shikimate + NAD⁺ (r) [1]
3 (-)-Dihydroshikimate + NAD⁺ (r) [1]
4 (-)-t-3,t-4-Dihydrocyclohexane-c-1-carboxylate + NAD⁺ (r) [1]

Product spectrum
1 (-)-3-Dehydroquinate + NADH [1]
2 (-)-3-Dehydroshikimate + NADH [1]
3 (-)-t-4,c-5-Dihydroxy-3-oxocyclohexane-c-1-carboxylate + NADH [1]
4 4-Hydroxy-3-oxocyclohexane-c-1-carboxylate + NADH [1]

Inhibitor(s)

Cofactor(s)/prosthetic group(s)/activating agents
NAD$^+$ (cannot be replaced by NADP$^+$) [1]

Metal compounds/salts

Turnover number (min^{-1})

Specific activity (U/mg)
58.9 [1]

K$_m$-value (mM)
0.26–0.45 (NAD$^+$, depending on substrate) [1]; 0.52 ((-)-dihydroshikimate) [1]; 0.74 ((-)t-3,t-4-dihydrocyclohexane-c-1-carboxylate) [1]; 0.75 ((-)-shikimate) [1]; 0.85 ((-)-quinate) [1]

pH-optimum
6.8–7.6 (reduction) [1]; 10.0–10.1 (oxidation) [1]

pH-range
6.0–8.0 (reduction) [1]; 6.9–11.4 (oxidation) [1]

Temperature optimum (°C)
25 (assay at) [1]

Temperature range (°C)

---

3 ENZYME STRUCTURE

Molecular weight

Subunits

Glycoprotein/Lipoprotein
–

---

4 ISOLATION/PREPARATION

Source organism
Lactobacillus plantarum [1]

Source tissue
Cell [1]

Localisation in source

Purification
Lactobacillus plantarum [1]

Crystallization

–

Cloned

–

Renaturated

–

5 STABILITY

pH

Temperature (°C)

Oxidation

Organic solvent

General stability information

Storage

6 CROSSREFERENCES TO STRUCTURE DATABANKS

PIR/MIPS code

Brookhaven code

7 LITERATURE REFERENCES

[1] Whiting G.C., Coggins R.: Biochem. J.,141,35–42 (1974)

1 NOMENCLATURE

EC number
1.1.1.167

Systematic name
Hydroxymalonate:NAD+ oxidoreductase

Recommended name
Hydroxymalonate dehydrogenase

Synonymes
Dehydrogenase, hydroxymalonate

CAS Reg. No.
58693-60-0

2 REACTION AND SPECIFICITY

Catalysed reaction
Hydroxymalonate + NAD+ →
→ oxomalonate + NADH

Reaction type
Redox reaction

Natural substrates

Substrate spectrum
1 Hydroxymalonate + NAD+ [1]

Product spectrum
1 Oxomalonate + NADH

Inhibitor(s)

Cofactor(s)/prosthetic group(s)/activating agents
NAD+ [1]

Metal compounds/salts

Turnover number (min^{-1})

Specific activity (U/mg)

K$_m$-value (mM)

pH-optimum

pH-range

Temperature optimum (°C)

Temperature range (°C)

---

3 ENZYME STRUCTURE

Molecular weight

Subunits

Glycoprotein/Lipoprotein

--

---

4 ISOLATION/PREPARATION

Source organism

Bombyx mori (silk worm) [1]

Source tissue

Localisation in source

Purification

Crystallization

--

Cloned

--

Renaturated

--

5 STABILITY

pH

Temperature (°C)

Oxidation

Organic solvent

General stability information

Storage

6 CROSSREFERENCES TO STRUCTURE DATABANKS

PIR/MIPS code

Brookhaven code

7 LITERATURE REFERENCES

[1] Zhukova, N.I.: Biol. Nauki (Mosc.) ,18,113–116 (1975)

1 NOMENCLATURE

EC number
1.1.1.168

Systematic name
(R)-Pantoyl-lactone:NADP+ oxidoreductase (A-specific)

Recommended name
2-Dehydropantoyl-lactone reductase (A-specific)

Synonymes
Reductase, 2-oxopantoyl lactone
Ketopantoyl lactone reductase
2-Ketopantoyl lactone reductase
2-Dehydropantoyl-lactone reductase [2]
More (cf. EC 1.1.1.214)

CAS Reg. No.
37211-75-9

2 REACTION AND SPECIFICITY

Catalysed reaction
2-Dehydropantoyl lactone + NADPH →
→ (R)-pantoyl lactone + NADP+

Reaction type
Redox reaction

Natural substrates
2-Dehydropantoyl lactone + NADPH [4]

Substrate spectrum
1 2-Dehydropantoyl lactone + NADPH (reverse reaction not detected [2])
  [1–7]
2 Quinone (e.g. camphoquinone, naphthoquinone [1]) + NADPH [1, 2]
3 Polyketone (e.g. ketopantoyl lactone and its derivatives [1], isatin and its
  derivatives [1, 2], ninhydrin [1]) + NADPH [1, 2]
4 Conjugated polyketones (e.g. 5-bromoisatin) + NADPH [1]
5 More (carbonyl reductase specifically catalyzing the reduction of conjugated polyketones [1, 2], keto-omega-methylpantoyl lactone is the only
  analog of ketopantoyl lactone tested that is a substrate [4, 7], not: keto-
  pantonic acid [4, 6, 7], not: several other 2-keto acids [4, 7]) [1, 2, 4, 6, 7]

Product spectrum
1 (R)-Pantoyl lactone + NADP+ [1–7]
2 ?
3 ?
4 ?
5 More (highly purified enzyme from Saccharomyces cerevisiae forms
D-(-)-pantoyl lactone from ketopantoyl lactone, whole or broken yeast
forms a mixture of D-(-)- and L-(+)-pantoyl lactone) [3]

Inhibitor(s)
Quercetin [1, 2]; Polyketones (e.g. parabanic acid [1, 2], cyclohexenedi-
ol-1,2,3,4-tetraone [1, 2]) [1, 2]; SH-ketopantoyl lactone (competitive to isatin) [1, 2]; 3,4-Dihydroxy-3-cyclobutene-1,2-dione [2]; Cu^{2+} [2]; Zn^{2+} [2]; Hg^{2+}
[2]; Cd^{2+} [2]; Al^{3+} [2]; p-Chloromercuribenzoate [2]; Iodoacetate [2];
2-Keto-4-hydroxy-3-methylbutyric acid-gamma-lactone [4]; 2-Keto-4-hydroxy-
butyric acid-gamma-lactone [4]; 1,2-Cyclopentanedione [4]; 2-Ketopantonic
acid [4]; 2-Ketoisovaleric acid [4]

Cofactor(s)/prosthetic group(s)/activating agents
NADPH (specific for [4], A-specific with respect to NADPH [4, 7]) [1–7];
More (not NADH) [6]

Metal compounds/salts

Turnover number (min⁻¹)

Specific activity (U/mg)
251 (isatin) [2]; 173 (ketopantoyl lactone) [2]; 69.2 [4]

K_m-value (mM)
0.17 (ketopantoyl lactone) [1]; 0.983 (dihydro-4-methyl-4-propyl-2,3-furane-
dione) [1]; 0.420 (dihydro-4,4-diethyl-2,3-furanedione) [1]; 0.0093
(dihydro-5-isopropyl-4,4-dimethyl-2,3-furanedione) [1]; 0.0098
(dihydro-5-(3-pentyl)-4,4-dimethyl-2,3-furanedione) [1]; 0.013
(dihydro-5-(2-pentyl)-4,4-dimethyl-2,3-furanedione) [1]; 0.090
(dihydro-5-(2-pentyl)-4-methyl-4-propyl-2,3-furanedione) [1]; 0.043
(dihydro-5-(2-butyl)-4-methyl-4-ethyl-2,3-furanedione) [1]; 0.0066 (isatin) [1];
0.0057 (1-methylisatin) [1]; 0.0031 (5-bromoisatin) [1]; 0.0083 (5-methylisa-
tin) [1]; 0.520 (ninhydrin) [1]; 0.830 (alloxan) [1]; 0.093 (beta-naphthoquino-
ne) [1]; 0.064 (DL-camphoquinone) [1]; 0.014 (isatin) [2]; 0.333 (ketopantoyl
lactone) [2]; 0.085 (1-methylisatin) [2]; 0.016 (5-methylisatin) [2]; 0.014 (ketopantoyl lactone, enzyme form A) [5]; 0.062 (NADPH, enzyme form A) [5];
0.031 (ketopantoyl lactone, enzyme form B) [5]; 0.039 (NADPH, enzyme
form B) [5]

pH-optimum
5.1–5.6 [5]; 7.0 [6]

pH-range
5.2–8.0 (at pH 5.2 and 8.0 about 50% of activity maximum) [6]

Temperature optimum (°C)
30 (assay at) [2]; 25 (assay at) [4]

Temperature range (°C)

---

3 ENZYME STRUCTURE

Molecular weight
39000 (Candida parapsilosis, gel filtration) [2]
27000 (Saccharomyces cerevisiae, enzyme form A and B, gel filtration) [4, 5]

Subunits
Monomer (1 × 41600, Candida parapsilosis, SDS-PAGE) [2]

Glycoprotein/Lipoprotein
–

---

4 ISOLATION/PREPARATION

Source organism
Saccharomyces cerevisiae (2 enzyme forms: A and B with similar properties [5]) [1, 3–7]; Candida parapsilosis (IFO 0708) [2]

Source tissue

Localisation in source

Purification
Candida parapsilosis (IFO 0708) [2]; Saccharomyces cerevisiae [3–5]

Crystallization
[2]

Cloned
–

Renaturated
–

5 STABILITY

pH
6.0–10.0 [2]; 4.0 (30°C, 30 min, 33% loss of activity) [2]

Temperature (°C)
40 (pH 7.0, 10 min, 58% loss of activity) [2]; 60 (pH 7.0, 10 min, 77% loss of activity) [2]

Oxidation

Organic solvent

General stability information
Glycerol stabilizes [4]

Storage
–20°C, 1 year, 15% loss of activity [4]

---

6 CROSSREFERENCES TO STRUCTURE DATABANKS

PIR/MIPS code

Brookhaven code

---

7 LITERATURE REFERENCES

[1] Hata, H., Shimizu, S., Hattori, S., Yamada, H.: FEMS Microbiol. Lett.,58,87–90 (1989)
[2] Hata, H., Shimizu, S., Hattori, S., Yamada, H.: Biochim. Biophys. Acta,990,175–181 (1989)
[3] Wilken, D.R., Dyar, R.E.: Arch. Biochem. Biophys.,189,251–255 (1978)
[4] Wilken, D.R., King, H.L., Dyar, R.E.: Methods Enzymol.,62,209–215 (1979) (Review)
[5] King, H.L., Dyar, R.E., Wilken, D.R.: J. Biol. Chem.,249,4689–4695 (1974)
[6] King, H.L., Wilken, D.R.: J. Biol. Chem.,247,4096–4105 (1972)
[7] Wilken, D.R., King, H.L., Dyar, R.E.: J. Biol. Chem.,250,2311–2314 (1975)

1 NOMENCLATURE

EC number
1.1.1.169

Systematic name
(R)-Pantoate:NADP+ 2-oxidoreductase

Recommended name
2-Dehydropantoate 2-reductase

Synonymes
Reductase, 2-oxopantoate
2-Ketopantoate reductase
2-Ketopantoic acid reductase
Ketopantoate reductase
Ketopantoic acid reductase

CAS Reg. No.
37211-74-8

2 REACTION AND SPECIFICITY

Catalysed reaction
2-Dehydropantoate + NADPH →
→ (R)-pantoate + NADP+

Reaction type
Redox reaction

Natural substrates
2-Dehydropantoate + NADPH (enzyme is responsible for the synthesis of
D-(-)-pantoic acid necessary for the biosynthesis of pantotheic acid in Pseudomonas maltophila 845) [2]

Substrate spectrum
1   2-Dehydropantoate + NADPH (r, reaction equilibrium greatly favors the di-
    rection of D-(-)-pantoic acid formation [2]) [1, 2]
2   2-Keto-3-hydroxyisovalerate + NADPH [2]
3   More (high specificity, among a variety of carbonyl compounds only keto-
    pantoic acid and 2-keto-3-hydroxyisovalerate can serve as substrate) [2]

Product spectrum
1   (R)-Pantoate + NADP+ [1, 2]
2   ?
3   ?

Inhibitor(s)
Diphenylhydantoin (slight) [2]; Barbital (slight) [2]; Quercetin (slight) [2];
Iodoacetate (slight) [2]: p-Chloromercuribenzoate (slight) [2]; $NaAsO_2$ [2];
Phenylhydrazine [2]; Semicarbazide [2]

Cofactor(s)/prosthetic group(s)/activating agents
NADPH (B-specific with respect to NADPH [4]) [1–4]; NADP+ [2]; More (no
activity with NADH) [2]

Metal compounds/salts
More (Mg^{2+} has no effect) [2]

Turnover number (min^{-1})

Specific activity (U/mg)
0.669 [2]

K_m-value (mM)
0.40 (ketopantoic acid) [2]; 8.55 (2-keto-3-hydroxyisovalerate) [2]; 0.0318
(NADPH) [2]

pH-optimum
5.0 (reduction of ketopantoic acid) [3]; 6.0 (reduction of ketopantoic acid)
[2]; 8.5 (oxidation of D-(-)pantoic acid) [2]

pH-range
4.3–6.5 (4.3: about 10% of activity maximum, 6.5: about 40% of activity maximum) [3]

Temperature optimum (°C)
35–40 (non-purified enzyme) [1]; 37 [2]

Temperature range (°C)
15–60 (15°C: about 70% of activity maximum, 60°C: about 85% of activity
maximum) [2]

3 ENZYME STRUCTURE

Molecular weight
140000–160000 (Pseudomonas maltophila, HPLC gel filtration) [2]
115000 (Pseudomonas maltophila, gel filtration) [2]
87000 (Pseudomonas maltophila, sedimentation equilibrium) [2]

Subunits
Oligomer (x × 30500, Pseudomonas maltophila, SDS-PAGE) [2]

Glycoprotein/Lipoprotein
–

4 ISOLATION/PREPARATION

Source organism
Saccharomyces cerevisiae [3, 4]; E. coli [4]; Agrobacterium sp. S-246 [1];
Agrobacterium tumefaciens [1]; Agrobacterium radiobacter [1]; Pseudomo-
nas maltophila [1, 2]

Source tissue

Localisation in source

Purification
Pseudomonas maltophila 845 [2]

Crystallization
[2]

Cloned
–

Renaturated
–

5 STABILITY

pH
6–10 (30°C, 30 min stable) [2]

Temperature (°C)
30 (30 min, pH 6–10, stable) [2]; 60 (10 min stable below 60°C) [2]; 70
(10 min, 70% loss of activity) [2]

Oxidation

Organic solvent

General stability information

Storage
4°C, pH 7.4, 0.1 mM dithiothreitol, 0.2 M NaCl, 6 months [2]

6 CROSSREFERENCES TO STRUCTURE DATABANKS

PIR/MIPS code

Brookhaven code

7 LITERATURE REFERENCES

[1] Kataoka, M., Shimizu, S., Yamada, H.: Agric. Biol. Chem.,54,177–182 (1990)
[2] Shimizu, S., Kataoka, M., Ching-Ming Chung, M., Yamada, H.: J. Biol. Chem.,263,12077–12084 (1988)
[3] King, H.L., Wilken, D.R.: J. Biol. Chem.,247,4096–4105 (1972)
[4] Wilken, D.R., King, H.L., Dyar, R.E.: J. Biol. Chem.,250,2311–2314 (1975)

1 NOMENCLATURE

EC number
1.1.1.170

Systematic name
3beta-Hydroxy-4beta-methyl-5alpha-cholest-7-ene-4alpha-carboxylate:NAD+ 3-oxidoreductase (decarboxylating)

Recommended name
3beta-Hydroxy-4beta-methylcholestenecarboxylate 3-dehydrogenase (decarboxylating)

Synonymes
Dehydrogenase, 3beta-hydroxy-4beta-methylcholestenoate
3beta-Hydroxy-4beta-methylcholestenoate dehydrogenase
Sterol 4alpha-carboxylic decarboxylase [1]

CAS Reg. No.
71822-23-6

2 REACTION AND SPECIFICITY

Catalysed reaction
3beta-Hydroxy-4beta-methyl-5alpha-cholest-7-ene-4alpha-carboxylate + NAD+ →
→ 4alpha-methyl-5alpha-cholest-7-ene-3-one + CO_2 + NADH (stereochemistry at C-4 is reversed during decarboxylation)

Reaction type
Redox reaction
Decarboxylation

Natural substrates

Substrate spectrum
1 3beta-Hydroxy-4beta-methyl-5alpha-cholest-7-ene-4alpha-oic acid + NAD+ (ir) [1]
2 3beta-Hydroxy-5alpha-cholest-7-ene-4alpha-carboxylate + NAD+ [1]

Product spectrum
1 4alpha-Methyl-5alpha-cholest-7-ene-3-one + CO_2 + NADH
2 5alpha-Cholest-7-ene-3-one-4alpha-carboxylate + NADH

Inhibitor(s)
  Zn^{2+} [1]; Fe^{2+} (1 mM, less than 25% inhibition) [1]

Cofactor(s)/prosthetic group(s)/activating agents
  NAD^+ [1]; $NADP^+$ (at 5% the rate of the reaction with NAD^+) [1]

Metal compounds/salts

Turnover number (min^{-1})

Specific activity (U/mg)
  More [1]

K_m-value (mM)
  More [1]

pH-optimum
  9.0 [1]

pH-range
  7.0–9.5

Temperature optimum (°C)
  37 (assay at) [1]

Temperature range (°C)

---

3 ENZYME STRUCTURE

Molecular weight

Subunits

Glycoprotein/Lipoprotein
  Lipoprotein (removal of bound phospholipid results in no loss of activity) [1]

---

4 ISOLATION/PREPARATION

Source organism
  Rat [1]

Source tissue
  Liver [1]

Localisation in source
  Microsomes [1]

Purification
  Rat (partial) [1]

Crystallization

–

Cloned

–

Renaturated

–

---

5 STABILITY

pH

Temperature (°C)

Oxidation

Organic solvent

General stability information

Storage

---

6 CROSSREFERENCES TO STRUCTURE DATABANKS

PIR/MIPS code

Brookhaven code

---

7 LITERATURE REFERENCES

[1] Rahimtula, A.D., Gaylor, J.L.: J. Biol. Chem.,247,9–15 (1972)

1 NOMENCLATURE

EC number
1.1.1.172

Systematic name
2-Hydroxyadipate:NAD+ 2-oxidoreductase

Recommended name
2-Oxoadipate reductase

Synonymes
Reductase, 2-ketoadipate
2-Ketoadipate reductase
alpha-Ketoadipate reductase

CAS Reg. No.
61116-21-0

2 REACTION AND SPECIFICITY

Catalysed reaction
2-Oxoadipate + NADH →
→ 2-hydroxyadipate + NAD+

Reaction type
Redox reaction

Natural substrates
2-Oxoadipate + NADH [1, 2]

Substrate spectrum
1 2-Oxoadipate + NADH (highly specific for 2-oxoadipate [1]) [1, 2]

Product spectrum
1 2-Hydroxyadipate + NAD+ [1, 2]

Inhibitor(s)
NADH (substrate inhibition at high concentrations) [1]; SDS (slight) [1];
p-Chloromercuribenzoate (slight) [1]

Cofactor(s)/prosthetic group(s)/activating agents
NADH [1]; More (not NADPH) [1]

Metal compounds/salts

Turnover number (min^{-1})

Specific activity (U/mg)
  More [1]

K$_m$-value (mM)
  0.011 (NADH) [1]; 2.6 (alpha-ketoadipate) [1]

pH-optimum
  6.3 [1]

pH-range

Temperature optimum (°C)
  37 (assay at) [1, 2]

Temperature range (°C)

---

3 ENZYME STRUCTURE

Molecular weight
  95000 (human, gel filtration) [1]

Subunits

Glycoprotein/Lipoprotein
  –

---

4 ISOLATION/PREPARATION

Source organism
  Human [1]; Rat [2]

Source tissue
  Placenta [1]; Leg muscle [2]; Liver [2]; Heart muscle [2]; Brain [2]; Small intestine [2]

Localisation in source
  More (principally localized in the supernatant fraction) [2]

Purification
  Human [1]

Crystallization
  –

Cloned
  –

Renaturated
  –

5 STABILITY

pH

Temperature (°C)

Oxidation

Organic solvent

General stability information
Glycerol stabilizes [1]

Storage
4°C, 20% glycerol, several months [1]

---

6 CROSSREFERENCES TO STRUCTURE DATABANKS

PIR/MIPS code

Brookhaven code

---

7 LITERATURE REFERENCES

[1] Suda, T., Robinson, J.C., Fjellstedt, T.A.: Arch. Biochem. Biophys.,176,610–620
    (1976)
[2] Suda, T., Robinson, J.C., Fjellstedt, T.A.: Biochem. Biophys. Res. Commun.,77,
    586–591 (1977)

Enzyme Handbook © Springer-Verlag Berlin Heidelberg 1995
Duplication, reproduction and storage in data banks are only
allowed with the prior permission of the publishers

1 NOMENCLATURE

EC number
1.1.1.173

Systematic name
L-Rhamnofuranose:NAD+ 1-oxidoreductase

Recommended name
L-Rhamnose 1-dehydrogenase

Synonymes

CAS Reg. No.
52227-67-5

2 REACTION AND SPECIFICITY

Catalysed reaction
L-Rhamnofuranose + NAD+ →
→ L-rhamno-1,4-lactone + NADH

Reaction type
Redox reaction

Natural substrates
L-Rhamnose + NAD+ [1, 2]

Substrate spectrum
1 L-Rhamnose + NAD+ (r) [1, 2]

Product spectrum
1 L-Rhamno-1,4-lactone + NADH [1, 2]

Inhibitor(s)
p-Substituted mercuribenzoate [2]; Iodoacetate [2]; Co^{2+} [2]; Zn^{2+} [2]; Cu^{2+} [2]; EDTA [2]

Cofactor(s)/prosthetic group(s)/activating agents
NAD+ [1, 2]; NADH [1, 2]

Metal compounds/salts

Turnover number (min^{-1})

Specific activity (U/mg)
2.0 [1]

K_m-value (mM)
  0.2 (L-rhamnose) [1, 2]; 0.02 (NAD⁺) [1, 2]

pH-optimum
  9.0 [1, 2]

pH-range

Temperature optimum (°C)

Temperature range (°C)

---

3 ENZYME STRUCTURE

Molecular weight

Subunits

Glycoprotein/Lipoprotein
  –

---

4 ISOLATION/PREPARATION

Source organism
  Pullularia pullulans [1, 2]

Source tissue

Localisation in source

Purification
  Pullularia pullulans [2]

Crystallization
  –

Cloned
  –

Renaturated
  –

5 STABILITY

pH

Temperature (°C)

Oxidation

Organic solvent

General stability information

Storage
   4°C, 1 week [2]

6 CROSSREFERENCES TO STRUCTURE DATABANKS

PIR/MIPS code

Brookhaven code

7 LITERATURE REFERENCES

[1] Pittner, F., Turecek, P.L.: Appl. Biochem. Biotechnol.,16,15–24 (1987)
[2] Rigo, L.U., Nakano, M., Veiga, L.A., Feingold, D.S.: Biochim. Biophys. Acta,445,286–293 (1976)

Enzyme Handbook © Springer-Verlag Berlin Heidelberg 1995
Duplication, reproduction and storage in data banks are only
allowed with the prior permission of the publishers

1 NOMENCLATURE

EC number
1.1.1.174

Systematic name
trans-Cyclohexane-1,2-diol:NAD+ 1-oxidoreductase

Recommended name
Cyclohexane-1,2-diol dehydrogenase

Synonymes

CAS Reg. No.
62628-27-7

2 REACTION AND SPECIFICITY

Catalysed reaction
trans-Cyclohexane-1,2-diol + NAD+ →
→ 2-hydroxycyclohexan-1-one + NADH

Reaction type
Redox reaction

Natural substrates
trans-Cyclohexane-1,2-diol + NAD+ [1]

Substrate spectrum
1 trans-Cyclohexane-1,2-diol + NAD+ (r) [1]

Product spectrum
1 2-Hydroxycyclohexan-1-one + NADH [1]

Inhibitor(s)

Cofactor(s)/prosthetic group(s)/activating agents
NAD+ [1]; NADH [1]

Metal compounds/salts

Enzyme Handbook © Springer-Verlag Berlin Heidelberg 1995
Duplication, reproduction and storage in data banks are only
allowed with the prior permission of the publishers

Turnover number (min^{-1})

Specific activity (U/mg)

K_m-value (mM)

pH-optimum

pH-range

Temperature optimum (°C)

Temperature range (°C)

---

3 ENZYME STRUCTURE

Molecular weight

Subunits

Glycoprotein/Lipoprotein

–

---

4 ISOLATION/PREPARATION

Source organism
  Acinetobacter sp. [1]

Source tissue

Localisation in source

Purification
  Acinetobacter sp. (partial) [1]

Crystallization

–

Cloned

–

Renaturated

–

5 STABILITY

pH

Temperature (°C)

Oxidation

Organic solvent

General stability information

Storage

6 CROSSREFERENCES TO STRUCTURE DATABANKS

PIR/MIPS code

Brookhaven code

7 LITERATURE REFERENCES

[1] Davey, J.F., Trudgill, P.W.: Eur. J. Biochem.,74,115–127 (1977)

1 NOMENCLATURE

EC number
1.1.1.175

Systematic name
D-Xylose:NAD$^+$ 1-oxidoreductase

Recommended name
D-Xylose 1-dehydrogenase

Synonymes
Dehydrogenase, D-xylose
NAD-D-xylose dehydrogenase
D-Xylose dehydrogenase
(NAD)-linked D-xylose dehydrogenase [2]

CAS Reg. No.
62931-20-8

2 REACTION AND SPECIFICITY

Catalysed reaction
D-Xylose + NAD$^+$ →
→ D-xylonolactone + NADH

Reaction type
Redox reaction

Natural substrates
D-Xylose + NAD$^+$ (metabolism of D-xylose in Pseudomonades) [3]

Substrate spectrum
1 D-Xylose + NAD$^+$ [1–3]
2 L-Arabinose + NAD$^+$ (not [2, 3]) [1]
3 D-Glucose + NAD$^+$ (not [2, 3]) [1]
4 D-Galactose + NAD$^+$ (not [2, 3]) [1]

Product spectrum
1 D-Xylonolactone + NADH (product is an unstable delta-lactone that spontaneously hydrolyzes to D-xylonate) [3]
2 ?
3 ?
4 ?

Inhibitor(s)

Cofactor(s)/prosthetic group(s)/activating agents
NAD$^+$ (specific for NAD$^+$) [1–3]

Metal compounds/salts

Turnover number (min^{-1})

Specific activity (U/mg)
More [1]; 24.12 [2]

K$_m$-value (mM)
30 (D-galactose, Pseudomonas sp., strain 89B) [1]; 2.9 (D-xylose, Pseudomonas sp., strain 89B) [1]; 0.77 (D-xylose, D-xylose-dissimilating bacterium,
strain 90A) [1]; 5.9 (L-arabinose, Pseudomonas sp., strain 89B) [1]; 100
(L-arabinose, D-xylose, D-xylose-dissimilating bacterium, strain 90A) [1];
0.35 (NAD$^+$, Pseudomonas sp., strain 89B) [1]; 0.39 (NAD$^+$, D-xylose, D-xylose-dissimilating bacterium, strain 90A) [1]; 17.4 (D-xylose) [2]; 0.27 (NAD$^+$)
[2]; 0.5 (D-xylose) [2]; 0.2 (NAD$^+$) [2]

pH-optimum
8.0 [3]; 9.0 (Pseudomonas sp., strain 89B) [1]; 10.0 (D-xylose dissimilating
bacterium, strain 90A) [1]; 10.4 [2]

pH-range
9–10.8 (9: about 15% of activity maximum, 10.8: about 75% of activity maximum) [1]

Temperature optimum (°C)
40–55 [1]; 30–35 [2]

Temperature range (°C)
10–45 (10°C: about 25% of activity maximum, 45°C: about 40% of activity
maximum) [2]

3 ENZYME STRUCTURE

Molecular weight
62000 (Arthrobacter sp., gel filtration) [2]
67000 (Pseudomonas sp. (strain 89B) and D-xylose-dissimilating bacterium
(strain 90A), gel filtration) [1]

Subunits

Glycoprotein/Lipoprotein
–

4 ISOLATION/PREPARATION

Source organism
Pseudomonas sp. (MSU-1 (ATCC 27855) [3], strain 89B [1]) [1, 3];
D-Xylose-dissimilating bacterium (strain 90A) [1]; Arthrobacter sp. [2]

Source tissue
Cell [1–3]

Localisation in source

Purification
Pseudomonas sp. (MSU-1 (ATCC 27855) [3], strain 89B [1]) [1, 3];
D-Xylose-dissimilating bacterium (strain 90A) [1]

Crystallization
–

Cloned
–

Renaturated
–

---

5 STABILITY

pH
6 (30°C, 30 min, 78% loss of activity) [2]; 6.5 (30°C, 30 min, about 20% loss of activity) [2]; 7 (30°C, 30 min, no loss of activity) [2]; 7.5 (30°C, 30 min, about 10% loss of activity) [2]; 8 (30°C, 30 min, about 35% loss of activity) [2]; 8.5 (30°C, 30 min, about 85% loss of activity) [2]; 9.5 (30°C, 30 min, 98% loss of activity) [2]

Temperature (°C)
25 (half-life: 25 min) [3]; 30 (15 min, pH 7.0, stable below) [2]; 50 (15 min, pH 7.0, complete loss of activity) [2]; 32 (half-life: 2 min) [3]

Oxidation

Organic solvent

General stability information
Stabilized by substrates individually or by high ionic strength [3]

Storage
–20°C [3]; Frozen state, stable for several weeks [3]

6 CROSSREFERENCES TO STRUCTURE DATABANKS

PIR/MIPS code

Brookhaven code

---

7 LITERATURE REFERENCES

[1] Yamanaka, K., Gino, M.: Hakko Kogaku Kaishi,57,322–331 (1979)
[2] Yamanaka, K., Gino, M., Kaneda, R.: Agric. Biol. Chem.,41,1493–1499 (1977)
[3] Dahms, A.S., Russo, J.: Methods Enzymol.,89,226–229 (1982) (Review)

1 NOMENCLATURE

EC number
1.1.1.176

Systematic name
12alpha-Hydroxysteroid:NADP$^+$ 12-oxidoreductase

Recommended name
12alpha-Hydroxysteroid dehydrogenase

Synonymes
Dehydrogenase, 12alpha-hydroxy steroid
12alpha-Hydroxy steroid dehydrogenase
NAD-dependent 12alpha-hydroxysteroid dehydrogenase
NADP-12alpha-hydroxysteroid dehydrogenase

CAS Reg. No.
61642-40-8

2 REACTION AND SPECIFICITY

Catalysed reaction
3alpha,7alpha,12alpha-Trihydroxy-5beta-cholanate + NADP$^+$ →
→ 3alpha,7alpha-dihydroxy-12-oxo-5beta-cholanate + NADPH

Reaction type
Redox reaction

Natural substrates

Substrate spectrum
1 Cholic acid + NAD(P)$^+$ (i.e. 3alpha,7alpha,12alpha-trihydroxy-5beta-cho-
  lan-24-oic acid, r [1]) [1, 2, 4, 5, 8]
2 Deoxycholic acid + NAD(P)$^+$ (i.e.
  3alpha,12alpha-dihydroxy-5beta-cholan-24-oic acid) [1, 2, 4, 5, 8]
3 Glycocholic acid + NAD$^+$ (i.e.
  3alpha,7alpha,12alpha-trihydroxy-5beta-cholanoyl glycine) [2, 4, 5, 8]
4 Taurocholic acid + NAD(P)$^+$ (i.e.
  3alpha,7alpha,12alpha-trihydroxy-5beta-cholanoyl taurine) [2, 4, 8]
5 Glycodeoxycholic acid + NAD$^+$ (i.e. 3alpha,12alpha-dihydroxy-5beta-cho-
  lanoyl glycine) [2, 4–6]
6 Taurodeoxycholic acid + NAD$^+$ [2, 5]
7 Allocholate + NAD$^+$ [4]
8 Allodeoxycholate + NAD$^+$ [4]
9 7alpha,12alpha-Dihydroxy-5beta-cholanoate + NAD$^+$ [7]

Product spectrum

1 3alpha,7alpha-Dihydroxy-12-oxo-5beta-cholan-24-oic acid + NAD(P)H [2]
2 3alpha-Hydroxy-12-oxo-5beta-cholan-24-oic acid + NAD(P)H
3 3alpha,7alpha-Dihydroxy-12-oxo-5beta-cholanoyl glycine + NADH
4 3alpha,7alpha-Dihydroxy-12-oxo-5beta-cholanoyl taurine + NAD(P)H
5 3alpha-Hydroxy-12-oxo-5beta-cholanoyl glycine + NADH
6 3alpha-Hydroxy-12-oxo-5beta-cholanoyl taurine + NADH
7 ?
8 ?
9 7alpha-Hydroxy-12-oxo-5beta-cholanoate + NADH

Inhibitor(s)

Cu^{2+} [1]; Ag^+ [1]; Hg^{2+} [1]; p-Chloromercuribenzoate [1]; Pyridoxal 5'-phosphate (protection by $NADP^+$ or NADPH) [1]

Cofactor(s)/prosthetic group(s)/activating agents

$NADP^+$ (low activity [7]) [1, 5–7, 9]; NADPH (B-stereospecific [9]) [1, 9]; NAD^+ (10% of $NADP^+$-activity [6]) [1, 4, 6–8]

Metal compounds/salts

Turnover number (min^{-1})

Specific activity (U/mg)

K_m-value (mM)

0.0012 (12-oxochenodeoxycholic acid) [1]; 0.0085 (NADPH) [1]; 0.024 (NAD$^+$) [7]; 0.028 (3alpha,12alpha-dihydroxy-5beta-cholanoate) [4]; 0.03 (3alpha,7alpha,12alpha-trihydroxy-5beta-cholanoyl methyl ester, 3alpha,12alpha-dihydroxy-5beta-cholanoate) [4]; 0.035 (NADP$^+$) [1]; 0.045 (deoxycholic acid) [1]; 0.059 (3alpha,7alpha,12alpha-trihydroxy-5beta-cholanoate) [4]; 0.072 (cholic acid) [1]; 0.1 (7alpha,12alpha-dihydroxy-5beta-cholanoate) [7]; 0.17 (3alpha,12alpha-dihydroxy-5beta-cholanoyl glycine) [4]; 0.25 (3alpha,7alpha,12alpha-trihydroxy-5beta-cholanoyl glycine) [4]; 0.44 (3alpha,12alpha-dihydroxy-5beta-cholanoyl taurine) [5]; 0.57 (3alpha,12alpha-dihydroxy-5beta-cholanoyl glycine) [5]; 0.8 (3alpha, 12alpha-dihydroxy-5beta-cholanoate) [8]; 1.6 (3alpha,12alpha-dihydroxy-5beta-cholanoate) [5]; 2.7 (3alpha,7alpha,12alpha-trihydroxy-5beta-cholanoyl glycine) [5]; 4.0 (3alpha,7alpha,12alpha-trihydroxy-5beta-cholanoyl taurine) [5]; 7.0 (3alpha,7alpha,12alpha-trihydroxy-5beta-cholanoate) [5]

pH-optimum

8–10.5 (oxidation of cholic acid) [4]; 8.5–9.0 (oxidation of cholic acid) [5]; 8.5–9.5 (oxidation of cholic acid) [1]; 8.5–10.5 (oxidation of bile acids) [7]; 10.5 (oxidation of bile acids) [8]; More (reduction of 12-oxochenodeoxycholic acid optimal below 5.0, not measurable due to insolubility of substrate) [1]

pH-range
   6–11 [4]

Temperature optimum (°C)
   More (optimum above 55°C, not measurable because enzyme is inactivated
   above 55°C) [1]

Temperature range (°C)

---

3 ENZYME STRUCTURE

Molecular weight
   102000–108000 (Clostridium sp. group P, native PAGE, analytical ultracentrifugation, FPLC gel filtration) [1]

Subunits
   Tetramer (4 × 26000, Clostridium sp. group P, SDS-PAGE) [1]

Glycoprotein/Lipoprotein
   –

---

4 ISOLATION/PREPARATION

Source organism
   Clostridium sp. group P (strain 48–50 [1]) [1, 3, 6, 9]; Brevibacterium fuscum [2]; Eubacterium lentum [4, 7]; Clostridium leptum [5]; Clostridium per-
   fringens [8]

Source tissue

Localisation in source

Purification
   Clostridium sp. group P [1, 3]; Brevibacterium fuscum [2]; Clostridium leptum [5]

Crystallization
   –

Cloned
   –

Renaturated
   –

5 STABILITY

pH

Temperature (°C)
-20 (half-life: 20 days) [5]; 0–4 (half-life: 11 days) [5]; 25 (half-life: 4 days, increases to 10 days by addition of NADP⁺) [1]; 30 (half-life: 2 days, increases to 5 days by addition of NADP⁺) [1]; 50 (30 min, inactivation) [7]; 43 (slow inactivation) [8]; 56 (rapid inactivation) [8]

Oxidation

Organic solvent

General stability information
Glycerol, 50% v/v, stabilization [1]

Storage
-30°C, concentrated form, at least 6 months [1]; -20°C, 3 months, 5% loss of activity [4]; -20°C, lyophilized, 4 months [7]

6 CROSSREFERENCES TO STRUCTURE DATABANKS

PIR/MIPS code

Brookhaven code

7 LITERATURE REFERENCES

[1] Braun, M., Lünsdorf, H., Bückmann, A.F.: Eur. J. Biochem.,196,439–450 (1991)
[2] Kinoshita, S., Kadota, K., Inoue, T., Sawada, H., Taguchi, H.: J. Ferment. Technol., 66,145–152 (1988)
[3] MacDonald, I.A., Rochon, Y.P.: J. Chromatogr.,259,154–158 (1983)
[4] MacDonald, I.A., Jellett, J.F., Mahony, D.E., Holdeman, L.V.: Appl. Environ. Microbiol.,37,992–1000 (1979)
[5] Harris, J.N., Hylemon, P.B.: Biochim. Biophys. Acta,528,148–157 (1978)
[6] Mahony, D.E., Meier, C.E., MacDonald, I.A., Holdeman, L.V.: Appl. Environ. Microbiol.,34,419–423 (1977)
[7] MacDonald, I.A., Mahony, D.E., Jellet, J.F., Meier, C.E.: Biochim. Biophys. Acta,489,466–476 (1977)
[8] MacDonald, I.A., Meier, E.C., Mahony, D.E., Costain, G.A.: Biochim. Biophys. Acta,450,142–153 (1976)
[9] Ottolina, G., Riva, S., Carrea, G., Danieli, B., Buckmann, A.F.: Biochim. Biophys. Acta,998,173–178 (1989)

1 NOMENCLATURE

EC number
1.1.1.177

Systematic name
sn-Glycerol-3-phosphate:NADP⁺ 1-oxidoreductase

Recommended name
Glycerol-3-phosphate 1-dehydrogenase (NADP⁺)

Synonymes
Dehydrogenase, glycerol phosphate (nicotinamide adenine dinucleotide phosphate)
L-Glycerol 3-phosphate:NADP oxidoreductase
Glycerin-3-phosphate dehydrogenase [1]
NADPH-dependent glycerin-3-phosphate dehydrogenase [1]

CAS Reg. No.
37213-46-0

---

2 REACTION AND SPECIFICITY

Catalysed reaction
D-Glyceraldehyde 3-phosphate + NADPH →
→ sn-glycerol 3-phosphate + NADP⁺

Reaction type
Redox reaction

Natural substrates
D-Glyceraldehyde 3-phosphate + NADPH (specific role in a hexose mono-phosphate shunt) [1]

Substrate spectrum
1 D-Glyceraldehyde 3-phosphate + NADPH [1]

Product spectrum
1 sn-Glycerol 3-phosphate + NADP⁺

Inhibitor(s)

Cofactor(s)/prosthetic group(s)/activating agents
NADPH [1]

Metal compounds/salts

Turnover number (min⁻¹)

Specific activity (U/mg)

K_m-value (mM)

pH-optimum

pH-range

Temperature optimum (°C)
  37 (assay at) [1]

Temperature range (°C)

---

3 ENZYME STRUCTURE

Molecular weight

Subunits

Glycoprotein/Lipoprotein
  –

---

4 ISOLATION/PREPARATION

Source organism
  Rat (white) [1]

Source tissue
  Skeletal muscle [1]; Myocardium [1]

Localisation in source
  Soluble [1]

Purification

Crystallization
  –

Cloned
  –

Renaturated
  –

5 STABILITY

pH

Temperature (°C)

Oxidation

Organic solvent

General stability information

Storage

6 CROSSREFERENCES TO STRUCTURE DATABANKS

PIR/MIPS code

Brookhaven code

7 LITERATURE REFERENCES

[1] Glushankov, E.P., Epifanova, Y.E., Kolotilova, A.I. : Biokhimiya,41,1788–1790 (1976)

1 NOMENCLATURE

EC number

1.1.1.178

Systematic name

(2S,3S)-3-Hydroxy-2-methylbutanoyl-CoA:NAD⁺ oxidoreductase

Recommended name

3-Hydroxy-2-methylbutyryl-CoA dehydrogenase

Synonymes

Dehydrogenase, 2-methyl-3-hydroxybutyryl coenzyme A

2-Methyl-3-hydroxybutyryl coenzyme A dehydrogenase

2-Methyl-3-hydroxy-butyryl CoA dehydrogenase [1]

CAS Reg. No.

52227-66-4

2 REACTION AND SPECIFICITY

Catalysed reaction

(2S,3S)-3-Hydroxy-2-methylbutanoyl-CoA + NAD⁺ →

→ 2-methylacetoacetyl-CoA + NADH

Reaction type

Redox reaction

Natural substrates

2-Methyl-3-hydroxybutyryl-CoA + NAD⁺ (isoleucine metabolism in Pseudo-

monas putida) [1]

Substrate spectrum

1 2-Methyl-3-hydroxybutyryl-CoA + NAD⁺ [1]

2 3-Hydroxybutyryl-CoA + NAD⁺ [1]

3 2-Hydroxy-3-methylpentanoyl-CoA + NAD⁺ [1]

Product spectrum

1 2-Methylacetoacetyl-CoA + NADH [1]

2 ?

3 ?

Inhibitor(s)

Cofactor(s)/prosthetic group(s)/activating agents
NAD$^+$ (cannot be replaced by NADP$^+$) [1]

Metal compounds/salts

Turnover number (min^{-1})

Specific activity (U/mg)
16.2 (2-methyl-3-hydroxybutyryl-CoA) [1]; 21.7 (3-hydroxybutyryl-CoA) [1]

K$_m$-value (mM)
0.037 (2-methyl-3-hydroxybutyryl-CoA) [1]; 0.5 (3-hydroxybutyryl-CoA) [1]

pH-optimum
9.5 [1]

pH-range

Temperature optimum (°C)
30 (assay at) [1]

Temperature range (°C)

---

3 ENZYME STRUCTURE

Molecular weight
130000 (Pseudomonas putida, gel filtration, sucrose gradient centrifugation)
[1]

Subunits

Glycoprotein/Lipoprotein
–

---

4 ISOLATION/PREPARATION

Source organism
Pseudomonas putida ATCC 23287 [1]

Source tissue
Cell [1]

Localisation in source

Purification

Crystallization
–

Cloned

–

Renaturated

–

---

5 STABILITY

pH

Temperature (°C)

Oxidation

Organic solvent

General stability information

Storage

---

6 CROSSREFERENCES TO STRUCTURE DATABANKS

PIR/MIPS code

Brookhaven code

---

7 LITERATURE REFERENCES

[1] Conrad R.S., Massey L.K., Sokatch J.R.: J. Bacteriol.,4,103–111 (1974)

---

1 NOMENCLATURE

EC number
1.1.1.179

Systematic name
D-Xylose:NADP⁺ 1-oxidoreductase

Recommended name
D-Xylose 1-dehydrogenase (NADP⁺)

Synonymes
Dehydrogenase, D-xylose (nicotinamide adenine dinucleotide phosphate)
D-Xylose-NADP dehydrogenase
D-Xylose:NADP⁺ oxidoreductase [1]

CAS Reg. No.
83534-37-6

2 REACTION AND SPECIFICITY

Catalysed reaction
D-Xylose + NADP⁺ →
→ D-xylono-1,5-lactone + NADPH (mechanism [3])

Reaction type
Redox reaction

Natural substrates

Substrate spectrum
1 D-Xylose + NADP⁺ [1–3]
2 L-Arabinose + NADP⁺ [1, 3]
3 D-Ribose + NADP⁺ [1, 3]
4 2-Deoxy-D-glucose + NADP⁺ [3]
5 D-Glucose + NADP⁺ [3]
6 D-Mannose + NADP⁺ [3]
7 D-Galactose + NADP⁺ [1]
8 More (D-arabinose, D-ribose, D-glucose and D-galactose are oxidized at
   10–30% the rate of D-xylose) [1]

Enzyme Handbook © Springer-Verlag Berlin Heidelberg 1995
Duplication, reproduction and storage in data banks are only
allowed with the prior permission of the publishers

Product spectrum
1 D-Xylono-1,5-lactone + NADPH
2 ?
3 ?
4 ?
5 ?
6 ?
7 ?
8 ?

Inhibitor(s)
NAD⁺ [3]; L-Lyxose [3]

Cofactor(s)/prosthetic group(s)/activating agents
NADP⁺ (beta-NADP⁺ acts as coenzyme exclusively [1]) [1–3]; More (not: NAD⁺) [3]

Metal compounds/salts

Turnover number (min⁻¹)
960 (D-glucose) [3]; 3480 (D-xylose) [3]; 2400 (D-ribose) [3]; 2700 (L-arabinose) [3]; 1320 (2-deoxy-D-glucose) [3]; 720 (D-galactose) [3]; 540 (D-arabinose) [3]

Specific activity (U/mg)
7.1 [3]

K$_m$-value (mM)
130 (D-arabinose) [3]; 6.5 (D-xylose) [3]; 27 (D-ribose) [3]; 49 (D-glucose) [3]; 57 (L-arabinose) [3]; 54 (2-deoxy-D-glucose) [3]; 140 (D-galactose) [3]; 8.8 (D-xylose) [3]; 0.09 (NADP⁺) [3]

pH-optimum
8 [3]

pH-range

Temperature optimum (°C)
30 (assay at) [3]

Temperature range (°C)

3 ENZYME STRUCTURE

Molecular weight
48000 (mammalia, molecular sieve chromatography) [1]
62000 (pig, gel filtration) [3]

Subunits
   Dimer (2 × 32000, pig, SDS-PAGE) [3]

Glycoprotein/Lipoprotein
   –

---

4 ISOLATION/PREPARATION

Source organism
   Pig [1, 3]; Dog [1]; Bovine [1]; Pichia quercuum [2]

Source tissue
   Aorta [1]; Coronaries [1]; Liver [3]; Eye lens [1]

Localisation in source
   Cytosol [3]

Purification
   Pig [3]

Crystallization
   –

Cloned
   –

Renaturated
   –

---

5 STABILITY

pH

Temperature (°C)

Oxidation

Organic solvent

General stability information

Storage

---

6 CROSSREFERENCES TO STRUCTURE DATABANKS

PIR/MIPS code

Brookhaven code

---

7 LITERATURE REFERENCES

[1] Wissler, J.H.: Hoppe-Seyler's Z. Physiol. Chem.,358,1300–1301 (1977)
[2] Suzuki, T., Onishi, H.: Appl. Microbiol.,25,850–852 (1973)
[3] Zepeda, S., Monasterio, O., Ureta, T.: Biochem. J.,266,637–644 (1990)

1 NOMENCLATURE

EC number
1.1.1.181

Systematic name
Cholest-5-ene-3beta,7alpha-diol:NAD$^+$ 3-oxidoreductase

Recommended name
Cholest-5-ene-3beta,7alpha-diol 3beta-dehydrogenase

Synonymes
Dehydrogenase, 5-cholestene-3beta,7alpha-diol-3beta-
Hydroxy-DELTA^5-C$_{27}$-steroid oxidoreductase

CAS Reg. No.
56626-16-5

2 REACTION AND SPECIFICITY

Catalysed reaction
Cholest-5-ene-3beta,7alpha-diol + NAD$^+$ →
→ 7alpha-hydroxycholest-5-en-3-one + NADH

Reaction type
Redox reaction

Natural substrates

Substrate spectrum
1 5-Cholestene-3beta,7alpha-diol + NAD$^+$ (sole substrate) [1]

Product spectrum
1 7alpha-Hydroxy-5-cholesten-3-one + NADH [1]

Inhibitor(s)
FMN [1]; Quinacrine [1]

Cofactor(s)/prosthetic group(s)/activating agents
NAD$^+$ (specific for, no reaction with NADP$^+$) [1]

Metal compounds/salts

Turnover number (min⁻¹)

Specific activity (U/mg)
0.228 [1]

K_m-value (mM)
0.045 (5-cholesten-3beta,7alpha-diol) [1]

pH-optimum
7.0–7.5 [1]

pH-range

Temperature optimum (°C)

Temperature range (°C)

---

3 ENZYME STRUCTURE

Molecular weight
45000–50000 (rabbit, gel filtration) [1]

Subunits
Monomer (1 × 46000, rabbit, SDS-PAGE) [1]

Glycoprotein/Lipoprotein
–

---

4 ISOLATION/PREPARATION

Source organism
Rabbit [1]

Source tissue
Liver [1]

Localisation in source
Microsomes [1]

Purification
Rabbit [1]

Crystallization
–

Cloned
–

Renaturated
–

5 STABILITY

pH

Temperature (°C)

Oxidation

Organic solvent

General stability information

Storage

6 CROSSREFERENCES TO STRUCTURE DATABANKS

PIR/MIPS code

Brookhaven code

7 LITERATURE REFERENCES

[1] Wikvall, K.: J. Biol. Chem.,256,3376–3380 (1981)

Enzyme Handbook © Springer-Verlag Berlin Heidelberg 1995
Duplication, reproduction and storage in data banks are only
allowed with the prior permission of the publishers

1 NOMENCLATURE

EC number
1.1.1.183

Systematic name
Geraniol:NADP$^+$ oxidoreductase

Recommended name
Geraniol dehydrogenase

Synonymes

CAS Reg. No.
56802-96-1

2 REACTION AND SPECIFICITY

Catalysed reaction
Geraniol + NADP$^+$ →
→ geranial + NADPH

Reaction type
Redox reaction

Natural substrates
Geraniol + NADP$^+$ [1]

Substrate spectrum
1 Geraniol + NADP$^+$ (r) [1]
2 Nerol + NADP$^+$ [1]
3 Citronellol + NADP$^+$ [1]
4 Farnesol + NADP$^+$ [1]

Product spectrum
1 Geranial + NADPH [1]
2 Neral + NADPH [1]
3 Citronellal + NADPH [1]
4 Farnesal + NADPH [1]

Inhibitor(s)
Iodoacetamide [1]; 8-Hydroxyquinoline [1]; Zn^{2+} [1]; Oxidized glutathione [1]; (+)-Limonene [1]; Linalool [1]; alpha-Terpineol [1]; o-Phenanthroline [1]; 2,2'-Dipyridyl [1]; EDTA [1]

Cofactor(s)/prosthetic group(s)/activating agents
$NADP^+$ [1]; NADPH [1]

Metal compounds/salts

Turnover number (min^{-1})

Specific activity (U/mg)

K_m-value (mM)
0.0465 (geraniol) [1]; 0.454 ($NADP^+$) [1]

pH-optimum
9.0 (geraniol + $NADP^+$) [1]; 6.5 (geranial + NADPH) [1]

pH-range

Temperature optimum (°C)

Temperature range (°C)

3 ENZYME STRUCTURE

Molecular weight
92000 (orange, gel filtration) [1]

Subunits

Glycoprotein/Lipoprotein
−

4 ISOLATION/PREPARATION

Source organism
Orange [1]

Source tissue
Orange juice vesicles [1]

Localisation in source

Purification
Orange (partially) [1]

Crystallization
−

Cloned

–

Renaturated

–

---

5 STABILITY

pH
   5.0–11.0 [1]

Temperature (°C)

Oxidation

Organic solvent

General stability information

Storage
   –96°C, 6 months [1]

---

6 CROSSREFERENCES TO STRUCTURE DATABANKS

PIR/MIPS code

Brookhaven code

---

7 LITERATURE REFERENCES

[1] Potty, V.H., Bruemmer, J.H.: Phytochemistry,9,1003–1007 (1970)

---

1 NOMENCLATURE

EC number
1.1.1.184

Systematic name
Secondary-alcohol:NADP+ oxidoreductase

Recommended name
Carbonyl reductase (NADPH)

Synonymes
Aldehyde reductase I
Prostaglandin 9-ketoreductase
Xenobiotic ketone reductase
NADPH-dependent carbonyl reductase
ALR3
Reductase, carbonyl
Carbonyl reductase
Nonspecific NADPH-dependent carbonyl reductase [1]
Aldehyde reductase 1 [1]
More (cf. EC 1.1.1.2)

CAS Reg. No.
77106-95-7; 89700-37-8

2 REACTION AND SPECIFICITY

Catalysed reaction
R-CO-R' + NADPH →
→ R-CHOH-R' + NADP+ (mechanism [5, 6])

Reaction type
Redox reaction

Natural substrates

Substrate spectrum
1 R-CO-R' + NADPH (NADH also acts as cofactor [2, 5, 7, 11], no activity
with NADH [4], low reverse direction only detected with cyclohexanol [2],
reverse direction: dehydrogenase activity towards alcohols and aldehydes: (S)-(+)-1-indanol, (S)-(+)-tetralol [7], specificity of various enzyme
forms [4, 10], overview [1, 2, 4, 7, 10, 11], quinones [1, 2, 4], aldehydes
[1, 2, 4], ketones [1, 2, 4], methyl-1,4-benzoquinone [2], menadione [1, 2,
6, 7], ubiquinone [1], tocopherolquinone [1], 4-nitroacetophenone [2, 7],
4-benzoylpyridine [2, 7], pyridine-3-aldehyde [2, 7], 4-nitrobenzaldehyde
[1, 7], methylglyoxal [1], crotonaldehyde [2], acetaldehyde [2], acetone
[2, 5, 7], aromatic aldehydes [4], daunorubicin [1], 3-ketosteroids [1],
1-heptanal [7], 1-decanal [7], 1-octanal [7], 1-propanal [7], 1-pentanal [7],
1-hexanal [7], 1-nonanal [7], 1-butanal [7], cyclohexanone [7], 2-butanone
[7], 5alpha-androstane-3,17-dione [7], 5beta-androstane-3,17-dione [7],
5alpha-androstan-17beta-ol-3-one [7], 5beta-androstan-17beta-ol-3-one
[7], testosterone [7], prostaglandin of the E and A class [1],
4-(6-methoxy-2-benzoxatolyl)acetophenone [8], propan-2-ol [5]) [1, 2, 4–7,
10, 11]

Product spectrum
1 R-CHOH-R' + NADP+

Inhibitor(s)
Flavonoids [1]; Quercetin [1, 2, 4]; Quercitrin [10]; Retin [1]; Indomethacin
[1, 4, 10]; Ethacrynic acid (inhibition of enzyme forms V1, V2 and T3 [4]) [1,
4]; Dicoumarol (enzyme form CR2 inhibited, CR1 not [10]) [1, 4, 10]; p-Chloromercuribenzoate [10]; p-Hydroxymercuribenzoate [1, 2, 4]; Iodoacetate
[1]; Cibacron blue (3G-A [1]) [1, 5]; Chlorogenic acid [1]; 4-Oxo-4H-benzo-
pyran-2-carboxylic acid [1]; FAD [1]; NAD+ [2, 5]; NADP+ [1, 2, 5]; NADPH
(competitive to NADH) [5]; D-Catechin [1]; Chlorpromazine [1, 4, 10]; Tetra-
methyleneglutaric acid [1]; 2-Mercaptoethanol [2, 5, 11]; Dithiothreitol [2,
11]; N-Ethylmaleimide [2]; Phenylpyruvic acid [2]; Nicotinic acid (weak) [2];
Pyrazole [2, 5, 7, 11]; Phenylglyoxal (irreversible inactivation) [3]; 2,3-Buta-
nedione (irreversible inactivation) [3]; Disulfiram [4]; CuSO$_4$ [4]; Furosemide
(inhibition of V1, V2 and T3) [4]; Benzamide [5, 7]; Hg^{2+} [10]; AgNO$_3$ (enzyme form CR1 inhibited, CR2 not) [10]; 3,3-Tetramethyleneglutaric acid (enzyme form CR2 inhibited, CR1 slightly) [10]; Ammonium molybdate (enzyme
form CR1 inhibited, CR2 not) [10]; Sodium phenobarbitone (enzyme form
CR2 inhibited, CR1 slightly) [10]; Progesterone [10];
5beta-Pregnane-3,20-dione [10]; 21-Hydroxy-5alpha-pregnane-3,20-dione
[10]; Androsterone (CR2 inhibited, CR1 not) [10]; Testosterone [10];
5beta-Androstane-3alpha,17beta-diol [10]; FeSO$_4$ [2, 11]; Isobutyramide
[11]

Cofactor(s)/prosthetic group(s)/activating agents
NADPH (pro-4S hydrogen atom of the nicotinamide ring of NADPH is transferred to the substrate [1, 7]) [1, 2, 4–7, 10, 11]; NADH (no activity with NADH [4]) [2, 5, 7]; NADP+ [7, 10]; Fatty acids (with carbon chain length greater than nine at pH 7.0 activate, e.g. arachidonic acid) [9]; Anionic detergents (e.g. SDS and sarkosyl stimulate) [9]

Metal compounds/salts
Zinc (detected in enzyme by atomic absorption spectroscopy) [1]; Mg (detected in enzyme by atomic absorption spectroscopy) [1]; Fe (detected in enzyme by atomic absorption spectroscopy) [1]

Turnover number (min^{-1})

Specific activity (U/mg)
More (enantiospecific assay [8]) [2, 7, 8, 10, 11]; 8.3 [1]

K_m-value (mM)
More (overview [1, 4, 10]) [1, 2, 4–7, 10, 11]; 0.002 (pyridine-4-aldehyde (+ NADPH)) [2]; 0.004 (4-nitroacetophenone (+ NADPH)) [2]; 0.004–0.01 (NADPH (+ 4-nitrobenzaldehyde), K_m depending on enzyme form) [1]; 0.009 (menadione (+ NADPH)) [2]; 0.012 (methyl-1,4-benzoquinone (+ NADPH)) [2]; 0.015 (pyridine-3-aldehyde (+ NADPH)) [2]; 0.015–0.017 (ubiquinone-1, K_m depending on enzyme form) [1]; 0.031 (benzalacetone (+ NADPH)) [2]; 0.039 (acetaldehyde (+ NADPH)) [2]; 0.045–0.06 (menadione, K_m depending on enzyme form) [1]; 0.057 (methyl-1,4-benzoquinone (+ NADH)) [2]; 0.091 (acetone (+ NADPH)) [2]; 0.12 (crotonaldehyde (+ NADPH)) [2]; 0.13 (daunorubicin, enzyme form CR8 [1], pyridine-3-aldehyde (+ NADH) [2]) [1, 2]; 0.24 (4-benzoylpyridine (+ NADPH)) [2]; 0.33 (benzalacetone (+ NADH)) [2]; 0.45 (prostaglandin E$_1$, enzyme form CR8) [1]; 0.5 (pyridine-4-aldehyde (+ NADH)) [2]; 1.6–2 (4-nitrobenzaldehyde, K_m depending on enzyme form) [1]; 3.1 (acetone (+ NADH)) [2]; 3.3 (4-benzoylpyridine (+ NADH)) [2]; 4–5 (phenylglyoxal, K_m depending on enzyme form) [1]; 5.1 (acetaldehyde (+ NADH)) [2]; 250 (crotonaldehyde (+ NADH)) [2]

pH-optimum
4.8 (NADPH + 4-nitroacetophenone) [2]; 4.8–5.3 (4-nitroacetophenone, pyridine-4-aldehyde, 5beta-dihydrotestosterone, enzyme CR2) [10]; 5–6 (NADH + 4-nitroacetophenone) [11]; 5.2 (NADPH + 4-nitroacetophenone) [11]; 5.8 [7, 9]; 5.8–6.2 (4-nitroacetophenone, pyridine-4-aldehyde, enzyme form CR1) [10]; 6 (menadione, 4-nitrobenzaldehyde, daunorubicin, in sodium phosphate buffer) [1]; 6.2 (4-(6-methoxy-2-benzoxazolyl)acetophenone + NADPH) [8]; More (enzyme activation by arachidonic acid shifts pH-optimum from 5.8 to 6.5) [9]

pH-range

Temperature optimum (°C)
25 (assay at) [1, 11]

Temperature range (°C)

---

3 ENZYME STRUCTURE

Molecular weight
30000 (human, gel filtration, SDS-PAGE) [1]
34000 (guinea pig, gel filtration, enzyme form CR1) [10]
86000 (guinea pig, gel filtration) [2]
90000 (mouse, gel filtration) [11]
103000 (pig, gel filtration, SDS-PAGE) [7]
115000 (guinea pig, gel filtration, enzyme form CR2) [10]

Subunits
Monomer (1 × 37500, guinea pig, enzyme form CR1, SDS-PAGE [10],
1 × 30000, human, SDS-PAGE [1]) [1, 10]
Tetramer (4 × 32000, guinea pig, enzyme form CR2, SDS-PAGE [10],
4 × 23000, mouse, SDS-PAGE [11], 4 × 24000, pig, SDS-PAGE [7])
[7, 10, 11]
? (x × 23000, guinea pig, SDS-PAGE [2], x × 33000, rat, enzyme form T2, T3,
V1, V2, SDS-PAGE [4], x × 32000, rat, enzyme form T1, SDS-PAGE [4]) [2, 4]

Glycoprotein/Lipoprotein
–

---

4 ISOLATION/PREPARATION

Source organism
Human [1, 3, 6]; Guinea pig [2, 5, 10, 11]; Rat [4, 8]; Pig [7–9]; Mouse [11];
Rabbit [8]

Source tissue
Brain [1, 3, 6]; Lung [2, 5, 7, 9, 11]; Liver [8, 10]; Testis [4]; Vas deferens
(mucosal epithelium cells) [4]; Lung (ciliated cells, nonciliated bronchiolar
cells, Type II alveolar pneumocytes, epithelial cells of the ducts of the bron-
chial gland) [7]

Localisation in source
Cytoplasm [1]; Cytosol [2, 8]; Microsomes [10]

Purification
Human (3 molecular forms: CR7, CR8, CR8.5) [1]; Rat (3 forms from testis:
T1, T2, T3, 2 forms from vas deferens: V1, V2) [4]; Guinea pig (2 enzyme
species [2], 2 enzyme forms: major form CR2, minor form CR1 [10]) [2, 10];
Mouse [11]

Crystallization

–

Cloned

–

Renaturated

–

---

5 STABILITY

ph
4.4 (unstable) [2]; 6–11 (25°C, 15 min, stable) [2]

Temperature (°C)
25 (pH 6–11, 15 min, stable) [2]; 50 (pH 6–11, 15 min, about 20% loss of activity) [2]; 60 (rapid inactivation above) [2]; More (activation by arachidonic acid leads to a decrease in thermal stability) [9]

Oxidation

Organic solvent

General stability information
Glycerol, 20% v/v, stabilizes [2]; Activation by arachidonic acid leads to a decrease in thermal stability [9]

Storage
4°C, 0.02 mg/ml protein concentration, 0.15 M KCl, stable for at least several days, 90% loss of activity after 2 weeks [2]; –20°C or 4°C, 20% v/v, glycerol, stable for 1 month [2]

---

6 CROSSREFERENCES TO STRUCTURE DATABANKS

PIR/MIPS code
PIR1:RDHUCB (human); PIR3:JN0703 (Pig)

Brookhaven code

---

7 LITERATURE REFERENCES

[1] Wermuth, B.: J. Biol. Chem.,256,1206–1213 (1981)
[2] Nakayama, T., Hara, A., Sawada, H.: Arch. Biochem. Biophys.,217,564–573 (1982)
[3] Bohren, K.M., von Wartburg, J.-P., Wermuth, B.: Biochim. Biophys. Acta,916,185–192 (1987)
[4] Iwata, N., Inazu, N., Takeo, S., Satoh, T.: Eur. J. Biochem.,193,75–81 (1990)
[5] Matsuura, K., Nakayama, T., Nakagawa, M., Hara, A., Sawada, H.: Biochem. J.,252,17–22 (1988)
[6] Bohren, K.M., von Wartburg, J.-P., Wermuth, B.: Biochem. J.,244,165–171 (1987)
[7] Oritani, H., Deyashiki, Y., Nakayama, T., Hara, A., Sawada, H., Matsuura, K., Bunai, Y., Ohya, I.: Arch. Biochem. Biophys.,292,539–547 (1992)
[8] Naganuma, H., Kondo, J.-I., Kawahara, Y.: J. Chromatogr.,532,65–74 (1990)
[9] Hara, A., Oritani, H., Deyashiki, Y., Nakayama, T., Sawada, H.: Arch. Biochem. Biophys.,292,548–554 (1992)
[10] Usui, S., Hara, A., Nakayama, T., Sawada, H.: Biochem. J.,223,697–705 (1984)
[11] Nakayama, T., Yashiro, K., Inoue, Y., Matsuura, K., Ichikawa, H., Hara, A., Sawada, H.: Biochim. Biophys. Acta,882,220–227 (1986)

1 NOMENCLATURE

EC number
1.1.1.185

Systematic name
L-Glycol:NAD(P)+ oxidoreductase

Recommended name
L-Glycol dehydrogenase

Synonyms
Dehydrogenase, glycol (nicotinamide adenine dinucleotide (phosphate))
L-(+)-Glycol:NAD(P) oxidoreductase [1]
L-Glycol:NAD(P) dehydrogenase [1]

CAS Reg. No.
77967-75-0

---

2 REACTION AND SPECIFICITY

Catalysed reaction
A 2-hydroxycarbonyl compound + NAD(P)H →
→ an L-Glycol + NAD(P)+ (glyoxal reduction by enzyme form pI 7.2 follows
ordered bi-bi mechanism in which the coenzyme is the first substrate to bind
to the enzyme [2, 3])

Reaction type
Redox reaction

Natural substrates

Substrate spectrum

1 R_1-CO-CHOH-R_2 + NAD(P)H (r, much more efficient in dehydrogenase direction [1], alpha-hydroxycarbonyl compound [1, 2]: e.g. glycolaldehyde
  [1], glyceraldehyde [1], acetoin [1], acetylethylcarbinol [1], diacetylme-
  thylcarbinol) [1, 2]
2 R_1-CO-CO-R_2 + NAD(P)H (vicinal dicarbonyl: e.g. glyoxal [1, 3], methylglyoxal [1, 3], diacetyl [1, 3], 2,3-pentanedione [1, 3]) [1, 3]
3 Ethyl pyruvate + NAD(P)H [1]
4 Methyl pyruvate + NAD(P)H [1]
5 More (not: acetaldehyde, monoketones, non-vicinal diketones, free keto
  acids) [1]

Product spectrum
1 R_1-CHOH-CHOH-R_2 + NAD(P)$^+$ [1]
2 R_1-CO-CHOH-R_2 + NAD(P)$^+$ [2]
3 ?
4 ?
5 ?

Inhibitor(s)
Acetone (pI 7.2 form) [2, 3]

Cofactor(s)/prosthetic group(s)/activating agents
NADH [1–3]; NADPH [1–3]; More (no activity with alpha-NADH) [1]

Metal compounds/salts

Turnover number (min^{-1})

Specific activity (U/mg)
More [1]; 0.4 [2]

K_m-value (mM)
0.0022 (NADPH (+ 2,3-pentanedione)) [3]; 0.0024 (NADPH (+ glyoxal)) [3];
0.0028 (NADPH (+ methylglyoxal)) [3]; 0.0031 (NADPH (+ diacetyl)) [3];
0.026 (glyceraldehyde, pI 7.2 form) [1]; 0.29 (glyceraldehyde, pI 6.2 form)
[1]; 0.035 (2,3-pentanedione) [3]; 0.38 (diacetyl, pI 6.2 form) [1]; 0.053
(glyoxal, pI 6.2 form) [1]; 0.060 (2,3-pentanedione, pI 7.2 form) [1]; 0.090
(methylglyoxal, pI 7.2 form) [3]; 0.097 (2,3-pentanedione, pI 6.2 form) [1];
0.098 (methylglyoxal, pI 6.2 form) [1]; 0.102 (methylglyoxal, pI 7.2 form) [1];
0.109 (diacetyl, pI 7.2 form) [1]; 0.120 (diacetyl, pI 7.2 form) [3]; 0.410
(glyoxal, pI 7.2 form) [3]; 0.411 (glyoxal, pI 4.8 form) [1]; 0.522 (glyoxal, pI
7.2 form) [1]; 0.566 (glyceraldehyde, pI 4.8 form) [1]; 0.675 (2,3-pentanedio-
ne, pI 4.8 form) [1]; 2.013 (acetoin, pI 6. 2 form) [1]; 2.164 (acetoin, pI 4.8
form) [1]; 2.590 (acetoin, pI 7.2 form) [1]; More [1]

pH-optimum
5–6.6 [1, 2]

pH-range

Temperature optimum (°C)
25 (assay at) [1, 2]

Temperature range (°C)

3 ENZYME STRUCTURE

Molecular weight
28000 (chicken, gel filtration) [1]

Subunits
Glycoprotein/Lipoprotein
–

---

4 ISOLATION/PREPARATION

Source organism
Chicken (hen) [1–3]

Source tissue
Muscle [1–3]

Localisation in source

Purification
Chicken (hen [1, 2], 3 enzyme forms: pl 7.2, pl 6.2 and pl 4.8 [1, 2]) [1–3]

Crystallization
–

Cloned
–

Renaturated
–

---

5 STABILITY

pH
7 (very stable in both water and low molarity buffers of pH 7 at 0–4°C) [1, 2]

Temperature (°C)

Oxidation

Organic solvent

General stability information
Very stable in both water and low molarity buffers of pH 7 at 0–4°C [1, 2];
Easily inactivated in high molarity buffers [1, 2]

Storage
–18°C, enzyme at any stage of purification [1, 2]

6 CROSSREFERENCES TO STRUCTURE DATABANKS

PIR/MIPS code

Brookhaven code

7 LITERATURE REFERENCES

[1] Bernardo, A., Burgos, J., Martin, R.: Biochim. Biophys. Acta,659,189–198 (1981)
[2] Burgos, J., Sarmiento, R.M.: Methods Enzymol.,89,523–526 (1982) (Review)
[3] Prieto, J.G., Sarmiento, R.M., Burgos, J.: Arch. Biochem. Biophys.,224,372–377 (1983)

1 NOMENCLATURE

EC number
1.1.1.186

Systematic name
dTDP-D-galactose:NAD$^+$ 6-oxidoreductase

Recommended name
dTDPgalactose 6-dehydrogenase

Synonymes
Thymidine-diphosphate-galactose dehydrogenase

CAS Reg. No.

---

2 REACTION AND SPECIFICITY

Catalysed reaction
dTDP-D-galactose + 2 NAD$^+$ + H$_2$O →
→ dTDP-D-galacturonate + 2 NADH

Reaction type
Redox reaction

Natural substrates

Substrate spectrum
1 dTDP-D-galactose + NAD$^+$ + H$_2$O

Product spectrum
1 dTDP-D-galacturonate + NADH

Inhibitor(s)

Cofactor(s)/prosthetic group(s)/activating agents

Metal compounds/salts

Turnover number (min^{-1})

Specific activity (U/mg)

K$_m$-value (mM)

pH-optimum

pH-range

Temperature optimum (°C)

Temperature range (°C)

---

3 ENZYME STRUCTURE

Molecular weight

Subunits

Glycoprotein/Lipoprotein

–

4 ISOLATION/PREPARATION

Source organism

Source tissue

Localisation in source

Purification

Crystallization

–

Cloned

–

Renaturated

–

5 STABILITY

pH

Temperature (°C)

Oxidation

Organic solvent

General stability information

Storage

6 CROSSREFERENCES TO STRUCTURE DATABANKS

PIR/MIPS code

Brookhaven code

7 LITERATURE REFERENCES

1 NOMENCLATURE

EC number
    1.1.1.187

Systematic name
    GDP-6-deoxy-D-mannose:NAD(P)+ 4-oxidoreductase

Recommended name
    GDP-4-dehydro-D-rhamnose reductase

Synonymes
    GDP-4-keto-6-deoxy-D-mannose reductase
    GDP-4-keto-D-rhamnose reductase
    Guanosine diphosphate-4-keto-D-rhamnose reductase [1]

CAS Reg. No.

---

2 REACTION AND SPECIFICITY

Catalysed reaction
    GDP-6-deoxy-D-mannose + NAD(P)+ →
    → GDP-4-dehydro-6-deoxy-D-mannose + NAD(P)H

Reaction type
    Redox reaction

Natural substrates
    GDP-4-dehydro-6-deoxy-D-mannose + NAD(P)H [2]

Substrate spectrum
    1 GDP-4-dehydro-6-deoxy-D-mannose + NAD(P)H [1, 2]
    2 More (a single enzyme protein is responsible for the formation of both
      C-4-epimers, GDP-D-rhamnose and GDP-D-talomethylose, no precise ste-
      reoselectivity with respect to its hydrogen receptor carbonyl function) [1]

Product spectrum
    1 GDP-6-deoxy-D-mannose + NAD(P)+
    2 ?

Inhibitor(s)
    Dithiothreitol (NADP+ protects) [1]; GDP-mannose [1]

Cofactor(s)/prosthetic group(s)/activating agents
    NADH [1]; NADPH [1]

Metal compounds/salts

Turnover number (min^{-1})

Specific activity (U/mg)
    More [1]

K$_m$-value (mM)
    0.027 (NADPH) [1]; 0.028 (NADH) [1]; 0.05 (GDP-4-dehydro-6-deoxy-D-man-
    nose (+ NADH)) [1]; 0.025 (GDP-4-dehydro-6-deoxy-D-mannose (+ NADPH))
    [1]

pH-optimum
    6–7 (GDP-4-dehydro-6-deoxy-D-mannose + NADPH) [1]; 8
    (GDP-4-dehydro-6-deoxy-D-mannose + NADH) [1]

pH-range
    5–9 (pH 5: about 50% of activity maximum (GDP-4-dehydro-6-deoxy-D-man-
    nose + NADPH), about 35% of activity maximum
    (GDP-4-dehydro-6-deoxy-D-mannose + NADH), pH 9: about 35% of activity
    maximum (GDP-4-dehydro-6-deoxy-D-mannose + NAD(P)H)) [1]

Temperature optimum (°C)

Temperature range (°C)

---

3 ENZYME STRUCTURE

Molecular weight
    50000–70000 (soil bacterium GS (ATCC 19241), gel filtration) [1].

Subunits

Glycoprotein/Lipoprotein
    –

---

4 ISOLATION/PREPARATION

Source organism
    Soil bacterium GS (ATCC 19241) [1]; Leucaena glauca [2]

Source tissue

Localisation in source

Purification
    Soil bacterium GS (ATCC 19241) [1]

Crystallization
    –

Cloned

–

Renaturated

–

5 STABILITY

pH

Temperature (°C)
23 (room temperature, 3 h, 51% loss of activity) [1]; 56 (5 min, complete loss of activity) [1]

Oxidation

Organic solvent

General stability information
NADPH protects against inactivation by dithiothreitol and heat [1]

Storage
–20°C, half-life: 5 months [1]

6 CROSSREFERENCES TO STRUCTURE DATABANKS

PIR/MIPS code

Brookhaven code

7 LITERATURE REFERENCES

[1] Winkler, N.W., Markovitz, A.: J. Biol. Chem.,246,5868–5876 (1971)
[2] Barber, G.A.: Biochim. Biophys. Acta,165,68–75 (1968)

1 NOMENCLATURE

EC number
1.1.1.188

Systematic name
(5Z,13E)-(15S)-9alpha,11alpha,15-Trihydroxyprosta-5,13-dienoate:NADP+
11-oxidoreductase

Recommended name
Prostaglandin-F synthase

Synonymes
Reductase, 15-hydroxy-11-oxoprostaglandin
PGD_2 11-ketoreductase
PGF_{2alpha} synthetase
Prostaglandin 11-ketoreductase
Prostaglandin D_2 11-ketoreductase
Prostaglandin D_2-ketoreductase
Prostaglandin F synthase
Prostaglandin F synthetase
Synthetase, prostaglandin F_{2alpha}
Prostaglandin-D_2 11-reductase
PGF synthetase [8]
NADPH-dependent prostaglandin D_2 11-keto reductase [2]
Prostaglandin 11-keto reductase [3]

CAS Reg. No.
55976-95-9

---

2 REACTION AND SPECIFICITY

Catalysed reaction
(5Z,13E)-(15S)-9alpha,15-Dihydroxy-11-oxoprosta-5,13-dienoate + NADPH
$\rightarrow$
$\rightarrow$ (5Z,13E)-(15S)-9alpha,11alpha,15-trihydroxyprosta-5,13-dienoate +
NADP+

Reaction type
Redox reaction

Natural substrates
Prostaglandin D_2 + NADPH (may be the major enzymatic pathway for metabolism of endogenous prostaglandin D_2) [1, 3]

Enzyme Handbook © Springer-Verlag Berlin Heidelberg 1995
Duplication, reproduction and storage in data banks are only
allowed with the prior permission of the publishers

Substrate spectrum
1 (5Z,13E)-(15S)-9alpha,15-Dihydroxy-11-oxoprosta-5,13-dienoate + NADPH (i.e. prostaglandin D_2, ir [8]) [1–5, 8–10]
2 Prostaglandin H_2 + NADPH (prostaglandin D_2 is not an intermediate in the reduction of prostaglandin H_2) [4, 8, 9]
3 9,10-Phenanthrenequinone + NADPH [8, 9]
4 4-Nitroacetophenone + NADPH [8]
5 Phenylglyoxal + NADPH [8]
6 Menadione + NADPH [8, 9]
7 4-Nitrobenzaldehyde + NADPH [8, 9]
8 Hydrindantin + NADPH [8, 9]
9 Daunorubicine + NADPH [8, 9]
10 2,6-Dichloroindophenol + NADPH [8]
11 Prostaglandin D_1 + NADPH [9]
12 Prostaglandin D_3 + NADPH [9]
13 More (broad specificity [8, 9], overview [8, 9, 11]) [8, 9, 11]

Product spectrum
1 (5Z,13E)-(15S)-9alpha,11alpha,15-Trihydroxyprosta-5,13-dienoate + NADP+ (i.e. prostaglandin F_{2alpha} [1–3, 8], 9alpha,11alpha-prostaglandin F_2 is the product [4, 5, 9], 11-epi-prostaglandin F_{2alpha} is the product [10]) [1–5, 8–10]
2 Prostaglandin F_{2alpha} + NADP+ [4, 8, 9]
3 ?
4 ?
5 ?
6 ?
7 ?
8 ?
9 ?
10 ?
11 ?
12 ?
13 ?

Inhibitor(s)
Phenanthrenequinone (inhibition of prostaglandin D_2 reductase activity, not prostaglandin H_2 reductase activity) [8, 9]; Prostaglandin D_2 (inhibition of phenanthrenequinone reductase activity, not prostaglandin H_2 reductase activity) [8]; $CuSO_4$ [9]; $HgCl_2$ [9]; NaF [9]; NaBr (below 0.5 M inhibition of liver enzyme, above 0.5 M inhibition of lung enzyme) [9]; NaI (below 0.2 M inhibition of liver enzyme, above 0.2 M inhibition of lung enzyme) [9]; $AgNO_3$ [9]; $ZnSO_4$ [9]; $NiSO_4$ (100 mM inhibition of phenanthrenequinone reductase activity and nitrobenzaldehyde reductase activity of liver enzyme, inhibition of prostaglandin D_2 reductase activity of lung enzyme) [9]; $MnSO_4$ (inhibition of prostaglandin D_2 reductase activity of lung enzyme) [9]; $LiSO_4$ (inhibition of prostaglandin reductase activity of lung enzyme) [9]; Quercitrin [9]; Quercetin [9]; Ethacrynic acid [9]; Fenbufen [9]; Mefenamic acid [9]; Indomethacin [9]; Disulfiram [9]; Rutin [9]; Clemastine [9]; Benzoic anhydride [9]; Acetylsalicylic acid [9]; Valproic acid [9]; Catechin [9]; Cromolyn [9]; Flurbiprofen [9]; Ketoprefen [9]; Cibacron blue 3GA [9]

Cofactor(s)/prosthetic group(s)/activating agents
NADPH [1–5, 8–10]; NADH (10% [8], 3% [9] activity of that with NADPH) [8, 9]

Metal compounds/salts
NaCl (stimulation of prostaglandin D_2 11-ketoreductase activity, phenanthrenequinone and nitrobenzaldehyde reductase activity of liver enzyme, maximal at 0.5–1.0 M [9], no effect on prostaglandin H_2 reductase activity) [9]; NaBr (stimulation of prostaglandin D_2 11-ketoreductase activity, phenanthrenequinone and nitrobenzaldehyde reductase activity of liver enzyme, maximal at 0.5 M) [9]; NaI (stimulation of prostaglandin D_2 11-ketoreductase activity, phenanthrenequinone and nitrobenzaldehyde reductase activity of liver enzyme, maximal at 0.2 M) [9]; $MnSO_4$ (100 mM, stimulation of prostaglandin D_2 11-ketoreductase activity and phenanthrenequinone and nitrobenzaldehyde reductase activity of liver enzyme) [9]; $NiSO_4$ (100 mM, stimulation of prostaglandin D_2 11-ketoreductase activity of liver enzyme) [9]

Turnover number (min^{-1})

Specific activity (U/mg)
More [1, 2, 8, 9]

K_m-value (mM)
  0.0007 (9,10-phenanthrenequinone) [8]; 0.002 (9,10-phenanthrenequinone)
  [9]; 0.0029 (NADPH (+ 9,10-phenanthrenequinone)) [9]; 0.0031 (hydrindantin) [9]; 0.0033 (NADPH (+ prostaglandin D_2)) [9]; 0.0054 (NADPH (+ p-nitroacetophenone)) [9]; 0.0056 (prostaglandin D_3) [9]; 0.007 (prostaglandin D_1)
  [9]; 0.010 (prostaglandin H_2 [8], prostaglandin D_2 [9]) [8, 9]; 0.0119 (p-nitroacetophenone) [9]; 0.013 (4-nitroacetophenone) [8]; 0.018 (hydrindantin)
  [8]; 0.020 (phenylglyoxal) [9]; 0.025 (prostaglandin H_2) [9]; 0.053 (cyclohexanone) [9]; 0.080 (phenylglyoxal) [8]; 0.120 (prostaglandin D_2) [8]; 0.125
  (4-nitrobenzaldehyde) [8]; 0.143 (D,L-glyceraldehyde) [9]; 0.2 (prostaglandi-
  n D_2) [2, 3]; 0.230 (prostaglandin J_2) [8]; 0.250 (prostaglandin A_2) [8]; 0.4
  (NADH (+ p-nitroacetophenone)) [9]; 0.670 (DELTA^{12}-prostaglandin J_2) [8];
  0.80 (NADH (+ 9,10-phenanthrenequinone)) [9]; 1.0 (NADH (+ menadione
  or phenylglyoxal)) [9]; 135 (D-xylose) [9]; More [9]

pH-optimum
  6–7 [9]; 7.5 [1, 2]

pH-range

Temperature optimum (°C)
  37 (assay at) [3, 8]

Temperature range (°C)

---

3 ENZYME STRUCTURE

Molecular weight
  30500 (bovine, gel filtration, SDS-PAGE) [8]
  36666 (bovine, nucleotide sequence analysis of cloned cDNA) [6]

Subunits
  ? (x × 66000, rabbit, SDS-PAGE) [1, 2]
  Monomer (1 × 30500, bovine, SDS-PAGE) [8]

Glycoprotein/Lipoprotein
  –

---

4 ISOLATION/PREPARATION

Source organism
  Rabbit [1–3]; Human [5]; Bovine (enzyme cloned and expressed in E. coli
  [11]) [4, 6, 8, 9, 11]; Rat [10]; European common frog (epsilon-crystallin of
  European common frog is identical to bovine lung PGF synthase) [7]

Source tissue
  Cerebral neocortex [5]; Liver [1–3, 5, 9]; Lung [4–6, 8]; Lens [7]; Ocular system (sclereal complex, anterior uveal complex, not retinal complex) [10]

Localisation in source
Cytoplasm [2]; Cytosol [5]

Purification
Rabbit [1, 2]; Bovine (enzyme cloned and expressed in E. coli [11]) [8, 9, 11]

Crystallization
–

Cloned
(expression of bovine enzyme in E. coli [11]) [4, 6, 11]

Renaturated
–

---

5 STABILITY

pH

Temperature (°C)
100 (5 min, complete loss of activity) [9]

Oxidation

Organic solvent

General stability information
Freezing and thawing: several times, less than 20% loss of activity [9]

Storage
–80°C, 1.5 mg/ml enzyme concentration, 10 mM potassium phosphate buffer, pH 7.0, stable for several months [9]; 4°C or at room temperature, 1.5 mg/ml enzyme concentration, 10–20% loss of activity after 1 month [9]

---

6 CROSSREFERENCES TO STRUCTURE DATABANKS

PIR/MIPS code
PIR2:A28396 (bovine); PIR2:JH0575 (II hepatic bovine)

Brookhaven code

7 LITERATURE REFERENCES

[1] Wong, P.Y.-K.: Methods Enzymol.,86,117–125 (1982) (Review)

[2] Wong, P.Y.-K.: Biochim. Biophys. Acta,659,169–178 (1981)

[3] Reingold, D.F., Kawasaki, A., Needleman, P.: Biochim. Biophys. Acta,659,179–188 (1981)

[4] Watanabe, K., Fujii, Y., Nakayama, K., Ohkubo, H., Kuramitsu, S., Hayashi, H., Kagamiyama, H., Nakanishi, S., Hayaishi, O.: Adv. Prostaglandin Thromboxane Leukotriene Res.,19 (Taipei Conf. Prostaglandin Leukotriene Res.) 462–465 (1988)

[5] Wolfe, L.S., Rostworowski, K., Pellerin, L., Sherwin, A.: J. Neurochem.,53,64–70 (1989)

[7] Watanabe, K., Fujii, Y., Nakayama, K., Ohkubo, H., Karamitsu, S., Kagamiyama, H., Nakanishi, S., Hayaishi, O. : Proc. Natl. Acad. Sci. USA,85,11–15 (1988)

[6] Wermuth, B., Omar, A., Forster, A., Francesco, C., Wolf, M., Wartburg, J.P., Bullock, B., Gabbay, K.H. in " Enzymology and Molecular Biology of Carbonyl Metabolism, Aldehyde Dehydrogenase Aldo-Keto Reductase, and Alcohol Dehydrogenase" (Weiner, H., Flynn, T.G., eds.) 297–307, Liss, New York (1987)

[8] Watanabe, K., Yoshida, R., Shimizu, T., Hayaishi, O.: J. Biol. Chem.,260,7035–7041 (1985)

[9] Chen, L.-Y., Watanabe, K., Hayaishi, O.: Arch. Biochem. Biophys.,296,17–26 (1992)

[10] Goh, Y., Urade, Y., Fujimoto, N., Hayaishi, O.: Biochim. Biophys. Acta,921,302–311 (1987)

[11] Watanabe, K., Fujii, Y., Ohkubo, H., Kuramitsu, S., Kagamiyama, H., Nakanishi, S., Hayaishi, O.: Biochem. Biophys. Res. Commun.,181,272–278 (1991)

1 NOMENCLATURE

EC number
1.1.1.189

Systematic name
(5Z,13E)-(15S)-9alpha,11alpha,15-Trihydroxyprosta-5,13-dienoate:NADP+ 9-oxidoreductase

Recommended name
Prostaglandin-E$_2$ 9-reductase

Synonymes
PGE$_2$-9-OR [5]
Reductase, 15-hydroxy-9-oxoprostaglandin
9-Keto-prostaglandin E$_2$ reductase
9-Ketoprostaglandin reductase
PGE-9-ketoreductase
PGE$_2$ 9-oxoreductase
PGE$_2$-9-ketoreductase
Prostaglandin 9-ketoreductase
Prostaglandin E 9-ketoreductase
Prostaglandin E$_2$-9-oxoreductase
More (may be identical with EC 1.1.1.197)

CAS Reg. No.
42613-35-4

---

2 REACTION AND SPECIFICITY

Catalysed reaction
(5Z,13E)-(15S)-11alpha,15-Dihydroxy-9-oxoprosta-5,13-dienoate + NADPH
→ (5Z,13E)-(15S)-9alpha,11alpha,15-trihydroxyprosta-5,13-dienoate +
NADP+

Reaction type
Redox reaction

Natural substrates
Prostaglandin E$_2$ + NADPH (enzyme might be responsible for the control of the prostaglandin E$_2$/F$_{2alpha}$ ratio in human decidua vera [2]) [2, 5]

Substrate spectrum

1 (5Z,13E)-(15S)-11,15-Dihydroxy-9-oxoprosta-5,13-dienoate + NADPH (i.e. prostaglandin E$_2$, r [2, 3, 5], low reverse reaction [3]) [1–10]
2 15-Keto-prostaglandin E$_2$ + NADPH [1]
3 15-Ketoprostaglandin F$_{2alpha}$ + NADPH [1]
4 13,14-Dihydro-15-ketoprostaglandin F$_{2alpha}$ + NADPH [1]
5 Prostaglandin A$_1$ + NADPH [1]
6 Prostaglandin E$_1$ + NADPH [8]
7 More (conversion of both the 9-keto group and the 15-keto group of 15-ketoprostaglandin E$_2$ to the corresponding hydroxyl group (the 15-keto group is reduced faster than the 9-keto group), conversion of 15-keto group of 15-ketoprostaglandin F$_{2alpha}$ to the corresponding hydroxyl group, not: cyclohexanone, prostaglandin D$_2$ [1], 2 forms: I and II, each of them possesses reversible prostaglandin 9-ketoreductase and 15-hydroxyprostaglandin dehydrogenase activity [9]) [1, 9]

Product spectrum

1 (5Z,13E)-(15S)-9,11,15-Trihydroxyprosta-5,13-dienoate + NADP$^+$ [1, 2] .
2 Prostaglandin F$_{2alpha}$ + prostaglandin E$_2$ + 15-ketoprostaglandin F$_{2alpha}$ + NADP$^+$ [1]
3 Prostaglandin F$_{2alpha}$ + NADP$^+$ [1]
4 13,14-Dihydro-prostaglandin F$_{2alpha}$ + NADP$^+$ [1]
5 9-Hydroxyprostaglandin A$_1$ + NADP$^+$ [1]
6 Prostaglandin F$_{1alpha}$ + NADP$^+$ [8]
7 ?

Inhibitor(s)

5,5'-Dithiobis(2-nitrobenzoate) [1]; p-Chloromercuribenzoate [1]; Indomethacin (not [3, 4]) [2, 5, 9]; Fentiazac [2]; Progesterone [2, 5]; Cu^{2+} [4]; NADP$^+$ [3, 4, 8]; NAD$^+$ [3, 4]; FAD [4]; Oestradiol-17beta [5]; Cortisol [5]; Acetylsalicylic acid [5]; Ionazolac (Irritren, Byk Gulden, Konstanz, FRG) [5]; Palmitic acid [6]; More (not: myristic acid, stearic acid) [6]

Cofactor(s)/prosthetic group(s)/activating agents

NADPH (obligatory [10]) [1–10]; NADH (NADPH is more effective [4]) [3, 4]; NADP$^+$ [2, 3, 5]; Sodium diphosphate (stimulation, effect blocked in presence of 3': 5-cAMP, AMP or several other ribonucleotides) [1]; Angiotensin I (stimulation, effect blocked in presence of 3': 5-cAMP, AMP or several other ribonucleotides) [1]; N^6,O$_2$'-Dibutyrylcyclic 3':5'-AMP (slight stimulation) [1]; 3':5'-AMP (slight stimulation) [1]; CMP (slight stimulation) [1]; CDP (slight stimulation) [1]; ATP (slight stimulation) [1]; CTP (slight stimulation) [1]; GTP (slight stimulation) [1]; Oxytocin (activates) [5]

Metal compounds/salts

Ca^{2+} (activates) [5]; More (no metal requirement) [4]

Turnover number (min^{-1})

Specific activity (U/mg)
5.7 [1]

K$_m$-value (mM)
0.0016 (NADPH) [2, 5]; 0.080 (prostaglandin E$_2$) [3]; 0.093 (prostaglandin E$_2$) [2]; 0.153 (prostaglandin E$_2$) [4]; 0.304 (prostaglandin E$_1$) [8]

pH-optimum
7.4–8.8 [1]; 7.4–9.0 [3]; 7.5 [10]

pH-range

Temperature optimum (°C)
37 (assay at) [1]

Temperature range (°C)

---

3 ENZYME STRUCTURE

Molecular weight
21800 (rabbit, gel filtration) [10]
29000 (human, FPLC) [2]
33000 (human, gel filtration) [5]
45000–55000 (chicken, gel filtration) [1]

Subunits
Monomer (1 × 30000, human, SDS-PAGE of denatured enzyme, FPLC) [1]

Glycoprotein/Lipoprotein
–

---

4 ISOLATION/PREPARATION

Source organism
Chicken [1]; Human [2, 3, 5]; Rat [4, 8]; Rabbit [7, 8, 10]; Pig [9]

Source tissue
Heart [1]; Uterine decidua vera [2, 5]; Erythrocytes [3]; Kidney (renal cortex [10]) [7, 10]; Testis [4]; Ovary [7]; Adrenal cortex [8]

Localisation in source
Cytoplasm [3, 4]; Cytosol [8]; Microsomes [4]

Purification
Chicken [1]; Human (partial [3]) [2, 3, 5]; Rabbit (partial) [10]; Pig (2 forms: I and II, each of them possesses reversible prostaglandin 9-ketoreductase and 15-hydroxyprostaglandin dehydrogenase activity) [9]

Crystallization

–

Cloned

–

Renaturated

–

5 STABILITY

pH

Temperature (°C)
45 (10 min, 75% loss of activity) [3]; 60 (half-life: 13.5 min (form I), 10 min (form II)) [9]

Oxidation

Organic solvent

General stability information

Storage

6 CROSSREFERENCES TO STRUCTURE DATABANKS

PIR/MIPS code

Brookhaven code

7 LITERATURE REFERENCES

[1] Lee, S.-C., Levine, L.: J. Biol. Chem.,250,4549–4555 (1975)
[2] Krüger, S., Schlegel, W.: Eur. J. Biochem.,157,481–485 (1986)
[3] Kaplan, L., Lee, S.-C., Levine, L.: Arch. Biochem. Biophys.,167,287–293 (1975)
[4] Thuy, L.P., Carpenter, M.P.: Biochem. Biophys. Res. Commun.,81,322–328 (1978)
[5] Schlegel, W., Krüger, S., Korte, K.: FEBS Lett.,171,141–144 (1984)
[6] Oshige, T., Ohtsuka, T., Mibe, M., Nagai, K., Yamaguchi, M., Ikenoue, T., Mori, N.: Prostaglandins, Leukotrienes Essent. Fatty Acids,45,245–247 (1992)
[7] Schlegel, W., Daniels, D., Krüger, S.: Clin. Physiol. Biochem.,5,336–342 (1987)
[8] Levasseur, S., Friedman, Y., Burke, G.: Biochem. Biophys. Res. Commun.,95, 236–242 (1980)
[9] Chang, D. G.-B., Sun, M., Tai, H.-H.: Biochem. Biophys. Res. Commun.,99,745–751 (1981) (Review)
[10] Stone, K.J., Hart, M.: Prostaglandins,10,273–288 (1975)

1 NOMENCLATURE

EC number
1.1.1.190

Systematic name
Indole-3-ethanol:NAD+ oxidoreductase

Recommended name
Indole-3-acetaldehyde reductase (NADH)

Synonymes
Indoleacetaldehyde reductase
Reductase, indoleacetaldehyde

CAS Reg. No.
58875-06-2

2 REACTION AND SPECIFICITY

Catalysed reaction
Indole-3-acetaldehyde + NADH →
→ indole-3-ethanol + NAD+

Reaction type
Redox reaction

Natural substrates
Indole-3-acetaldehyde + NADH (possible role in regulation of auxin biosynthesis) [4]

Substrate spectrum
1 Indole-3-acetaldehyde + NADH (no reverse reaction in vitro [3]) [1–4]

Product spectrum
1 Indole-3-ethanol + NAD+

Inhibitor(s)
Ionic strength (slight decrease of activity above 0.04 M NaCl) [3, 4]; Sodium azide [4]; EDTA [4]; Cyanide [4]; Iodoacetate [4]; N-Ethylmaleimide [4]

Cofactor(s)/prosthetic group(s)/activating agents
NADH [1–4]

Metal compounds/salts
No activation by divalent metal ions [4]

Turnover number (min⁻¹)

Specific activity (U/mg)

Kₘ-value (mM)
0.40 (indole-3-acetaldehyde) [3]; 0.180 (NADH) [3]; 0.5 (indole-3-acetaldehyde) [4]; 0.14 (NADH) [4]

pH-optimum
7.0 [3]; 7.6 [1]

pH-range
4.5–8 [3]

Temperature optimum (°C)

Temperature range (°C)

---

3 ENZYME STRUCTURE

Molecular weight
32000 (Cucumis sativus, gel filtration) [4]

Subunits

Glycoprotein/Lipoprotein
–

---

4 ISOLATION/PREPARATION

Source organism
Nicotiana glauca [1]; Nicotiana langsdorffii [1]; Cucumis sativus [2–4]

Source tissue
Callus tissue [1]; Seedlings [2–4]

Localisation in source
Cytosol [2]

Purification
Cucumis sativus (partial, no separation from NADPH-specific enzyme) [3]

Crystallization
–

Cloned
–

Renaturated
–

5 STABILITY

pH

Temperature (°C)

Oxidation

Organic solvent

General stability information

Storage

6 CROSSREFERENCES TO STRUCTURE DATABANKS

PIR/MIPS code

Brookhaven code

7 LITERATURE REFERENCES

[1] Liu, S.-T., Katz, C.D., Knight, C.A.: Plant Physiol.,61,743–747 (1978)
[2] Bower, P.J., Brown, H.M., Purves, W.K.: Plant Physiol.,57,850–854 (1976)
[3] Brown, H.M., Purves, W.K.: J. Biol. Chem.,251,907–913 (1976)
[4] Brown, H.M., Purves, W.K.: Plant Physiol.,65,107–113 (1980)

1 NOMENCLATURE

EC number
1.1.1.191

Systematic name
Indole-3-ethanol:NADP+ oxidoreductase

Recommended name
Indole-3-acetaldehyde reductase (NADPH)

Synonyms
Reductase, indoleacetaldehyde (reduced nicotinamide adenine dinucleotide phosphate)

CAS Reg. No.
58875-05-1

2 REACTION AND SPECIFICITY

Catalysed reaction
Indole-3-acetaldehyde + NADPH →
→ indole-3-ethanol + NADP+

Reaction type
Redox reaction

Natural substrates
Indole-3-acetaldehyde + NADPH (possible role in regulation of auxin biosynthesis [6]) [5, 6]

Substrate spectrum
1 Indole-3-acetaldehyde + NADPH (no reverse reaction in vitro [5], enzymes from Brassica campestris, Brassica oleracea, Arabidopsis thaliana can use either NADPH or NADH [1], activity with NADH 10 times lower compared to NADPH [2]) [1, 2, 5, 6]
2 Phenylacetaldehyde + NADPH [1, 2, 5]
3 Benzaldehyde + NADPH [1, 2]
4 Acetaldehyde + NADPH [1, 5]
5 Dihydroxyacetone + NADPH [2]
6 trans-Cinnamaldehyde + NADPH [5]
7 Butyraldehyde + NADPH [5]
8 More (not: indole-3-aldehyde, indole-3-carbaldehyde) [1]

Enzyme Handbook © Springer-Verlag Berlin Heidelberg 1995
Duplication, reproduction and storage in data banks are only
allowed with the prior permission of the publishers

Product spectrum

1 Indole-3-ethanol + NADP+
2 Phenylethanol + NADP+ [5]
3 Benzyl alcohol + NADP+
4 Ethanol + NADP+
5 ?
6 3-Phenyl-2-propenol + NADP+
7 Butanol + NADP+
8 ?

Inhibitor(s)

NADPH (more than 200 mM [1], non competitive [2]) [1, 2]; Iodoacetate (more than 0.05 mM [1]) [1, 6]; Tryptophan [1]; Acetaldehyde [1]; p-Chloromercuriphenylsulfonate [1]; p-Chloromercuribenzoate [1]; $MnCl_2$ [2]; $ZnSO_4$ [2]; $HgCl_2$ [2]; NaN_3 [6]; EDTA [6]; NaF [6]; N-Ethylmaleimide [6]

Cofactor(s)/prosthetic group(s)/activating agents

NADPH (specific for [3], enzymes from Brassica rapa and Sinapis alba specific for [1]) [1–6]; NADH (enzymes from Brassica campestris, Brassica oleracea, Arabidopsis thaliana can use either NADPH or NADH [1], activity with NADH 10 times lower compared to NADPH [2]) [1, 2]

Metal compounds/salts

Fe-citrate (activation) [2]; $FeCl_3$ (activation) [2]; Mn^{2+} (activation) [5]; Mg^{2+} (activation) [5]; Ca^{2+} (activation) [5]

Turnover number (min^{-1})

Specific activity (U/mg)

More [2, 3]

K_m-value (mM)

0.125 (indole-3-acetaldehyde, similar value [2, 5], (+ NADPH [1])) [1, 2, 5, 6]; 0.4 (indole-3-acetaldehyde (+ NADH)) [1]; 0.036 (NADPH, similar values [2, 5, 6]) [1, 2, 5, 6]; 0.5 (phenylacetaldehyde) [1]; 0.715 (acetaldehyde) [1]; 0.645 (dihydroxyacetone, light-grown organism) [2]; 0.910 (dihydroxyacetone, dark-grown organism) [2]

pH-optimum

5.2 [5]; 6–7 [1]; 6–8 [2]

pH-range

4.5–8 [5]; 5–7.5 [1]

Temperature optimum (°C)

30 [2]

Temperature range (°C)

3 ENZYME STRUCTURE

Molecular weight
38000 (Phycomyces blakesleeanus, gel filtration) [2]
32000 (Brassica sp., gel permeation HPLC [1], Cucumis sativus, gel filtration [6], SDS-PAGE [3]) [1, 3, 6]

Subunits

Glycoprotein/Lipoprotein
–

4 ISOLATION/PREPARATION

Source organism
Brassica campestris spp. pekinensis [1]; Brassica oleracea [1]; Arabidopsis thaliana [1]; Brassica rapa [1]; Sinapis alba [1]; Phycomyces blakesleeanus [2]; Cucumis sativus [3–6]

Source tissue
Seedlings [1, 4–6]; Hypocotyls [3]; Roots [3]; Fruits [3]

Localisation in source
Cytoplasm [3]; Microsomes [4]

Purification
Phycomyces blakesleeanus [2]; Cucumis sativus (partially, no separation from NADH specific enzyme) [5]

Crystallization
–

Cloned
–

Renaturated
–

5 STABILITY

pH

Temperature (°C)

Oxidation

Organic solvent

General stability information

Storage

6 CROSSREFERENCES TO STRUCTURE DATABANKS

PIR/MIPS code

Brookhaven code

---

7 LITERATURE REFERENCES

[1] Ludwig-Müller, J., Hilgenberg, W.: Physiol. Plant.,80,541–548 (1990)
[2] Ludwig-Müller, J., Schramm, P., Hilgenberg, W.: Physiol. Plant.,80,472–478 (1990)
[3] Ludwig-Müller, J., Hilgenberg, W.: Physiol. Plant.,77,613–619 (1989)
[4] Bower, P.J., Brown, H.M., Purves, W.K.: Plant Physiol.,57,850–854 (1976)
[5] Brown, H.M., Purves, W.K.: J. Biol. Chem.,251,907–913 (1976)
[6] Brown, H.M., Purves, W.K.: Plant Physiol.,65,107–113 (1980)

1 NOMENCLATURE

EC number
1.1.1.192

Systematic name
Long-chain-alcohol:NAD+ oxidoreductase

Recommended name
Long-chain-alcohol dehydrogenase

Synonymes
Dehydrogenase, long-chain alcohol
Long-chain alcohol dehydrogenase
Fatty alcohol oxidoreductase [1]

CAS Reg. No.
76774-36-2

2 REACTION AND SPECIFICITY

Catalysed reaction
A long-chain alcohol + 2 NAD+ + H_2O →
→ a long-chain carboxylate + 2 NADH

Reaction type
Redox reaction

Natural substrates
Long-chain alcohol + NAD+ + H_2O (enzyme may play an important role in regulating the cellular levels of ether-linked lipids [1], microsomal and mitochondrial enzyme may have an indispensable role in lipid biosynthesis [2, 3], peroxisomal enzyme is participating in fatty acid degradation [2, 3])

Substrate spectrum
1 Long-chain alcohol + NAD+ + H_2O (rat: C_{10}-C_{18} [1], Candida tropicalis: C_8-C_{16} [2], C_{17}-C_{19} alcohols do not serve as a good substrate [2], Candida lipolytica: C_6-C_{18} alcohols [2], maximal activity with decanol (rat [1], Candida lipolytica [2])) [1–3]

Product spectrum
1 Long-chain carboxylate + NADH

Inhibitor(s)
Cyanide [1]; N-Ethylmaleimide [1]; More (not: pyrazole, sodium azide, AMP, ATP, quinacrine, iodoacetate) [1]

Cofactor(s)/prosthetic group(s)/activating agents
NAD^+ (absolute requirement) [1, 2]; Albumin (0.5 mg/ml increases reaction rate) [1]

Metal compounds/salts

Turnover number (min^{-1})

Specific activity (U/mg)

K_m-value (mM)
0.00067 (hexadecanol) [1]

pH-optimum
8.4 (hexadecanol, diphosphate buffer) [1]; 8.8–9.0 (hexadecanol, glycine or barbital buffer) [1]

pH-range

Temperature optimum (°C)
30 (assay at) [2]; 37 (assay at) [1]

Temperature range (°C)

---

3 ENZYME STRUCTURE

Molecular weight

Subunits

Glycoprotein/Lipoprotein
–

---

4 ISOLATION/PREPARATION

Source organism
Rat [1]; Candida tropicalis [2, 3]; Candida lipolytica [2]

Source tissue
Liver [1]; Cell [2]

Localisation in source
Microsomes [1–3]; Mitochondria [2, 3]; Peroxisomes [2, 3]; Membrane (bound) [2]

Purification

Crystallization
–

Cloned
–

Renaturated
–

5 STABILITY

pH

Temperature (°C)

Oxidation

Organic solvent

General stability information
Solubilization with 1% Tween 80 or Triton X-100 at 4°C stimulates inactivation, crude extract [2]; Dithiothreitol, 2 mM, no stabilization, crude extract [2]; Ethylene glycol, 20% w/w, no stabilization, crude extract [2]

Storage

6 CROSSREFERENCES TO STRUCTURE DATABANKS

PIR/MIPS code

Brookhaven code

7 LITERATURE REFERENCES

[1] Lee, T.-C.: J. Biol. Chem.,254,2892–2896 (1979)
[2] Ueda, M., Tanaka, A.: Methods Enzymol.,188,171–175 (1990) (Review)
[3] Yamada, T., Nawa, H., Kawamoto, S., Tanaka, A., Fukui, S.: Arch. Microbiol.,128, 145–151 (1980)

1 NOMENCLATURE

EC number
1.1.1.193

Systematic name
5-Amino-6-(5-phosphoribitylamino)uracil:NADP$^+$ 1'-oxidoreductase

Recommended name
5-Amino-6-(5-phosphoribosylamino)uracil reductase

Synonymes
Aminodioxyphosphoribosylaminopyrimidine reductase

CAS Reg. No.
69020-28-6

---

2 REACTION AND SPECIFICITY

Catalysed reaction
5-Amino-6-(5-phosphoribitylamino)uracil + NADP$^+$ →
→ 5-amino-6-(5-phosphoribosylamino)uracil + NADPH

Reaction type
Redox reaction

Natural substrates
5-Amino-6-(5-phosphoribosylamino)uracil + NADPH [1]

Substrate spectrum
1 5-Amino-6-(5-phosphoribosylamino)uracil + NAD(P)H [1]

Product spectrum
1 5-Amino-6-(5-phosphoribitylamino)uracil + NAD(P)$^+$ [1]

Inhibitor(s)
Riboflavin [1]; FMN [1]; FAD [1]

Cofactor(s)/prosthetic group(s)/activating agents
NADPH [1]; NADH [1]

Metal compounds/salts

Turnover number (min⁻¹)

Specific activity (U/mg)

K_m-value (mM)
   0.005 (NADPH) [1]

pH-optimum
   7.5 [1]

pH-range

Temperature optimum (°C)

Temperature range (°C)

---

3 ENZYME STRUCTURE

Molecular weight
   37000 (E. coli, gel filtration) [1]

Subunits

Glycoprotein/Lipoprotein
   –

---

4 ISOLATION/PREPARATION

Source organism
   E. coli [1]

Source tissue

Localisation in source

Purification
   E. coli (partially) [1]

Crystallization
   –

Cloned
   –

Renaturated
   –

5 STABILITY

pH

Temperature (°C)

Oxidation

Organic solvent

General stability information

Storage
–6°C, 1 month [1]

6 CROSSREFERENCES TO STRUCTURE DATABANKS

PIR/MIPS code

Brookhaven code

7 LITERATURE REFERENCES

[1] Burrows, R.B., Brown, G.M.: J. Bacteriol.,136,657–667 (1978)

1 NOMENCLATURE

EC number
1.1.1.194

Systematic name
Coniferyl-alcohol:NADP$^+$ oxidoreductase

Recommended name
Coniferyl-alcohol dehydrogenase

Synonymes
CAD [1]

CAS Reg. No.

2 REACTION AND SPECIFICITY

Catalysed reaction
Coniferyl alcohol + NADP$^+$ →
→ coniferyl aldehyde + NADPH

Reaction type
Redox reaction

Natural substrates
Coniferyl aldehyde + NADPH (lignin biosynthesis) [1, 3]

Substrate spectrum
1 Coniferyl alcohol + NADP$^+$ (r [4], the only substrate found, specific for co-niferyl alcohol, does not act on cinnamyl alcohol, 4-coumaryl alcohol or sinapyl alcohol [5]) [1–5]
2 More (not: benzyl alcohol, anisic alcohol, ethanol) [5]

Product spectrum
1 Coniferyl aldehyde + NADPH [1–5]
2 ?

Inhibitor(s)
More (no substrate inhibition) [5]

Cofactor(s)/prosthetic group(s)/activating agents
NADP$^+$ [1–5]; NADPH [4, 5]; More (no reaction with NAD$^+$ or NADH) [5]

Metal compounds/salts

Turnover number (min⁻¹)

Specific activity (U/mg)
   0.36 [5]

K_m-value (mM)
   0.830 (coniferyl alcohol) [5]; 0.0074 (NADP⁺) [5]

pH-optimum
   9.2 (coniferyl alcohol oxidation) [5]

pH-range

Temperature optimum (°C)

Temperature range (°C)

---

3 ENZYME STRUCTURE

Molecular weight
   43000 (Glycine max, gel filtration) [5]

Subunits

Glycoprotein/Lipoprotein
   –

---

4 ISOLATION/PREPARATION

Source organism
   Metasequoia glyptostroboides [2]; Pinus palustris [2]; Zea mays [2]; Medi-
   cago sativa L. [1, 2]; Schinus terebinthifolius [2]; Petunia hybrida [3]; Acer
   rubrum [2]; Prunus padus [2]; Salix purpurea [2]; Salix alba [4]; Glycine max
   (L. var. Mandarin) [5]

Source tissue
   Internode [3]; Cell suspension cultures [5]

Localisation in source

Purification
   Glycine max (var. Mandarin) [5]

Crystallization
   –

Cloned
   –

Renaturated
   –

5 STABILITY

pH

Temperature (°C)

Oxidation

Organic solvent

General stability information

Storage
−20°C, 42 mM 2-mercaptoethanol, 10% ethyleneglycol, several weeks [5]

6 CROSSREFERENCES TO STRUCTURE DATABANKS

PIR/MIPS code

Brookhaven code

7 LITERATURE REFERENCES

[1] Dalkin, K., Edwards, R., Edington, B., Dixon, R.A.: Plant Physiol.,92,440–446 (1990)
[2] Mansell, R.L., Babbel, G.R., Zenk, M.H.: Phytochemistry,15,1849–1853 (1976)
[3] Ranjeva, R., Boudet, A.M., Harada, H., Marigo, G.: Biochim. Biophys. Acta,399,23–30 (1975)
[4] Gross, G.G., Stöckigt, J., Mansell, R.L., Zenk, M.H.: FEBS Lett.,31,283–286 (1973)
[5] Wyrambik, D., Grisebach, H.: Eur. J. Biochem.,59,9–15 (1975)

1 NOMENCLATURE

EC number
1.1.1.195

Systematic name
Cinnamyl-alcohol:NADP+ oxidoreductase

Recommended name
Cinnamyl-alcohol dehydrogenase

Synonymes
Dehydrogenase, cinnamyl alcohol
Cinnamyl alcohol dehydrogenase
CAD [8]
More (cf. EC 1.1.1.194)

CAS Reg. No.
55467-36-2

2 REACTION AND SPECIFICITY

Catalysed reaction
Cinnamyl alcohol + NADP+ →
→ cinnamaldehyde + NADPH (A-specific enzyme [11, 12], ordered bi-bi
mechanism [12])

Reaction type
Redox reaction

Natural substrates
More (enzyme of lignification [3], final step in a branch of phenylpropanoid
synthesis specific for production of lignin monomers [4], one of the regulating enzymes which controls the formation of guaiacyl and syringyl lignins
[8]) [3, 4, 8]

Substrate spectrum

1 Cinnamyl alcohol + NADP$^+$ (r [1]) [1, 11]
2 Coniferyl alcohol + NADP$^+$ (r [1, 8, 9]) [1, 7–9, 11]
3 p-Coumaryl alcohol + NADP$^+$ (r [1, 9]) [1, 7, 9, 11]
4 Sinapyl alcohol + NADP$^+$ (r [1, 8, 9]) [1, 7–9]
5 3,4-Dimethoxycinnamyl alcohol + NADP$^+$ (r [1]) [1]
6 4-Methoxycinnamyl alcohol + NADP$^+$ [11]
7 3,4-Methylenedioxycinnamyl alcohol + NADP$^+$ [11]
8 4-Methylcinnamyl alcohol + NADP$^+$ [11]
9 2,4-Dimethylcinnamyl alcohol + NADP$^+$ [11]
10 4-Chlorocinnamyl alcohol + NADP$^+$ [11]
11 4-Bromocinnamyl alcohol + NADP$^+$ [11]
12 More (not: methanol, ethanol, n-propanol, n-butanol, isobutanol, geraniol, various aromatic alcohols (e.g. salicyl alcohol, vanillyl alcohol, piperonyl alcohol)) [11]

Product spectrum

1 Cinnamaldehyde + NADPH [1]
2 Coniferyl aldehyde + NADPH [1, 7]
3 p-Coumaraldehyde + NADPH [1, 7]
4 Sinapaldehyde + NADPH [1, 7]
5 3,4-Dimethoxycinnamaldehyde + NADPH [1]
6 4-Methoxycinnamaldehyde + NADPH [11]
7 3,4-Methylenedioxycinnamaldehyde + NADPH [11]
8 4-Methylcinnamaldehyde + NADPH [11]
9 2,4-Dimethylcinnamaldehyde + NADPH [11]
10 4-Chlorocinnamaldehyde + NADH [11]
11 4-Bromocinnamaldehyde + NADPH [11]
12 ?

Inhibitor(s)

Cofactor(s)/prosthetic group(s)/activating agents

NADP$^+$ (A-specific [11, 12]) [1, 7, 11, 12]; NADPH [1, 8, 9]

Metal compounds/salts

Zn^{2+} (restores activity after incubation in buffer containing a divalent-cation exchange resin) [7]

Turnover number (min^{-1})

Specific activity (U/mg)

3.86 [1]; 1.71 [7]; 0.6 [8]; 64.56 [9]; 3.5 [11]; 56.4 [12]

K_m-value (mM)

0.0038 ($NADP^+$) [1]; 0.057 (p-coumaryl alcohol) [1]; 0.011 (coniferyl alcohol) [1]; 0.050 (sinapyl alcohol) [1]; 0.022 (3,4-dimethoxy-cinnamyl alcohol) [1]; 0.026 (cinnamyl alcohol) [1]; 0.0072 (NADPH) [1]; 0.0091 (p-coumaraldehyde) [1]; 0.0017 (coniferaldehyde) [1]; 0.0043 (sinapaldehyde) [1]; 0.0069 (3,4-dimethoxycinnamaldehyde) [1]; 0.0025 (cinnamaldehyde) [1]; 0.00077 (coniferaldehyde) [7]; 0.008 (NADPH) [7]; 0.0012 (p-coumaraldehyde) [7]; 0.0048 (sinapaldehyde) [7]; 0.0068 (NADPH) [8]; 0.0091 (coniferaldehyde) [8]; 0.014 (cinnamaldehyde) [8]; 0.030 (p-coumaraldehyde) [8]; 0.043 (3,4-dimethoxycinnamaldehyde) [8]; More [9, 11]

pH-optimum

6.0 (reduction of coniferyl aldehyde, KH_2PO_4/Na_2HPO_4 buffer) [7]; 6.1 [9]; 6.5 (coniferyl aldehyde reduction) [1]; 7.0 (reduction of coniferyl aldehyde, Tris maleate/NaOH buffer) [7]; 7.6 (aldehyde reduction) [11]; 8.8 (coniferyl alcohol oxidation [1], alcohol oxidation [11]) [1, 11]

pH-range

4.5–8.5 (4.5: about 55% of activity maximum, 8.5: about 30% of activity maximum) [9]; 6.0–8.0 (coniferyl aldehyde reduction, 6.0: about 70% of activity maximum, 8.0: about 30% of activity maximum) [1]; 7.2–9.7 (coniferyl alcohol oxidation, 7.2: about 40% of activity maximum, 9.7: about 35% of activity maximum) [1]

Temperature optimum (°C)

40 [11]

Temperature range (°C)

3 ENZYME STRUCTURE

Molecular weight

69000 (Glycine max, gel filtration) [1]
70000 (Populus X euramericana, gel filtration) [7]
80000 (Forsythia suspensa) [11]

Subunits

Dimer (2 × 40000, Populus X euramericana, SDS-PAGE [7], Glycine max, SDS-PAGE [12]) [7, 12]; ?(x × 65000, Phaseolus vulgaris, SDS-PAGE) [5]

Glycoprotein/Lipoprotein

–

4 ISOLATION/PREPARATION

Source organism
More (overview: occurence in plants) [11]; Wheat (induced with elicitor from Puccinia graminis f. sp. tritici germ tube walls) [6]; Populus X euramericana [7, 8]; Pinus thunbergii (Japanese black pine) [8]; Ginkgo biloba [8]; Thuja orientalis [8, 10]; Cryptomeria japonica [8]; Metasequoia glyptostroboides [8, 10]; Liridendron tulipifera [8]; Robinia pseudoacacia [8, 10]; Erythrina crista-galli [8]; Prunus yedoensis [8]; Prunus persica [8]; Picea abies [9]; Forsythia suspensa [11]; Glycine max L. (var. Mandarin) [1, 12]; Hevea brasiliensis [2]; Phaseolus mungo (neoformed calli induced by Agrobacterium rhizogenes on Phaseolus mungo hypocotyl) [3]; Phaseolus vulgaris L. (with a high-molecular-mass elicitor preparation heat-released from mycelial cell walls [5]) [4, 5]

Source tissue
Cell suspension culture [1, 12]; Laticifer [2]; Leaf (low activity) [2]; Cambial sap [9]; Wood [10]; Young stem sections [11]; More (neoformed calli induced by Agrobacterium rhizogenes on Phaseolus hypocotyl) [3]

Localisation in source

Purification
Glycine max L. (var. Mandarin) [1, 12]; Forsythia suspensa [11]; Populus X euramericana [7]; Pinus thunbergii [8]

Crystallization
[4]

Cloned
–

Renaturated
–

5 STABILITY

pH
7.5 (4°C, 1 week, 50% loss of activity) [7]; 6.8–8.4 (maximum stability) [11]; 6.8 (unstable below) [11]

Temperature (°C)
40 (rapid inactivation above) [11]

Oxidation

Organic solvent

General stability information
2-Mercaptoethanol stabilizes [7]; 2-Mercaptoethanol: stable in presence of high concentration [9]; Ethylene glycol stabilizes [7]; Ethylene glycol stabilizes dilute solutions [9]; Dithiothreitol stabilizes [7]; Freezing of glycerol containing fractions causes 50% loss of activity after 24 h [11]; Glycerol stabilizes [11]; 60% Loss of activity after 20 h incubation in buffer containing a divalent cation-exchange resin [7]

Storage
-20°C, 42 mM 2-mercaptoethanol, 10% ethylene glycol, 5 months [1]; 4°C, mercaptoethanol, ethylene glycol, 1 month [7]; -20°C, NADP+, mercaptoethanol, ethylene glycol, several months [9]; -18°C, 20 or 50% glycerol, several months [11]

6 CROSSREFERENCES TO STRUCTURE DATABANKS

PIR/MIPS code
PIR3:S31572 (alfalfa); PIR3:S31571 (cottonwood (fragment)); PIR2:S33534 (cider tree); PIR3:S23525 (common tobacco); PIR3:S23526 (common tobacco); PIR1:DEFBC (kidney bean)

Brookhaven code

7 LITERATURE REFERENCES

[1] Wyrambik, D., Grisebach, H.: Eur. J. Biochem.,59,9–15 (1975)
[2] Kush, A., Goyvaerts, E., Chye, M.-L., Chua, N.-H.: Proc. Natl. Acad. Sci. USA,87,1787–1790 (1990)
[3] Grima-Pettenati, J., Chriqui, D., Sarni Manchado, P., Prinsen, E.: Plant Sci.,61,179–188 (1989)
[4] Walter, M.H., Grima-Pettenati, J., Grand, C., Boudet, A.M., Lamb, C.J.: Proc. Natl. Acad. Sci. USA,85,5546–5550 (1988)
[5] Grand, C., Sarni, F., Lamb, C.J.: Eur. J. Biochem.,169,73–77 (1987)
[6] Moerschbacher, B., Heck, B., Kogel, K. H., Obst, O., Reisener, H.J.: Z. Naturforsch.,41c,839–844 (1986)
[7] Sarni, F., Grand, C., Boudet, A.M.: Eur. J. Biochem.,139,259–265 (1984)
[8] Kutsuki, H., Shimada, M., Higuchi, T.: Phytochemistry,21,19–23 (1982)
[9] Lüderitz, T., Grisebach, H.: Eur. J. Biochem.,119,115–124 (1981)
[10] Kutsuki, H., Higuchi, T.: Planta,152,365–368 (1981)
[11] Mansell, R.L., Gross, G.G., Stöckigt, J., Franke, H., Zenk, M.H.: Phytochemistry,13, 2427–2435 (1974)
[12] Wyrambik, D., Grisebach, H.: Eur. J. Biochem.,97,503–509 (1979)

1 NOMENCLATURE

EC number
1.1.1.196

Systematic name
(5Z,13E)-(15S)-9alpha,15-Dihydroxy-11-oxoprosta-5,13-dienoate:NADP⁺
15-oxidoreductase

Recommended name
15-Hydroxyprostaglandin-D dehydrogenase (NADP⁺)

Synonymes
Dehydrogenase, prostaglandin D_2
NADP-PGD$_2$ dehydrogenase
Prostaglandin-D 15-dehydrogenase (NADP⁺)
Dehydrogenase, 15-hydroxyprostaglandin (nicotinamide adenine dinucleoti-
de phosphate)
15-Hydroxy PGD$_2$ dehydrogenase
15-Hydroxyprostaglandin dehydrogenase (NADP)
NADP-dependent 15-hydroxyprostaglandin dehydrogenase
Prostaglandin D_2 dehydrogenase
NADP-linked 15-hydroxyprostaglandin dehydrogenase
NADP-specific 15-hydroxyprostaglandin dehydrogenase
NADP-linked prostaglandin D_2 dehydrogenase [3]

CAS Reg. No.
84399-95-1; 54989-39-8 (identical with CAS Reg. No. of EC 1.1.1.197)

2 REACTION AND SPECIFICITY

Catalysed reaction
(5Z,13E)-(15S)-9alpha,15-Dihydroxy-11-oxoprosta-5,13-dienoate + NADP⁺ →
→ (5Z,13E)-9alpha-hydroxy-11,15-dioxoprosta-5,13-dienoate + NADPH

Reaction type
Redox reaction

Natural substrates
Prostaglandin D_2 + NADP⁺ (enzyme is responsible for specific interaction of
prostaglandin D_2 which is the major prostaglandin in the central nervous system) [1]

Substrate spectrum

1 (5Z,13E)-(15S)-9alpha,15-Dihydroxy-11-oxoprosta-5,13-dienoate + NADP+
(i.e. prostaglandin D_2, specific for) [1, 2]
2 Prostaglandin D_1 + NADP+ (25% of the activity with prostaglandin D_2) [2]
3 More (poor substrates with less than 10% of the reaction rate with prostaglandin D_2: prostaglandin A_2 (not [3]) [1, 2], B_2 (not [3]) [1], D_3 [1, 2], E_2
[1, 2], F_{2alpha} [1, 2], A and B series of prostaglandins cannot serve as substrates [2], not: E_1 [3]) [1–3]

Product spectrum

1 (5Z,13E)-9alpha-Hydroxy-11,15-dioxoprosta-5,13-dienoate + NADPH (i.e.
15-ketoprostaglandin D_2)
2 ?
3 ?

Inhibitor(s)

N-Ethylmaleimide [3]; Arachidonic acid [3]; 11,14-Eicosadienoic acid [3];
p-Chloromercuribenzoate [3]; Linoleic acid [3]; Indomethacin [3]; Flufena-
mic acid [3]; L-Thyroxine [3]; 3,3',5-Triiodothyroacetic acid [3]; More (not:
arachidic acid, stearic acid, capric acid, phenylbutazone, 2-hexadecenoic
acid) [3]

Cofactor(s)/prosthetic group(s)/activating agents

NADP+ (exclusively [2, 3]) [1–3]; NAD+ (less than 1% of the activity with
NADP+) [1, 3]

Metal compounds/salts

Turnover number (min⁻¹)

Specific activity (U/mg)

More [1–3]

K_m-value (mM)

0.0004 (NADP+) [3]; 0.005 (NADP+) [1, 2]; 0.052 (prostaglandin D_2) [3];
0.070 (prostaglandin D_2) [1, 2]; 0.250 (prostaglandin D_1) [3]; 0.350 (prostaglandin E_2) [3]; 0.4 (prostaglandin F_{2alpha}) [3]; 0.54 (prostaglandin I_2) [3]

pH-optimum

9.0 (assay at) [1, 2]; 9.5 [3]

pH-range

Temperature optimum (°C)

24 (assay at) [1–3]

Temperature range (°C)

3 ENZYME STRUCTURE

Molecular weight
30000 (pig, sedimentation equilibrium analysis) [3]
52000–62000 (pig, gel filtration) [1, 2]

Subunits
Monomer (1 × 28000, pig, SDS-PAGE) [3]

Glycoprotein/Lipoprotein
–

---

4 ISOLATION/PREPARATION

Source organism
Pig [1–3]

Source tissue
Brain [1–3]

Localisation in source
Cytosol [1]

Purification
Pig (partial) [1]

Crystallization
–

Cloned
–

Renaturated
–

---

5 STABILITY

pH
5 (4°C, 6 h, 100% loss of activity) [3]; 6 (4°C, 6 h, 70% loss of activity) [3]; 7 (4°C, 6 h, 54% loss of activity) [3]; 8 (4°C, 6 h, 76% loss of activity) [3]; 9 (4°C, 6 h, 100% loss of activity) [3]

Temperature (°C)

Oxidation

Organic solvent

General stability information
Unstable at protein concentrations of less than 10 mg/ml [2]; Sulfhydryl reagents, such as dithiothreitol, 2-mercaptoethanol and reduced glutathione, 0.1 mM, protect against inactivation [2, 3]

Storage

---

6 CROSSREFERENCES TO STRUCTURE DATABANKS

PIR/MIPS code

Brookhaven code

---

7 LITERATURE REFERENCES

[1] Watanabe, K., Shimizu, T., Iguchi, S., Wakutsuka, H., Hayashi, M., Hayaishi, O.: J. Biol. Chem.,255,1779–1782 (1980)
[2] Shimizu, T., Watanabe, K., Tokumoto, H., Hayaishi, O.: Methods Enzymol.,86, 147–152 (1982) (Review)
[3] Tokumoto, H., Watanabe, K., Fukushima, D., Shimizu, T., Hayaishi, O.: J. Biol. Chem.,257,13576–13580 (1982)

1 NOMENCLATURE

EC number

1.1.1.197

Systematic name

(13E)-(15S)-11alpha,15-Dihydroxy-9-oxoprost-13-enoate:NADP⁺ 15-oxidore-
ductase

Recommended name

15-Hydroxyprostaglandin dehydrogenase (NADP⁺)

Synonymes

Dehydrogenase, 15-hydroxyprostaglandin (nicotinamide adenine dinucleotide phosphate)

NADP-dependent 15-hydroxyprostaglandin dehydrogenase

NADP-linked 15-hydroxyprostaglandin dehydrogenase

NADP-specific 15-hydroxyprostaglandin dehydrogenase

Type II 15-hydroxyprostaglandin dehydrogenase [3]

More (may be identical with EC 1.1.1.189)

CAS Reg. No.

54989-39-8 (identical with one of the two CAS Reg. No. of EC 1.1.1.196)

2 REACTION AND SPECIFICITY

Catalysed reaction

(13E)-(15S)-11alpha,15-Dihydroxy-9-oxoprost-13-enoate + NADP⁺ →
→ (13E)-11alpha-hydroxy-9,15-dioxoprosta-13-enoate + NADPH

Reaction type

Redox reaction

Natural substrates

Prostaglandin + NADP⁺ (enzyme may play a significant role in regulation of
intracellular levels of prostaglandins of the E and F series in blood vessels) [5]

Substrate spectrum

1 (13E)-(15S)-11alpha,15-Dihydroxy-9-oxoprost-13-enoate + NADP⁺ (i.e.
prostaglandin E_1) [6]

2 Prostaglandin E_2 + NADP⁺ [1, 3, 5]

3 Prostaglandin F_{2alpha} + NADP⁺ (r) [3, 4]

4 More (2 enzyme forms: I and II, each of them possesses reversible prostaglandin 9-ketoreductase and NADP-linked 15-hydroxyprostaglandin
dehydrogenase activity, compare EC 1.1.1.189 [3], not: prostaglandin B_2
[1], prostaglandin D_2) [1–3]

Product spectrum
  1 (13E)-11alpha-Hydroxy-9,15-dioxoprosta-13-enoate + NADPH (i.e. 15-ke-
    toprostaglandin E$_1$)
  2 15-Ketoprostaglandin E$_2$ + NADPH [3]
  3 15-Ketoprostaglandin F$_{2alpha}$ + NADPH
  4 ?

Inhibitor(s)
  NADPH [1, 2]; Indomethacin (not [4]) [1, 3]; Prostaglandin B$_2$ (inhibition of
  oxidation of 15-hydroxyl group of prostaglandin F$_{2alpha}$) [1]; More (not: thyro-
  id hormones [1], NADH [1, 2]) [1, 2]

Cofactor(s)/prosthetic group(s)/activating agents
  NADP$^+$ [1, 3–6]; NADPH [3, 4]

Metal compounds/salts

Turnover number (min^{-1})

Specific activity (U/mg)
  More [1]

K$_m$-value (mM)
  0.025 (prostaglandin F$_{2alpha}$) [1]; 0.035 (prostaglandin F$_{2alpha}$) [4]; 0.157
  (prostaglandin E$_1$) [6]; 0.200 (prostaglandin E$_2$) [1]

pH-optimum
  7.2–8.5 [1]

pH-range

Temperature optimum (°C)
  37 (assay at) [1, 4]

Temperature range (°C)

---

3 ENZYME STRUCTURE

Molecular weight

Subunits

Glycoprotein/Lipoprotein
  –

4 ISOLATION/PREPARATION

Source organism
Monkey [1]; Human [1, 4]; Pig [1–3]; Bovine [5]; Rat [6]

Source tissue
Brain [1]; Erythrocytes [1, 4]; Mesenteric arteries [5]; Adrenal cortex [6]; Kidney (medulla, cortex [2]) [1–3]

Localisation in source
Cytoplasm [5]; Cytosol [6]

Purification
Monkey [1]; Pig (partial [2], 2 enzyme forms: I and II, each of them posses-ses reversible prostaglandin 9-ketoreductase and NADP-linked 15-hydroxy-prostaglandin dehydrogenase activity, compare EC 1.1.1.189 [3]) [2, 3]; Human (partial [1]) [1, 4]

Crystallization
–

Cloned
–

Renaturated
–

---

5 STABILITY

pH

Temperature (°C)
45 (10 min, 75% loss of activity) [4]; 60 (half-life: 13.5 min (form I), 10 min (form II)) [3]

Oxidation

Organic solvent

General stability information

Storage
–80°C [3]

---

6 CROSSREFERENCES TO STRUCTURE DATABANKS

PIR/MIPS code

Brookhaven code

---

7 LITERATURE REFERENCES

[1] Lee, S.-C., Levine, L.: J. Biol. Chem.,250,548–552 (1975)
[2] Lee, S.-C., Pong, S.-S., Katzen, D., Wu, K.-Y., Levine, L.: Biochemistry,14,142–145 (1975)
[3] Chang, D.G.-B., Sun, M., Tai, H.-H.: Biochem. Biophys. Res. Commun.,99,745–751 (1981)
[4] Kaplan, L., Lee, S.-C., Levine, L.: Arch. Biochem. Biophys.,167,287–293 (1975)
[5] Wong, P.Y.-K., McGiff, J.C.: Biochim. Biophys. Acta,500,436–439 (1977)
[6] Levasseur, S., Friedman, Y., Burke, G.: Biochem. Biophys. Res. Commun.,95, 236–242 (1980)

1 NOMENCLATURE

EC number
1.1.1.198

Systematic name
(+)-Borneol:NAD+ oxidoreductase

Recommended name
(+)-Borneol dehydrogenase

Synonymes
Bicyclic monoterpenol dehydrogenase

CAS Reg. No.
67185-75-5

2 REACTION AND SPECIFICITY

Catalysed reaction
(+)-Borneol + NAD+ →
→ (+)-camphor + NADH

Reaction type
Redox reaction

Natural substrates
(+)-Borneol + NAD+ [1, 2]
(-)-Thujol + NAD+ [2]

Substrate spectrum
1 (+)-Borneol + NAD(P)+ (r) [1, 2]
2 (-)-Thujol + NAD(P)+ [2]

Product spectrum
1 (+)-Camphor + NAD(P)H [1, 2]
2 (-)-Thujone + NAD(P)H [2]

Inhibitor(s)
p-Hydroxymercuribenzoate [2]; NADH [2]; (+)-Camphor [2]; Borate [2]; 8-Hydroxyquinoline [2]; Cu^{2+} [2]; Hg^{2+} [2]; Pb^{2+} [2]; N-Ethylmaleimide [2]; Iodoacetamide [2]

Cofactor(s)/prosthetic group(s)/activating agents
NAD+ [1, 2]; NADH [1, 2]; NADP+ [1, 2]; NADPH [1, 2]

Metal compounds/salts

Turnover number (min⁻¹)

Specific activity (U/mg)

K_m-value (mM)
    0.07 (NAD⁺) [2]; 0.03 ((+)-borneol) [2]; 0.041 ((-)-thujol) [2]

pH-optimum
    8.0 [2]

pH-range
    5.0 (not active below) [2]

Temperature optimum (°C)

Temperature range (°C)

---

3 ENZYME STRUCTURE

Molecular weight
    91000 (Salvia officinalis, gel filtration) [2]

Subunits

Glycoprotein/Lipoprotein
    –

---

4 ISOLATION/PREPARATION

Source organism
    Salvia officinalis (sage) [1, 2]

Source tissue
    Leaf [1, 2]

Localisation in source

Purification
    Salvia officinalis (partial) [2]

Crystallization
    –

Cloned
    –

Renaturated
    –

5 STABILITY

pH
    8.0 [2]

Temperature (°C)

Oxidation

Organic solvent

General stability information

Storage
    0°C, 15% glycerol, 20 mM mercaptoethanol, 3 days [2]

6 CROSSREFERENCES TO STRUCTURE DATABANKS

PIR/MIPS code

Brookhaven code

7 LITERATURE REFERENCES

[1] Dehal, S.S., Croteau, R.: Arch. Biochem. Biophys.,258,287–291 (1987)
[2] Croteau, R., Hooper, C.L., Felton, M.: Arch. Biochem. Biophys.,188,182–193 (1978)

1 NOMENCLATURE

EC number
1.1.1.199

Systematic name
Reduced-(S)-usnate:NAD+ oxidoreductase (ether-bond-forming)

Recommended name
(S)-Usnate reductase

Synonymes
L-Usnic acid dehydrogenase

CAS Reg. No.

2 REACTION AND SPECIFICITY

Catalysed reaction
L-Usnic acid + NADH →
→ NAD+ +
2-acetyl-6-(3-acetyl-2,4,6-trihydroxy-5-methyl-phenyl)-3-hydroxy-6-methyl-cyclohexa-2,4-dienone

Reaction type
Redox reaction

Natural substrates
L-Usnic acid + NADH (catabolism of usnic acid in starvation conditions, re-mobilization of L-usnic acid (lichen acid, storage metabolite)) [1]

Substrate spectrum
1 L-Usnic acid + NADH (D- and L-form of usnic acid appear to be reduced by the same enzyme) [1]

Product spectrum
1 NAD+ +
2-acetyl-6-(3-acetyl-2,4,6-trihydroxy-5-methyl-phenyl)-3-hydroxy-6-methyl-cyclohexa-2,4-dienone (arising from reductive cleavage on the ether bond in L-usnic acid molecule to form a 7-hydroxy group [1]

Inhibitor(s)

Cofactor(s)/prosthetic group(s)/activating agents
  NADH (strict requirement for NADH, not replaceable by NADPH) [1]

Metal compounds/salts

Turnover number (min^{-1})

Specific activity (U/mg)
  More [1]

K_m-value (mM)

pH-optimum
  8.0 [1]

pH-range
  6.8–9.2 (6.8: about 20% of activity maximum, 9.2: about 50% of activity maximum) [1]

Temperature optimum (°C)
  30 [1]

Temperature range (°C)
  22–37 (22°C: about 30% of activity maximum, 37°C: about 35% of activity maximum) [1]

3 ENZYME STRUCTURE

Molecular weight
  450000 (Evernia prunastri, gel filtration) [1]

Subunits

Glycoprotein/Lipoprotein
  –

4 ISOLATION/PREPARATION

Source organism
  Evernia prunastri [1]

Source tissue
  Thallus [1]

Localisation in source

Purification
  Evernia prunastri [1]

Crystallization
–

Cloned
–

Renaturated
–

5 STABILITY

pH

Temperature (°C)
More (thermolabile) [1]

Oxidation

Organic solvent

General stability information
Thermolabile [1]

Storage

6 CROSSREFERENCES TO STRUCTURE DATABANKS

PIR/MIPS code

Brookhaven code

7 LITERATURE REFERENCES

[1] Estevez, M.P., Legaz, E., Olmeda, L., Perez, F.J., Vicente, C.: Z. Naturforsch.,36c,35–39 (1981)

1 NOMENCLATURE

EC number
1.1.1.200

Systematic name
D-Aldose-6-phosphate:NADP$^+$ 1-oxidoreductase

Recommended name
Aldose-6-phosphate reductase (NADPH)

Synonymes
Reductase, aldose 6-phosphate
NADP-dependent aldose 6-phosphate reductase
A6PR [1]
Aldose-6-P reductase [2]
Aldose-6-phosphate reductase [2]
Alditol 6-phosphate:NADP 1-oxidoreductase [2]

CAS Reg. No.
76901-04-7

2 REACTION AND SPECIFICITY

Catalysed reaction
D-Sorbitol 6-phosphate + NADP$^+$ →
→ D-glucose 6-phosphate + NADPH

Reaction type
Redox reaction

Natural substrates
More (important role in sorbitol synthesis) [2]

Substrate spectrum
1 D-Sorbitol 6-phosphate + NADP$^+$ (r [2]) [1, 2]
2 D-Galactose 6-phosphate + NADPH (reduced at a higher rate than D-glucose 6-phosphate) [2]
3 D-Mannose 6-phosphate + NADPH (reduced at low rate) [2]
4 2-Deoxy-D-glucose 6-phosphate + NADPH (reduced at low rate) [2]
5 More (not: D-glucose 1-phosphate, D-fructose 6-phosphate, D-ribose 5-phosphate, D-glucose, sorbitol) [2]

Product spectrum
1 D-Glucose 6-phosphate + NADPH [2]
2 ?
3 ?
4 ?
5 ?

Inhibitor(s)
Na_2EDTA [2]; Cysteine [2]; p-Chloromercuribenzoate [2]; $AgNO_3$ [2]; $ZnSO_4$ [2]; Iodoacetate [2]

Cofactor(s)/prosthetic group(s)/activating agents
$NADP^+$ [1, 2]; NADPH [2]; More (no activity with NAD^+ or NADH) [2]

Metal compounds/salts

Turnover number (min^{-1})

Specific activity (U/mg)
628 [2]

K_m-value (mM)
3.9 (D-sorbitol 6-phosphate) [2]; 0.20 (D-glucose 6-phosphate) [2]

pH-optimum
9.5 (both directions) [2]

pH-range
8.5–10.1 (about 50% of activity maximum at pH 8.5 and 10.1) [2]

Temperature optimum (°C)
35 (D-glucose 6-phosphate reduction) [2]; 30–35 (D-sorbitol 6-phosphate oxidation) [2]

Temperature range (°C)
More (no activity at 50°C) [2]

---

3 ENZYME STRUCTURE

Molecular weight

Subunits

Glycoprotein/Lipoprotein
–

4 ISOLATION/PREPARATION

Source organism
Pyrus communis (pear, Conference, Bartlett) [2]; Prunus persica (Redhaven peach) [2]; Prunus armenica (Perfection apricot) [2]; Malus domestica (apple, cv. Starkrimson [1, 2], Golden Delicious [2], Antonovka [2]) [1, 2]

Source tissue
Green tissues [1]; Leaf [1, 2]

Localisation in source

Purification
Malus domestica cv. Starkrimson [2]

Crystallization
--

Cloned
--

Renaturated
--

5 STABILITY

pH

Temperature (°C)

Oxidation

Organic solvent

General stability information

Storage
1°C, 1 week, partially purified enzyme [2]

6 CROSSREFERENCES TO STRUCTURE DATABANKS

PIR/MIPS code

Brookhaven code

7 LITERATURE REFERENCES

[1] Loescher, W.H., Marlow, G.C., Kennedey, R.A.: Plant Physiol.,70,335–339 (1982)
[2] Negm, F.B., Loescher, W.H.: Plant Physiol.,67,139–142 (1981)

1 NOMENCLATURE

EC number
1.1.1.201

Systematic name
7beta-Hydroxysteroid:NADP⁺ 7-oxidoreductase

Recommended name
7beta-Hydroxysteroid dehydrogenase (NADP⁺)

Synonymes
Dehydrogenase, 7beta-hydroxy steroid (nicotinamide adenine dinucleotide phosphate)
NADP-dependent 7beta-hydroxysteroid dehydrogenase

CAS Reg. No.
79393-83-2

2 REACTION AND SPECIFICITY

Catalysed reaction
7beta-Hydroxysteroid + NADP⁺ →
→ 7-oxosteroid + NADPH

Reaction type
Redox reaction

Natural substrates
Ursodeoxycholanoic acid + NADP⁺ (pathway in metabolism of bile acids or salts with 7beta-hydroxy-group) [3]

Substrate spectrum

1 Ursodeoxycholic acid + NADP⁺ (i.e. 3alpha,7beta-dihydroxy-5beta-cholanoic acid, r [1, 5, 6], ir [3], traces of activity, reverse reaction favored [5]) [1–6]

2 Ursocholic acid + NADP⁺ (i.e. 3alpha,7beta,12alpha-trihydroxy-5beta-cholanoic acid, oxidation at 20% the rate of ursodeoxycholanoic acid) [1]

3 3alpha,7beta-Dihydroxy-5beta-cholanoyl taurine + NADP⁺ (ir [3], oxidation at 20% the rate of ursodeoxycholanoic acid [1], traces of activity [5]) [1, 3–5]

4 3alpha,7beta-Dihydroxy-5beta-cholanoyl glycine + NADP⁺ (ir [3]) [3, 4]

5 3,7-Diketo-5beta-cholanoic acid + NADPH [5]

6 7,12-Diketo-5beta-cholanoic acid + NADPH [5]

7 3alpha,12alpha-Dihydroxy-7-keto-5beta-cholanoic acid + NADPH [5]

8 3alpha-Hydroxy-7,12-diketo-5beta-cholanoic acid + NADPH [5]

9 3,7,12-Triketo-5beta-cholanoic acid + NADPH [5]

10 3,7,12-Triketo-5beta-cholanoyl taurine + NADPH [5]

11 More (no substrate: chenodeoxycholic, cholic [1] or deoxycholic acids [1, 4], free bile acids are more readily oxidized than their conjugates [3]) [1, 3, 4]

Product spectrum

1 7-Ketolithocholic acid + NADPH (i.e. 3alpha-hydroxy-7-keto-5beta-cholanoic acid) [3, 5]

2 3alpha,12alpha-Dihydroxy-7keto-5beta-cholanoic acid + NADPH [1]

3 3alpha-Hydroxy-7-keto-5beta-cholanoyl taurine + NADPH

4 3alpha-Hydroxy-7-keto-5beta-cholanoyl glycine + NADPH

5 7beta-Hydroxy-3-keto-5beta-cholanoic acid + NADP⁺

6 7beta-Hydroxy-12-keto-5beta-cholanoic acid + NADP⁺

7 3alpha,7beta,12alpha-Trihydroxy-5beta-cholanoic acid + NADP⁺

8 3alpha,7beta-Dihydroxy-12-keto-5beta-cholanoic acid + NADP⁺

9 7beta-Hydroxy-3,12-diketo-5beta-cholanoic acid + NADP⁺

10 7beta-Hydroxy-3,12-diketo-5beta-cholanoyl taurine + NADP⁺

11 ?

Inhibitor(s)

Iodoacetate [3]; p-Chloromercuribenzoate [3]; More (Ca²⁺, Mg²⁺, EDTA do not inhibit) [3]

Cofactor(s)/prosthetic group(s)/activating agents

NADP⁺ [1–6]; NADPH [1, 5, 6]; More (NAD⁺/NADH no cofactor) [1, 5]

Metal compounds/salts

Turnover number (min⁻¹)

Specific activity (U/mg)
0.17 (Eubacterium aerofaciens) [3]; 1.059 (Peptostreptococcus productus)
[3]

K_m-value (mM)
0.0095 (ursodeoxycholanoic acid) [2]; 0.018 (ursodeoxycholanoic acid) [4];
0.022 (ursodeoxycholanoic acid, Peptostreptococcus) [3]; 0.055 (3alpha-hy-
droxy-7-keto-5beta-cholanoic acid, NADPH) [5]; 0.0703 (3,12-diketochola-
noic acid) [5]; 0.072 (NADP⁺) [2]; 0.076 (ursocholanoic acid) [4]; 0.0865
(3,7-diketocholanoic acid) [5]; 0.093 (3alpha,7beta-dihydroxy-5beta-chola-
noyl glycine) [4]; 0.1 (NADP⁺, Peptostreptococcus) [3]; 0.108 (ursodeoxycholanoic acid, Eubacterium) [3]; 0.123 (3,7,12-triketo-5beta-cholanoic acid)
[5]; 0.1239 (3alpha-hydroxy-7,12-diketo-5beta-cholanoic acid) [5]; 0.14
(3alpha,7beta-dihydroxy-5beta-cholanoyl taurine) [4]; 0.1414
(3alpha,12alpha-dihydroxy-7-keto-5beta-cholanoic acid) [5]; 0.1445
(3,7,12-triketo-5beta-cholanoyl taurine) [5]; 0.238 (3alpha,7beta-dihy-
droxy-5beta-cholanoyl glycine, Peptostreptococcus) [3]; 0.268
(3alpha,7beta-dihydroxy-5beta-cholanoyl taurine, Peptostreptococcus) [3];
0.4 (NADP⁺, Eubacterium) [3]; 0.909 (3alpha,7beta-dihydroxy-5beta-chola-
noyl glycine, Eubacterium) [3]; 1.639 (3alpha,7beta-dihydroxy-5beta-chola-
noyl taurine, Eubacterium) [3]

pH-optimum
9.8 (Peptostreptococcus productus) [3]; 9.8–10.2 (oxidation) [1]; 9.8–10.5
[4]; 10.0 [5]; 10.5 (Eubacterium aerofaciens) [3]

pH-range
8.5–10.8 (half-maximal activity at pH 8.5 and 10.8) [4]; 8.9–10.6 (half-maximal activity at pH 8.9 and 10.6, Peptostreptococcus) [3]; 9.0–11.2 (half-maximal activity at pH 9.0 and 11.2, Eubacterium aerofaciens) [3]; 9.5–10.4
(half-maximal activity at pH 9.5 and 10.4) [5]

Temperature optimum (°C)
25 (assay at) [1–6]

Temperature range (°C)

3 ENZYME STRUCTURE

Molecular weight
45000 (Eubacterium aerofaciens, gel filtration) [3]
53000 (Peptostreptococcus productus, gel filtration) [3]
82000 (Peptostreptococcus productus, gel filtration) [5]
115000 (Clostridium sp., gel filtration) [6]

Subunits

Glycoprotein/Lipoprotein

–

---

4 ISOLATION/PREPARATION

Source organism
  Eubacterium aerofaciens (major species of intestinal microflora) [1, 3]; Peptostreptococcus productus [3, 5]; Clostridium absonum (enzyme synthesis is induced by dihydroxy bile salts, i.e. chenodeoxycholate or deoxycholate, induction is repressed by ursodeoxycholanoic acid) [2]; Clostridium limosum (strain F-14) [4]; Clostridium sp. (strain 25.11.c) [6]

Source tissue
  Cell (crude [1, 2, 4]) [1–6]

Localisation in source
  Cytoplasm [3–5]; Membranes [5, 6]

Purification
  Peptostreptococcus productus (partial) [3, 5]

Crystallization
  –

Cloned
  –

Renaturated
  –

---

5 STABILITY

pH

Temperature (°C)
  4 (6 h stable, complete loss of activity after 48 h) [5]; 20 ($t_{1/2}$: 60 min) [5]; 30 (complete inactivation after 120 min [2], $t_{1/2}$: 25 min [4]) [2, 4]; 37 ($t_{1/2}$: 15 min, complete loss of activity after 4 h) [5]; 40 ($t_{1/2}$: 12 min) [3]; 50 ($t_{1/2}$: about 90 s) [3]

Oxidation

Organic solvent

General stability information
  Thioglycolate stabilizes [3]; DTT stabilizes Peptostreptococcus enzyme [3, 5]

Storage
  –70°C, stable under nitrogen [4]; –20°C, stable [3, 5]; 4°C, stable [3]

6 CROSSREFERENCES TO STRUCTURE DATABANKS

PIR/MIPS code

Brookhaven code

7 LITERATURE REFERENCES

[1] Macdonald, I.A., Rochon, Y.P., Hutchinson, D.M., Holdeman, L.V.: Appl. Environ. Microbiol.,44,1187–1195 (1982)
[2] Macdonald, I.A., Roach, P.D.: Biochim. Biophys. Acta,665,262–269 (1981)
[3] Hirano, S., Masuda, N.: Appl. Environ. Microbiol.,43,1057–1063 (1982)
[4] Sutherland, J.D., Williams, C.N.: J. Lipid Res.,26,344–350 (1985)
[5] Edenharder, R., Pfützner, A., Hammann, R.: Biochim. Biophys. Acta,1004,230–238 (1989)
[6] Edenharder, R., Pfützner, A., Hammann, R.: Biochim. Biophys. Acta,1002,37–44 (1989)

1 NOMENCLATURE

EC number
1.1.1.202

Systematic name
Propane-1,3-diol:NAD$^+$ 1-oxidoreductase

Recommended name
1,3-Propanediol dehydrogenase

Synonymes
3-Hydroxypropionaldehyde reductase
1,3-PD:NAD$^+$ oxidoreductase [3]
Dehydrogenase, 1,3-propanediol
1,3-Propanediol:NAD$^+$ oxidoreductase

CAS Reg. No.
81611-70-3

2 REACTION AND SPECIFICITY

Catalysed reaction
Propane-1,3-diol + NAD$^+$ →
→ 3-hydroxypropanal + NADH

Reaction type
Redox reaction

Natural substrates
3-Hydroxypropanal + NADH (enzyme in fermentative utilization of glycerol) [1–3]

Substrate spectrum
1 Propane-1,3-diol + NAD$^+$ (r [1, 3], high specificity, at pH 7.4 in 50 mM HEPES buffer, the enzyme catalyzes the reduction of propionaldehyde with an activity of 2.1 U/mg) [1–3]

Product spectrum
1 3-Hydroxypropanal + NADH (r [1, 3])

Inhibitor(s)
2,2'-Dipyridyl (reactivation by Mn^{2+} or Fe^{2+}) [1]; 8-Hydroxyquinoline [1]; 1,10-Phenanthroline [1]; More (with 100 mM Na^+, Li^+, Mg^{2+} or NH_4^+ in addition to K^+: 13–23% decrease in activity) [3]

Cofactor(s)/prosthetic group(s)/activating agents
NAD^+ [1, 3]; NADH [1, 3]; More ($NADP^+$ has no coenzyme activity) [1]

Metal compounds/salts
K^+ (with respect to cation effects: highest levels of activity in presence of 100 mM K^+) [3]

Turnover number (min^{-1})

Specific activity (U/mg)
37 [1]; 7.3 [3]

K_m-value (mM)
18 (propane-1,3-diol) [1]; 0.31 (NAD^+) [1]; 0.15 (NAD^+) [3]

pH-optimum
6.2 (3-hydroxypropanal + NADH) [3]; 7 (propane-1,3-diol + NAD^+) [3]; 9.0 (assay at) [1]

pH-range

Temperature optimum (°C)
25 (assay at) [1]; 37 (assay at) [3]

Temperature range (°C)

---

3 ENZYME STRUCTURE

Molecular weight
328000 (Klebsiella pneumoniae, gel filtration) [1]
180000 (Lactobacillus reuteri, gel filtration) [3]

Subunits
Tetramer (4 × 42000, Lactobacillus reuteri, SDS-PAGE) [3]
Hexamer or octamer (6 × or 8 × 45000, Klebsiella pneumoniae, SDS-PAGE) [1]

Glycoprotein/Lipoprotein
–

4 ISOLATION/PREPARATION

Source organism
    Klebsiella pneumoniae (i.e. Aerobacter aerogenes [4], all genes of the dha
    regulon, including the gene for 1,3-propanediol oxidoreductase of Klebsiella
    pneumoniae mobilized by the plasmid RP4:mini Mu and transferred to E.
    coli [6]) [1, 4, 6]; Clostridium butyricum (LMG 1212t2) [2]; Clostridium past-
    eurianum (LMG 3285) [2]; Lactobacillus reuteri [3]; Lactobacillus brevis [5]

Source tissue
    Cells [1]

Localisation in source

Purification
    Klebsiella pneumoniae [1]; Lactobacillus reuteri [3]

Crystallization
    –

Cloned
    (all genes of the dha regulon, including the gene for 1,3-propanediol oxi-
    doreductase of Klebsiella pneumoniae mobilized by the plasmid RP4:mini
    Mu and transferred to E. coli) [6]

Renaturated
    –

5 STABILITY

pH

Temperature (°C)

Oxidation
    Inactivated by oxidation during aerobic metabolism [1]

Organic solvent

General stability information
    –20°C, Mn^{2+}-DTT-HEPES buffer, stable for 3 days, 82% loss of activity after 3
    weeks [1]

Storage

6 CROSSREFERENCES TO STRUCTURE DATABANKS

PIR/MIPS code

Brookhaven code

7 LITERATURE REFERENCES

[1] Johnson, E.A., Lin, E.C.C.: J. Bacteriol.,169,2050–2054 (1987)
[2] Heyndrickx, M., De Vos, P., Vancanneyt, M., De Ley, J.: Appl. Microbiol. Biotechnol., 34,637–642 (1991)
[3] Talarico, T.L., Axelsson, L.T., Novotny, J., Fiuzat, M., Dobrogosz, W.J.: Appl. Environ. Microbiol.,56,943–948 (1990)
[4] Abeles, R.H., Brownstein, A.M., Randles, C.H.: Biochim. Biophys. Acta,41,530–531 (1960)
[5] Schütz, H., Radler, F.: Syst. Appl. Microbiol.,5,169–178 (1984)
[6] Sprenger, G.A., Hammer, B.A., Johnson, E.A., Lin, E.C. C.: J. Gen. Microbiol.,135, 1255–1262 (1989)

1 NOMENCLATURE

EC number
1.1.1.203

Systematic name
Uronate:NAD$^+$ 1-oxidoreductase

Recommended name
Uronate dehydrogenase

Synonymes
EC 1.2.1.35 (formerly)
Dehydrogenase, uronate
Uronate: NAD-oxidoreductase
Uronic acid dehydrogenase [2]

CAS Reg. No.
37250-98-9

2 REACTION AND SPECIFICITY

Catalysed reaction
D-Galacturonate + NAD$^+$ + H$_2$O →
→ D-galactarate + NADH

Reaction type
Redox reaction

Natural substrates
D-Galacturonate + NAD$^+$ + H$_2$O (catabolism) [1]
D-Glucuronate + NAD$^+$ + H$_2$O (catabolism) [1]

Substrate spectrum
1 D-Galacturonate + NAD$^+$ + H$_2$O (ir [1], weak reverse reaction with glucaric acid lactones at acidic pH) [1, 2]
2 D-Glucuronate + NAD$^+$ + H$_2$O [1, 2]

Product spectrum
1 D-Galactarate +NADH
2 D-Glucaric acid + NADH

Inhibitor(s)
Hg^{2+} [1, 2]; NADH [2]; p-Chloromercuribenzoate [2]; N-Ethylmaleimide [2];
EDTA [2]; AsO$_4^{3-}$ [2]; NaN$_3$ [2]; K$_4$[Fe(CN)$_6$] [2]; KCN [2]

Cofactor(s)/prosthetic group(s)/activating agents
NAD$^+$ [1, 2]; More (not NADP$^+$ [1], no evidence for a prosthetic group [2])
[1, 2]

Metal compounds/salts
No metal requirement [2]

Turnover number (min^{-1})
1600 (galacturonate) [2]; 4800 (glucuronate) [2]

Specific activity (U/mg)
5.2 [1]; 81 [2]

K_m-value (mM)
0.11 (D-glucuronate, enzyme from cells grown in medium with D-glucuronate as carbon source) [1]; 0.056 (D-galacturonate, enzyme from cells grown in medium with D-glucuronate as carbon source) [1]; 0.058 (NAD$^+$ (+ 0.67 mM D-galacturonate), enzyme from cells grown in medium with D-glucuronate as carbon source) [1]; 0.1 (D-glucuronate, enzyme from cells grown in medium with D-galacturonate as carbon source) [1]; 0.059 (D-galacturonate, enzyme from cells grown in medium with D-galacturonate as carbon source) [1]; 0.057 (NAD$^+$ (+ 0.67 mM D-galacturonate), enzyme from cells grown in medium with D-galacturonate as carbon source) [1]; 0.083 (NAD$^+$ (+ 1.33 mM glucuronate), enzyme from cells grown in medium with D-galacturonate as carbon source) [1]; 0.085 (NAD$^+$ (+ 1.33 mM glucuronate), enzyme from cells grown in medium with D-glucuronate as carbon source) [1]; 0.37 (glucuronate) [2]; 0.54 (galacturonate) [2]; 0.080 (NAD$^+$ (+ glucuronate)) [2]

pH-optimum
7.0–8.4 [1]

pH-range
6.5–9.3 (at pH 6.5 and 9.4: about 50% of activity maximum) [1]

Temperature optimum (°C)
37 [2]

Temperature range (°C)

---

3 ENZYME STRUCTURE

Molecular weight
60000 (Pseudomonas syringae, gel filtration, ultracentrifugation) [2]

Subunits
Dimer (2 × 30000, Pseudomonas syringae, SDS-PAGE) [2]

Glycoprotein/Lipoprotein
–

4 ISOLATION/PREPARATION

Source organism
Pseudomonas syringae [1, 2]

Source tissue
Cell [1, 2]

Localisation in source

Purification
Pseudomonas syringae [1, 2]

Crystallization
–

Cloned
–

Renaturated
More (24 h dialysis: SDS-denatured enzyme remains inactive, urea-denatured enzyme regains 10% of its original activity) [2]

---

5 STABILITY

pH

Temperature (°C)
0–15 (24 h, stable) [2]; 25 (2 h, 15% loss of activity [2], 30 min, stable [1]) [1, 2]; 30 (30 min, 87% loss of activity) [1]; 35 (30 min, complete loss of activity) [1]; 37 (1 h, 90% loss of activity) [2]; More (NAD+, 0.58 mM, enhances heat stability) [1]

Oxidation

Organic solvent

General stability information
Reducing agents (e.g. 2-mercaptoethanol, dithiothreitol and dithioerythritol protect against thermal destruction) [2]; Storage under vacuum or N_2 has no effect on stability [1]; NAD+, 0.58 mM, enhances heat stability [1]

Storage
−20°C, freeze-dried, 6 weeks, 30–55% loss of activity [1]; 23°C, freeze-dried, 6 weeks, 85% loss of activity [1]; −10°C, 0.5 mg/ml protein, Tris/HCl buffer, pH 7.2, 2.9 mM NAD+, 6 months, less than 10% loss of activity [1]; 4°C, 1.5 mg/ml protein, 0.05 M sodium diphosphate-HCl buffer, pH 8.0, 30% glycerol, 1 month, 5% loss of activity [2]; −20°C, 0.1% bovine serum albumin, stable [2]

6 CROSSREFERENCES TO STRUCTURE DATABANKS

PIR/MIPS code

Brookhaven code

7 LITERATURE REFERENCES

[1] Bateman, D.F., Kosuge, T., Kilgore, W.W.: Arch. Biochem. Biophys.,136,97–105
    (1970)
[2] Wagner, G., Hollmann, S.: Eur. J. Biochem.,61,589–596 (1976)

1 NOMENCLATURE

EC number
1.1.1.204

Systematic name
Xanthine:NAD$^+$ oxidoreductase

Recommended name
Xanthine dehydrogenase

Synonymes
Dehydrogenase, xanthine
NAD-xanthine dehydrogenase
Xanthine–NAD oxidoreductase
Xanthine/NAD$^+$ oxidoreductase [4]
Xanthine oxidoreductase [11, 20]
EC 1.2.1.37 (formerly)

CAS Reg. No.
9054-84-6

---

2 REACTION AND SPECIFICITY

Catalysed reaction
Xanthine + NAD$^+$ + H$_2$O →
→ urate + NADH (kinetic mechanism [4])

Reaction type
Redox reaction

Natural substrates
Xanthine + NAD$^+$ + H$_2$O (liver xanthine dehydrogenase participates in the liberation of iron from liver ferritin stores [12], essential enzyme for ureide metabolism [13]) [12, 13]
Purines + NAD$^+$ + H$_2$O (first enzyme of degradative pathway by which fungi convert purines to ammonia) [15]
Uric acid + NADH (participates in uric acid degradation in Rhodopseudomonas capsulata (reduction of uric acid to xanthine by purified enzyme not detected)) [19]

Substrate spectrum

1 Xanthine + NAD$^+$ + H$_2$O (other electron acceptors: 2,6-dichloroindophenol (low activity [29]) [14, 19, 27–29, 33, 36, 37], 3-acetylpyridine adenine dinucleotide [10], methylene blue (low activity [29]) [14, 29, 36, 37], phenazine methosulfate [14, 28, 29, 33, 36, 37], trinitrobenzene sulfonate [14, 37], p-benzoquinone (low activity [28]) [18, 28], ferricyanide [19, 27, 33, 36], nitroblue tetrazolium (low activity [28]) [27, 28, 36], NADP$^+$ (not [15], slight [27]) [27, 29]) [1–40]

2 Hypoxanthine + NAD$^+$ + H$_2$O [8, 10, 15, 19, 27, 28, 33]

3 Purine derivatives + NAD$^+$ + H$_2$O (e.g. 2-hydroxypurine [19, 27], 6,8-dihydroxypurine [19, 27], 2,6-dithiopurine [28], purine is a poor substrate (2% of the activity with xanthine [8, 27])) [8, 19, 27, 28]

4 Pteridine derivatives + NAD$^+$ + H$_2$O (e.g. 2-amino-4-hydroxypteridine) [15]

5 Xanthine derivatives + NAD$^+$ + H$_2$O (e.g. 1-methylxanthine [19], 2-thioxanthine [28], 6-thioxanthine [28]) [19, 28]

6 8-Azahypoxanthine + NAD$^+$ + H$_2$O [19, 28]

7 5-Azacytosine + NAD$^+$ + H$_2$O [19]

8 5-Azauracil + NAD$^+$ + H$_2$O [19]

9 Guanine + NAD$^+$ + H$_2$O (not [8, 28, 33]) [27]

10 Xanthoperin + NAD$^+$ + H$_2$O [31]

11 Dihydropterin + NAD$^+$ + H$_2$O [31]

12 Allopurinol + NAD$^+$ + H$_2$ O [31]

13 NADH + electron acceptor (e.g. nitroblue tetrazolium [2, 10], 2,6-dichloroindophenol [14, 15], 3-acetylpyridine-adenine dinucleotide [10], methylene blue [14], phenazine methosulfate [14], trinitrobenzene sulfonate [14], NADH diaphorase activity with several acceptors [2, 14, 31]) [2, 10, 14, 15, 31, 32, 37]

14 More (reversibility: although reduction of uric acid to xanthine is not detected with purified enzyme, in vitro experiments indicate that the enzyme participates in uric acid degradation in Rhodopseudomonas capsulata [19], reoxidation of NADH is extremely slow [33], enzyme also catalyzes reversible dismutation of xanthine to hypoxanthine and urate [36], the animal enzyme can be interconverted to EC 1.1.3.22 (the oxidase form), liver enzyme exists in vivo mainly in the dehydrogenase form, but can be converted into EC 1.1.3.22 by storage at −20°C, by treatment with proteolytic agents or organic solvents, or by thiol reagents such as Cu^{2+}, N-ethylmaleimide or 4-hydroxymercuribenzoate, the effect of thiol reagents can be reversed by thiols such as 1,4-dithioerythritol, in other animal tissues the enzyme exists almost entirely as EC 1.1.3.22, but can be converted into the dehydrogenase form by 1,4-dithioerythritol, enzyme converted to oxidase by: limited proteolysis [5, 7, 16], storage at −20°C [34], sulfhydryl oxidation (by tetraethylthiooxydicarbonic diamide [11], p-hydroxymercuribenzoate [34], CuSO$_4$ [34], 5,5'-dithiobis(2-nitro-

benzoic acid) [34], N-ethylmaleimide [34]) [5, 7, 11, 16, 18, 34], EC
1.1.1.204 can also be converted into EC 1.1.3.22 by EC 1.8.4.7 in the
presence of oxidized glutathione [23], dehydrogenase is predominant
form in vivo [5], rat: enzyme appears as oxidase in supernatant of rat
heart, intestine, spleen, pancreas, lung and kidney [34], the enzyme of
all organs but intestine can be converted into dehydrogenase by dithioerythritol [34], pea: not possible to convert xanthine dehydrogenase to
oxidase form [29]) [5, 7, 11, 16, 18, 19, 23, 29, 33, 34, 36]

Product spectrum
  1 Urate + NADH [1–40]
  2 ?
  3 ?
  4 ?
  5 ?
  6 ?
  7 ?
  8 ?
  9 ?
10 Leucopterin + NADH [31]
11 Dihydroxyxanthopterin + NADH [31]
12 4,6-Dihydroxypyrazolo[3,4-d]pyrimidine + NADH [31]
13 NAD$^+$ + reduced electron acceptor
14 ?

Inhibitor(s)
Urate (product inhibition) [10, 15, 27, 31, 36]; NADH (product inhibition) [10,
20]; Methanol (xanthine oxidation inactivated, NADH diaphorase activity activated [14, 37]) [14, 32, 36, 37, 40]; Arsenite (NADH diaphorase activity
activated, xanthine diaphorase activity inhibited [10, 37]) [10, 14, 32, 37];
Cyanide [14, 15, 27–29, 32, 36, 37]; Ammeline [14, 37, 39]; Urea [14, 37];
NAD$^+$ (competitive to NADH oxidation) [14, 37]; Xanthine (substrate inhibition) [15]; Hypoxanthine (substrate inhibition) [15]; Pterine (substrate inhibition) [15]; p-Hydroxymercuribenzoate [15]; Iodosobenzoate [15]; NaN$_3$ [15];
Thiourea [15]; Adenine [15, 27, 33]; Guanine [15, 33]; Allopurinol (4-hy-
droxypyrazolo(3,4-d)pyrimidine [29]) [15, 29]; Purines (e.g. 8-azahypoxan-
thine, 1-methylhypoxanthine) [19, 33, 36]; EDTA (not [28]) [27]; o-Phenanthroline (not [28]) [27]; Tiron [27]; p-Chloromercuribenzoate [27, 28]; Dinitro-
benzoate [27]; Iodoacetate [27]; Hg^{2+} [28]; Ag^{2+} [28]; Cu^{2+} [28]; Salicylhy-
droxamic acid [29]; Leucopterin [31]; 7,8-Dihydroxanthopterin [31];
4,6-Dihydroxypyrazolo[3,4-d]pyrimidine (inhibits allopurinol oxidation) [31];
4-Amino-2,6-dihydroxypyrimidine [31]; 2-Amino-4-hydroxypteridine-6-carb-
oxaldehyde [39]; 8-Azaguanine [39]; Borate [36]

Cofactor(s)/prosthetic group(s)/activating agents
Flavin (flavoprotein [3, 13–15, 27, 36, 37], Glycine max: 89–100% of the flavin present in the form of FMN, 0–11% of the flavin is present in the form of FAD, 1.4–6.7 mol of FMN per mol of enzyme [13], Rhodopseudomonas capsulata: 2 mol of flavin per mol of enzyme [19], Neurospora crassa: 2.0 mol of flavin per mol of enzyme, no FMN present [15], Streptomyces cyanogenis: molar ratio of FAD:iron:labile sulfur is 2:14:2 [27], rat: 2 mol of flavin per mol of enzyme [32], Micrococcus lactilyticus: 2 mol of flavin per mol of enzyme [40]) [3, 13–15, 19, 27, 32, 36, 37, 40]; FMN (Glycine max: 89–100% of the flavin is present in the form of FMN, 1.4–6.7 mol of FMN per mol of enzyme [13], Neurospora crassa: no FMN present [15]) [13, 15]; FAD (Glycine max: 0–11% of the flavin is present in form of FAD [13], Neurospora crassa: 2.0 mol of FAD per mol of enzyme [15]) [13, 15]; NAD⁺ (binding of NAD⁺ to enzyme flavin, NAD⁺ and NADH share a common binding site [37]) [1–40]; NADH (NAD⁺ and NADH share a common binding site [37]) [2, 10, 14, 15, 32, 36, 37]

Metal compounds/salts
Iron (iron containing flavoprotein [3, 27, 36, 37], molybdenum-iron protein [13–15], Glycine max: 8.2 mol of iron per mol of enzyme [13], Neurospora crassa: 12.1 mol of iron per mol of enzyme [15], Rhodopseudomonas capsulata: 8 mol of iron per mol of enzyme [19], Streptomyces cyanogenes: molar ratio of FAD: iron:labile sulfur is 2:14:2 [27], Pisum sativum: presence of non-heme iron [29], rat: 7.6 mol of iron per mol of enzyme [32], Micrococcus lactilyticus: 8 mol of iron per mol of enzyme [36]) [3, 13–15, 19, 27, 29, 32, 36, 37]; Molybdenum (molybdenum-iron protein [13–15, 36, 37], Glycine max: 1.7 mol of molybdenum per mol of enzyme [13], Neurospora crassa: 1.01 mol of molybdenum per mol of enzyme [15], Rhodopseudomonas capsulata: 2 mol of molybdenum per mol of enzyme [19], Streptomyces cyanogenes: little molybdenum detected [27], rat: 1.4 mol of molybdenum per mol of enzyme [32], Micrococus lactilyticus: 2 mol of molybdenum per mol of enzyme [36]) [13–15, 19, 27, 32, 36, 37]; K⁺ (stimulates at 50 mM or more) [27]

Turnover number (min⁻¹)
12000 (hypoxanthine + NAD⁺) [19]; 1210 (xanthine + ferredoxin) [36]; 1090 (xanthine + ferricyanide) [36]; 573 (xanthine + nitroblue tetrazolium) [36]; 340 (xanthine + dichlorophenolindophenol) [36]; 43 (xanthine + cytochrome c) [36]

Specific activity (U/mg)
2.47 [3]; 80.2 [8]; 0.152 [15]; 2.5 [9]; 2.2 [10]; 20 [28]; More [19, 32, 33, 35, 36, 37]

K_m-value (mM)
More (K_m values for free and immobilized enzyme at various pH values [24],
K_m for hypoxanthine varies 2.5-fold between pH 6 and 10.7 [10]) [10, 15, 18,
21, 24, 27, 32, 33, 35–37, 40]; 0.040 (hypoxanthine) [8]; 0.064 (xanthine)
[8]; 0.052 (NAD$^+$) [8]; 0.05 (xanthine (+ NAD$^+$), pH 7.5) [10]; 0.0125 (NAD$^+$
(+ xanthine), pH 7.5) [10]; 0.052 (hypoxanthine (+ NAD$^+$)) [10, 19]; 0.020
(NAD$^+$ (+ hypoxanthine), pH 7.5) [10]; 0.0212 (3-acetylpyridine adenine
dinucleotide (+ xanthine)) [10]; 0.102 (3-acetylpyridine adenine dinucleotide
(+ NADH)) [10]; 0.061 (NAD$^+$) [19]; 0.032 (xanthine (+ NAD$^+$)) [19]; 0.055
(hypoxanthine) [27]; 0.015 (guanine) [27]; 0.020 (thionicotinamide adenine
dinucleotide) [35]; 0.02 (ferredoxin (+ xanthine)) [36]; 0.09 (ferricyanide (+
xanthine)) [36]; 0.05 (nitroblue tetrazolium (+ xanthine)) [36]; 0.005 (cytochrome c (+ xanthine)) [36]

pH-optimum
7.75 (30°C) [31]; 7.9 [33]; 8.3 (hypoxanthine + NAD$^+$) [19]; 8.5–9 (hypoxanthine + NAD$^+$) [10]; 8.7 [27]; 9–10 (or higher, xanthine + NAD$^+$) [10]

pH-range
More [10]

Temperature optimum (°C)
37 [19]; 80 [27, 28]

Temperature range (°C)

---

3 ENZYME STRUCTURE

Molecular weight
125000 (Streptomyces cyanogenes, gel filtration, sedimentation equilibrium
analysis) [27]
230000–300000 (Colias eurytheme, gel filtration) [31]
250000 (Micrococcus lactilyticus, gel filtration) [36]
275000 (Pseudomonas acidovorans, nondissociating gel electrophoresis)
[33]
285000 (Glycine max, nondissociating vertical gradient polyacrylamide gel
electrophoresis) [13]
300000 (rat, nondenaturating disc gel electrophoresis [3], chicken, sedimentation equilibrium measurement [14, 37], Drosophila melanogaster, gel
filtration [38]) [3, 14, 37, 38]
345000 (Rhodopseudomonas capsulata, gel filtration) [19]
350000 (Pseudomonas putida F1, HPLC gel filtration) [8]
357000 (Neurospora crassa, gel filtration, sucrose density gradient centrifugation) [15]
540000 (Pseudomonas synxantha, gel filtration) [28]

Subunits

Dimer (1 × 155000 + 1 × 135000, chicken liver, SDS-PAGE [9], 2 × 141000, Glycine max, SDS-PAGE [13], 2 × 155000, Neurospora crassa, SDS-PAGE [15], 2 × 67000, Streptomyces cyanogenus, SDS-PAGE [27], 1 × 130000 + 1 × 140000, Drosophila melanogaster, SDS-PAGE [38]) [9, 13, 15, 27, 38]
Tetramer (4 × 84000, Rhodopseudomonas capsulata, SDS-PAGE) [19]
Oligomer (x × 92000 + x × 46000, Pseudomonas putida F–1, SDS-PAGE [8], x × 54000 + x × 76000, Pseudomonas synxantha, SDS-PAGE [28], x × 81000 + x × 63000, Pseudomonas acidovorans, SDS-PAGE [33]) [8, 28, 33]

Glycoprotein/Lipoprotein

–

---

4 ISOLATION/PREPARATION

Source organism

Chicken [4, 9, 14, 17, 24, 26, 34, 37]; Calliphora vicina [1]; Chlamydomonas reinhardtii [2, 22]; Rat [3, 7, 11, 12, 20, 23, 30, 32]; Human [5, 6]; Rabbit [5, 6]; Pseudomonas putida F1 [8, 40]; Glycine max (L. Merr. c.v. Williams) [10, 13]; Neurospora crassa [15]; Turkey [16]; Bovine [18]; Rhodopseudomonas capsulata [19]; Drosophila melanogaster (4 electrophoretically different forms [39]) [21, 35, 38, 39]; Clostridium acidiurici [25]; Clostridium cylindrosporum [25]; Pseudomonas synxantha A3 [28]; Pisum sativum (pea) [29]; Streptomyces cyanogenus [27]; Colias eurytheme (Colias butterfly) [31]; Pseudomonas acidovorans [33]; Micrococcus lactilyticus [36]; Bacteria (overview, distribution among bacteria, e.g. Pseudomonas putida, Pseudomonas cepacia, Pseudomonas acidovorans, Escherichia sp., Klebsiella sp., Alcaligenes sp.) [40]; More (similar enzymes: 1. Pseudomonas putida 40: activity separate from NAD+ and oxygen-utilizing activities which utilize ferricyanide, enzyme lacks flavin but possesses heme and is resistant to cyanide treatment [41], 2. Clostridium acidiurici: methyl viologen is the most effective electron acceptor, NAD+, NADP+ and ferredoxin are inactive as acceptors, uric acid, xanthine, hypoxanthine, 6-mercaptopurine, allopurinol and purine can serve as substrates [42]) [41, 42]

Source tissue

Liver [3–6, 9, 11, 12, 14, 17, 20, 24, 26, 30, 32, 34, 37]; Brain (high activity) [5, 6]; Heart (high activity) [5, 6, 34]; Kidney [5, 6, 34]; Skeletal muscle [5, 6]; Root nodule [10, 13]; Mycelium [15]; Milk [18]; Cells [19]; Leaf [29]; Wing [31]; Intestine [34]; Spleen [34]; Pancreas [34]; Lung [34]

Localisation in source

Cytoplasm (enzyme is entirely cytosolic [17]) [13, 17, 29]; Soluble [29]

Purification

Rat [3, 32]; Pseudomonas putida F1 [8]; Chicken [9, 14, 24, 37]; Glycine max (L. Merr. c.v. Williams) [13]; Neurospora crassa [15]; Rhodopseudomonas capsulata [19]; Streptomyces cyanogenus [27]; Pseudomonas synxantha A3 [28]; Colias eurytheme (Colias butterfly, partial) [31]; Pseudomonas acidovorans [33]; Drosophila melanogaster [35, 38]; Micrococcus lactilyticus [36]

Crystallization

[28]

Cloned

(Calliphora vicina gene) [1]

Renaturated

–

---

5 STABILITY

pH

5.3 (4°C: 10% loss of activity after 1 h, 83% loss of activity after 60 h) [36]; 6.5 (4°C: 12% loss of activity after 60 h, 27% loss of activity after 2 weeks) [36]; 7.0 (4°C: 26% loss of activity after 2 weeks) [36]; 7–12 (4°C, 24 h, stable) [27]; 8.0 (4°C, 2 weeks, 18% loss of activity, stable above pH 8.0) [36]

Temperature (°C)

4 (pH 7.8, $t_{1/2}$: 423 h (free enzyme), 2114 h (immobilized enzyme)) [24]; 20 (pH 7.8, $t_{1/2}$: 292 h (free enzyme), 1167 h (immobilized enzyme)) [24]; 30 (pH 7.8, $t_{1/2}$: 71 h (free enzyme), 810 h (immobilized enzyme)) [24]; 37 (inactivation above) [19]; 40 (pH 7.8, $t_{1/2}$: 21 h (free enzyme), 33 h (immobilized enzyme) [24], pH 8.5, 10 min, stable up to 40°C [28]) [24, 28]; 50 (pH 7.8, $t_{1/2}$: 11 h (free enzyme), 15 h (immobilized enzyme)) [24]; 55 (pH 9, 10 min, stable) [27]; 65 (pH 9, 10 min, complete loss of activity) [27]

Oxidation

Instable to oxygen during turnover [26]

Organic solvent

General stability information

Immobilized enzyme, $t_{1/2}$ at 4°C: 2114 h, at 20°C: 1167 h, at 30°C: 810 h, at 40°C: 33 h, at 50°C: 15 h [24]; Freezing at −15°C, 10% loss of activity [27]; Does not tolerate freezing well [31]; Unstable as dehydrogenase, gradually converted to an oxidase [32]; More (can be converted into EC 1.1.3.22 by storage at −20°C, by treatment with proteolytic agents or organic solvents, or by thiol reagents such as Cu^{2+}, N-ethylmaleimide or 4-hydroxymercuribenzoate) [5, 7, 16, 34]

Enzyme Handbook © Springer-Verlag Berlin Heidelberg 1995
Duplication, reproduction and storage in data banks are only
allowed with the prior permission of the publishers

Storage

4°C, pH 7.8, $t_{1/2}$: 2114 h immobilized enzyme, 423 h, free enzyme [24]; 1°C, 0.001 M dithiothreitol, 0.1 M ammonium sulfate, 8–12 weeks, 60–70% loss of activity [31]; More (can be converted into EC 1.1.3.22 by storage at –20°C) [34]

6 CROSSREFERENCES TO STRUCTURE DATABANKS

PIR/MIPS code

PIR2:JQ0407 (bluebottle fly (Calliphora vicina)); PIR2:S03392 (bluebottle fly (Calliphora vicina)); PIR2:A29627 (bluebottle fly (Calliphora vicina) (fragment)); PIR3:S07244 (fruit fly (Drosophila melanogaster)); PIR3:S07245 (fruit fly (Drosophila melanogaster)); PIR3:S10132 (fruit fly (Drosophila melanogaster)); PIR2:A31946 (fruit fly (Drosophila pseudoobscura)); PIR3:JC2138 (Human); PIR3:S22419 (mouse); PIR2:A37810 (rat)

Brookhaven code

7 LITERATURE REFERENCES

[1] Rocher-Chambonnet, C., Berreur, P., Houde, M., Tiveron, M.C., Lepesant, J.A., Bregegere, F.: Gene,59,201–212 (1987)
[2] Perez-Vicente, R., Pineda, M., Cardenas, J.: FEMS Microbiol. Lett.,43,321–325 (1987)
[3] Suleiman, S.A., Stevens, J.B.: Arch. Biochem. Biophys.,258,219–225 (1987)
[4] Bruguera, P., Lopez-Cabrera, A., Canela, E.I.: Biochem. J.,249,171–178 (1988)
[5] Wajner, M., Harkness, R.A.: Biochem. Soc. Trans.,16,358–359 (1988)
[6] Wajner, M., Harkness, R.A.: Biochim. Biophys. Acta,991,79–84 (1989)
[7] Stark, K., Seubert, P., Lynch, G., Baudry, M.: Biochem. Biophys. Res. Commun., 165,858–864 (1989)
[8] Kim, J.M., Schmid, R.D. in "GBF Monogr. (Biosens. Appl. Med., Environ. Prot. Process Control) ",13,421–424 (1989)
[9] Irie, S.: J. Biochem.,95,405–412 (1984)
[10] Boland, M.J., Blevins, D.G., Randall, D.D.: Arch. Biochem. Biophys.,222,435–441 (1983)
[11] Kaminski, Z.W., Jezewska, M.M.: Biochem. J.,207,341–346 (1982)
[12] Topham, R.W., Walker, M.C., Calisch, M.P.: Biochem. Biophys. Res. Commun.,109, 1240–1246 (1982)
[13] Triplett, E.W., Blevins, D.G., Randall, D.D.: Arch. Biochem. Biophys.,219,39–46 (1982)
[14] Rajagopalan, K.V., Handler, P.: J. Biol. Chem.,242,4097–4107 (1967)
[15] Lyon, E.S., Garrett, R.H.: J. Biol. Chem.,253,2604–2614 (1978)
[16] De Ni Fhaolain, Coughlan, M.P.: Biochem. Soc. Trans.,5,1705–1707 (1977)
[17] Coolbear, K.P., Herzberg, G.R., Brosnan, J.T.: Biochem. Soc. Trans.,9,394–395 (1981)
[18] Nakamura, M., Yamazaki, I.: J. Biochem.,92,1279–1286 (1982)
[19] Aretz, W., Kaspari, H., Klemme, J.-H.: Z. Naturforsch.,36c,933–941 (1981)

[20] Kaminski, Z., Jezewska, M.M.: Biochem. J.,200,597–603 (1981)
[21] Edwards, T.C.R., Candido, E.P.M.: Mol. Gen. Genet.,154,1–6 (1977)
[22] Fernandez, E., Cardenas, J.: Planta,153,254–257 (1981)
[23] Batelli, M.G., Lorenzoni, E.: Biochem. J.,207,133–138 (1982)
[24] Tramper, J., Angelino, S.A.G.F., Muller, F., van der Plas, H.C.: Biotechnol.
      Bioeng.,21,1767–1786 (1979)
[25] Wagner, R., Andreesen, J.R.: Arch. Microbiol.,121,255–260 (1979)
[26] Coughlan, M.P., Johnson, D.B.: Biochem. Soc. Trans.,7,18–21 (1979)
[27] Ohe, T., Watanabe, Y.: J. Biochem.,86,45–53 (1979)
[28] Sakai, T., Jun, H.-K.: Agric. Biol. Chem.,43,753–760 (1979)
[29] Nguyen, J., Feierabend, J.: Plant Sci. Lett.,13,125–132 (1978)
[30] Waud, W.R., Rajagopalan, K.V.: Arch. Biochem. Biophys.,172,365–379 (1976)
[31] Watt, W.B.: J. Biol. Chem.,247,1445–1451 (1972)
[32] Waud, W.R., Rajagopalan, K.V.: Arch. Biochem. Biophys.,172,354–364 (1976)
[33] Sin, I.L.: Biochim. Biophys. Acta,410,12–20 (1975)
[34] Della Corte, E., Stirpe, F.: Biochem. J.,126,739–745 (1972)
[35] Parzen, S.D., Fox, A.S.: Biochim. Biophys. Acta,92,465–471 (1964)
[36] Smith, S.T., Rajagopalan, K.V., Handler, P.: J. Biol. Chem.,242,4108–4117 (1967)
[37] Rajagopalan, K.V., Handler, P.: J. Biol. Chem.,242,4097–4107 (1967)
[38] Seybold, W.D.: Biochim. Biophys. Acta,334,266–271 (1974)
[39] Yen, T.T.T., Glassman, E.: Biochim. Biophys. Acta,146,35–44 (1967)
[40] Woolfolk, C.A., Downard, J.S.: J. Bacteriol.,130,1175–1191 (1977)
[41] Woolfolk, C.A.: J. Bacteriol.,163,600–609 (1985)
[42] Wagner, R., Cammack, R., Andreesen, J.R.: Biochim. Biophys. Acta,791,63–74
      (1984)

1 NOMENCLATURE

EC number
1.1.1.205

Systematic name
IMP:NAD+ oxidoreductase

Recommended name
IMP dehydrogenase

Synonymes
EC 1.2.1.14 (formerly)
Dehydrogenase, inosinate
Inosine-5'-phosphate dehydrogenase
Inosinic acid dehydrogenase
Inosinate dehydrogenase
Inosine 5'-monophosphate dehydrogenase
Inosine monophosphate dehydrogenase
IMP oxidoreductase [30]
Inosine monophosphate oxidoreductase [33]

CAS Reg. No.
9028-93-7

---

2 REACTION AND SPECIFICITY

Catalysed reaction
Inosine 5'-phosphate + NAD+ + H_2O →
→ xanthosine 5'-phosphate + NADH (enzyme acts on the hydroxyl group of
the hydrated derivative of the substrate, ordered bi-bi mechanism [12, 25]:
IMP binds first, followed by NAD+, NADH dissociates from the ternary complex first, then XMP is released [12], ordered sequential mechanism [28, 37,
42]: sequential addition of IMP, NAD+ and K+ [42], IMP binds first, XMP is released last [28], mechanism is partially random [21, 22]: IMP and K+ can
bind randomly to the free enzyme, while NAD+ does not react unless K+ or
both K+ and IMP are present on the enzyme [21])

Reaction type
Redox reaction

Natural substrates
Inosine 5'-phosphate + NAD+ + H_2O (enzyme linked to proliferation and ma-
lignancy [29], IMP oxidation is the predominant metabolic route leading to
ureide synthesis [30], obligatory step in biosynthesis of nucleic acid guanine [41]) [29, 30, 41]

Substrate spectrum

1 Inosine 5'-phosphate + NAD$^+$ + H$_2$O (ir [41–43]) [1–43]
2 6-Thioinosine 5'-phosphate + NAD$^+$ + H$_2$O [5]
3 Inosine 5'-phosphorothioate + NAD$^+$ + H$_2$O [25]
4 5-Mercapto-5'-deoxyinosine 5'-S-phosphate + NAD$^+$ + H$_2$O [25]
5 5'-Amino-5'-deoxyinosine 5'-N-phosphate + NAD$^+$ + H$_2$O [25]
6 Inosine + NAD$^+$ + H$_2$O (8% of the rate with inosine 5'-phosphate) [40]

Product spectrum

1 Xanthine 5'-phosphate + NADH
2 ?
3 ?
4 ?
5 ?
6 ?

Inhibitor(s)

6-Chloropurine ribonucleoside 5'-phosphate (inosine 5'-phosphate and guanosine 5'-phosphate retard inactivation [5]) [5, 20]; 6-Thioinosine 5'-phos-
phate (in absence of glutathione) [5, 25]; K$^+$ (above 0.1 M) [5]; Glutathione
(above 4 mM) [5]; Thiazole-4-carboxamide adenine dinucleotide [7, 12];
Thiazofurin (i.e. 2-beta-D-ribofuranosyl-thiazole-4-carboxamide, overview: cli-
nical and molecular impact of inhibition [13]) [8, 9, 13]; Mycophenolic acid
[10, 11, 28, 37]; XMP [11, 15, 19, 20, 28, 34, 37, 42, 43]; Sesquiterpene lactones (a class of anti-neoplastic drugs, overview) [11]; Ribavirin 5'-mono-
phosphate (inhibitory mechanism [15]) [12, 15]; NADH [15, 20, 33, 37, 42];
GMP [5, 6, 18, 19, 27, 28, 33, 34, 36, 37, 43]; ATP (at high concentration of
IMP and KCl [19], 0.8 mM, not [27]) [19, 43]; AMP [28, 33, 34, 37, 43]; Mg^{2+}
(inhibits K$^+$ activation) [33, 42, 43]; Selenazofurin (inhibition (in vivo) by the
NAD analog formed intracellularly) [14]; Bredinin [14]; Thiazole-4-carboxa-
mide adenine dinucleotide (inhibitory mechanism [15]) [14, 15]; Inorganic
phosphate [17]; Ribose 5-phosphate (noncompetitive) [17]; Inosine (very
weak noncompetitive) [17]; Inosine 5'-methylphosphonate (very weak non-
competitive) [17]; Inosine 5'-phosphofluoridate (very weak noncompetitive)
[17]; Inosine 5'-phosphite (very weak noncompetitive) [17];
6-Chloro-9-beta-D-ribofuranosylpurine 5 '-phosphate (irreversible inactivation, retarded by ligands that bind at IMP-binding site: IMP > XMP > GMP >
AMP [26]) [26, 34]; 5,5'-Dithiobis(2-nitrobenzoic acid) [26]; Iodoacetate [26,
40]; Iodoacetamide [26]; Methyl methanethiosulfonate [26]; UMP [27]; ADP
[43]; CMP [27]; TMP [27]; GDP [27, 43]; GTP [27, 43]; 6-Mercaptopurine ribonucleotide [28]; Allopurinol ribonucleotide [28]; p-Chloromercuribenzoate
[34, 39, 40, 43]; 6-Thio-IMP [37]; N-Ethylmaleimide [39]; HgCl$_2$ [40]; NAD$^+$
(substrate inhibition at high concentration) [42]; Li$^+$ (inhibits stimulatory effect of K$^+$ [42], no effect [43]) [42]

Cofactor(s)/prosthetic group(s)/activating agents
NAD+ (B-sided stereospecificity: 2–3H of IMP is transferred to pro-S position
of carbon atom C-4 of the nicotinamide ring in NAD+ [16]) [1–43]; Acetylpyri-
dine-NAD+ (60% as effective as NAD+) [43]; Thionicotinamide-NAD+ (5% as
effective as NAD+) [43]; NADP+ (not [43], reduction less than 10% of that
with NAD+ [33]) [33]; Cysteine (cysteine or glutathione required) [41]; Glutathione (cysteine or glutathione required [41], level for maximum activity:
2–4 mM [5]) [5, 41]; ATP (0.8 mM, activation) [27]; Sulfhydryl-reducing
agents (required for maximal activity) [33]

Metal compounds/salts
K+ (activation [33, 40–43], requirement [5], full activity in presence of K+
[34], 0.1 M: maximum activity [5], maximum activity at 33–66 mM KCl [40],
0.1 M [42], monovalent cation required for maximal activity [28]) [5, 28, 33,
34, 40–43]; NH₄+ (required [5], activation [41, 43], K+ can be partially replaced by NH₄+ [28, 42], no activation [33]) [5, 28, 41–43]; Na+ (can partially
replace K+ in activation [34, 42], no effect [43]) [34, 42]

Turnover number (min⁻¹)

Specific activity (U/mg)
0.236 [11]; 1.1–1.5 [33]; 1.19 [34]; 1.7 [12]; 5.6 [35]; 9.2 [36]; 6.49 [37];
More [38, 39, 43]

K_m-value (mM)
0.021 (inosine 5'-phosphate, 6-thioinosine 5'-phosphate, constant in
pH-range 6.5–8.2, sharp decrease at higher pH) [5]; 1.1 (NAD+) [5]; 0.020
(6-thio-IMP, pH 8.1) [5]; 0.012 (IMP) [11]; 0.023 (IMP) [12]; 0.025 (NAD+)
[11]; 0.065 (NAD+) [12]; 0.21 (inosine 5'-phosphorothioate) [25]; 0.013
(5'-mercapto-5-deoxyinosine 5'-S-phosphate) [25]; 0.038 (5'-amino-5'-deoxy-
inosine 5'-N-phosphate) [25]; More [15, 28, 30, 33, 34, 36–40, 42]

pH-optimum
7.8–8.0 [34]; 8.0 [40]; 8.1 [37]; 8.2 [5]; 8.2–9.5 [27]; 8.4 [39]; 8.5–9.5 [30];
9.4 [33]

pH-range
6.7–9.0 (6.7: about 40% of activity maximum, 9.0: about 55% of activity maximum) [34]; 7–9 (7: about 45% of activity maximum, 9: about 85% of activity maximum) [37]; 7.5–8.7 (7.5: about 60% of activity maximum, 8.7: about
40% of activity maximum) [40]; 7.5–9.5 (7.5: about 55% of activity maximum, 9.5: about 70% of activity maximum) [40]; 8.5–10 (8.5: about 60% of
activity maximum, 10: about 75% of activity maximum) [33]

Temperature optimum (°C)
25 (assay at) [6, 39, 43]; 35 (assay at) [35]; 23 (assay at) [36]

Temperature range (°C)

Enzyme Handbook © Springer-Verlag Berlin Heidelberg 1995
Duplication, reproduction and storage in data banks are only
allowed with the prior permission of the publishers

3 ENZYME STRUCTURE

Molecular weight

86000 (Aerobacter aerogenes, smallest active species (MW 86000), polymerization in absence of reducing agents, largest species (MW 248000), gel filtration) [32]

90000–100000 (Aerobacter aerogenes, basic catalytic molecular species, the species of this MW are of more than one type, under following conditions 2 species of indicated MW are present in approximately equal amount: 56000 and 95000 (0.02 M phosphate, pH 7.4), 185000 and 300000 (0.1 M Tris-citrate, pH 8.1 alone or 0.02 M buffer, pH 7.4 or 8.1 containing 0.1 M KCl, 0.1 M NaCl or 0.25 M KCl), 90000 and 180000 (0.38 M Tris-HCl), ultracentrifugation, sucrose density sedimentation) [23]

245000 (rat hepatoma 3924 A cells, gel filtration) [12]

127000 (rat, Yoshida sarcoma ascites tumor cells, molecular sieve chromatography in presence of 10% $(NH_4)_2SO_4$) [34]

165000 (Bacillus subtilis, sedimentation equilibrium) [39]

200000 (Vigna unguiculata, gel filtration) [33]

232000 (E. coli, gel filtration) [36]

249000 (E. coli, gel filtration, smallest enzymatically active species) [31]

More (nucleotide sequence of IMP dehydrogenase gene [4], various molecular species [43], enzyme tends to aggregate owing to its own physicochemical characteristics [34]) [4, 34, 43]

Subunits

Dimer (2 × 38000, Aerobacter aerogenes, amino acid analysis [32], 2 × 68000, rat, Yoshida sarcoma ascites tumor cells, SDS-PAGE [34]) [32, 34]

Tetramer (4 × 58000, E. coli, SDS-PAGE [36], 4 × 60000, rat hepatoma 3924 A cells, SDS-PAGE [12], 4 × 50000, Vigna unguiculata, SDS-PAGE [33]) [12, 33, 36]

? (x × 62000 + x × 44000, E. coli, gel filtration in presence of guanidine) [31]

Glycoprotein/Lipoprotein

–

4 ISOLATION/PREPARATION

Source organism
Leishmania donovania [1]; Human [2, 13, 15, 28]; Chinese hamster [2, 7];
Bacillus subtilis [3, 4, 19, 39]; Aerobacter aerogenes [5, 17, 20–23, 25, 32,
41]; E. coli (guaA strain PL 1068 [6], derepressed mutant [43]) [6, 8, 16, 26,
31, 35, 36, 38, 43]; Eimeria tenella [10]; Mouse [11, 16, 42]; Salmonella ty-
phimurium [18]; Chicken [24]; Schizosaccharomyces pombe [27]; Vigna unguiculata L. (Walp., infected with Rhizobium strain CB756 [30]) [30, 33]; Rat
(hepatoma [29]) [9, 12, 29, 34, 37]; Pisum sativum [40]; African green monkey (vero cell-line infected with vaccinia virus and parainfluenza virus type 3
strain C243) [14]

Source tissue
Cultured cells (rat hepatoma 3924 A cells [12, 37], Yoshida sarcoma ascites
tumor cells [34], MOLT 4F human T-lymphoblasts [15], CHO-cells (Chinese
hamster ovary cells) [7], hepatoma 3924 A cells [9], P-388 lymphocytic leu-
kemia tumor cells [11]) [7, 9, 11, 12, 15, 34, 37, 42]; Cells [5]; Lymphoblasts
[16]; Liver [24, 37]; Placenta [28]; Root nodules [30, 33]; Thymus (highest
activity of rat tissues tested) [37]; Spleen (highest activity of rat tissues tested) [37]; Seeds [40]

Localisation in source
Microsomes (most of the activity is associated with microsomal fraction)
[24]; Cytosol [28, 30]; Soluble [33]

Purification
Aerobacter aerogenes [5, 41]; Vigna unguiculata [33]; Rat (hepatoma cells
3924 A [12, 37], Yoshida sarcoma ascites tumor cells [34]) [12, 34, 37]; Bacillus subtilis [38]; E. coli (affinity chromatography [35, 36, 38], large scale
[36], derepressed mutant [43]) [35, 36, 38, 43]; Mouse (sarcoma 180 ascites cells [42], P-388 lymphocytic leucemia tumor cells [11]) [11, 42]; Human
(MOLT 4F human T-lymphoblasts) [15]

Crystallization
–

Cloned
(genes from: Leishmania donovania [1], human [2], Chinese hamster [2],
Bacillus subtilis [3]) [1–3]

Renaturated
(readily renatured following denaturation with urea, guanidine-HCl and SDS)
[38]

5 STABILITY

pH

Temperature (°C)

Oxidation

Organic solvent

General stability information
Purified enzyme very unstable [12]; Bovine serum albumin stabilizes during
dialysis [12]; Glycerol stabilizes [30, 33]; NAD$^+$ stabilizes [30]; IMP stabilizes [30]; p-Chloromercuribenzoate stabilizes [39]; Dithiothreitol stabilizes
[43]

Storage
−70°C, 10 mg/ml bovine serum albumin, 3 weeks, 15% loss of activity [12];
4°C, + NAD$^+$, + IMP, 20% loss of activity after 18 h [30]; −20°C [37]; 0°C, +
p-chloromercuribenzoate, several weeks without loss of activity [39]; 4°C,
24 h, 13% loss of activity [43]; −17°C, 24 h, 10% loss of activity [43]; Frozen
in 5 mM mercaptoethanol, 3 months, 50% loss of activity [43]

6 CROSSREFERENCES TO STRUCTURE DATABANKS

PIR/MIPS code
PIR3:S23226 (Acinetobacter calcoaceticus); PIR1:DEBSMP (Bacillus subti-
lis); PIR2:B31997 (Chinese hamster); PIR1:DEECIP (Escherichia coli);
PIR2:S20017 (Escherichia coli (fragment)); PIR2:A31997 (human);
PIR2:A38668 (Leishmania donovani); PIR2:JT0565 (mouse); PIR2:A35566 (I
human); PIR2:B35566 (II human)

Brookhaven code

7 LITERATURE REFERENCES

[1] Wilson, K., Collart, F.R., Huberman, E., Stringer, J. R., Ullman, B.: J. Biol.
    Chem.,266,1665–1671 (1991)
[2] Collart, F.R., Huberman, E.: J. Biol. Chem.,263,15769–15772 (1988)
[3] Miyagawa, K., Kimura, H., Nakahama, K., Kikuchi, M., Doi, M., Akiyama, S., Nakao,
    Y.: Bio/Technology,4,225–228 (1986)
[4] Kanzaki, N., Miyagawa, K.: Nucleic Acids Res.,18,6710 (1990)
[5] Hampton, A., Nomura, A.: Biochemistry,6,679–689 (1967)
[6] Lambden, P.R., Drabble, W.T.: Biochem. J.,133,607–608 (1973)
[7] Kuttan, R., Robins, R.K., Saunders, P.P.: Biochem. Biophys. Res. Commun.,107,
    862–868 (1982)
[8] Cooney, D.A., Jayaram, H.N., Glazer, R.I., Kelley, J. A., Marquez, V.E., Gebeyehu,
    G., Van Cott, A.C., Zwelling, L.A., Johns, D.G.: Adv. Enzyme Regul.,21,271–303
    (1983)

[9] Lui, M.S., Faderan, M.A., Liepnieks, J.J., Natsumeda, Y., Olah, E., Jayaram, H.N., Weber, G.: J. Biol. Chem.,259,5078–5082 (1984)
[10] Hupe, D.J., Azzolina, B.A., Behrens, N.D.: J. Biol. Chem.,261,8363–8369 (1986)
[11] Page, J.D., Chaney, S.G., Hall, I.H., Lee, K.H., Holbrook, D.J.: Biochim. Biophys. Acta,926,186–194 (1987)
[12] Yamada, Y., Natsumeda, Y., Weber, G.: Biochemistry,27,2193–2196 (1988)
[13] Weber, G., Yamaji, Y., Olah, E., Natsumeda, Y., Jayaram, H.N., Lapis, E., Zhen, W., Prajda, N., Hoffman, R., Tricot, G.J.: Adv. Enzyme Regul.,28,335–356 (1989)
[14] Robins, R.K., Revankar, G.R., McKernan, P.A., Murray, B.K., Kirsi, J.J., North, J.A.: Adv. Enzyme Regul.,24,29–43 (1985)
[15] Yamada, Y., Goto, H., Yoshino, M., Ogasawara, N.: Biochim. Biophys. Acta,1051,209–214 (1990)
[16] Cooney, D., Hamel, E., Cohen, M., Kang, G.J., Dalal, M., Marquez, V.: Biochim. Biophys. Acta,916,89–93 (1987)
[17] Nichol, A.W., Nomura, A., Hampton, A.: Biochemistry,6,1008–1015 (1967)
[18] Buzzee, D.H., Levin, A.P.: Biochem. Biophys. Res. Commun.,30,673–677 (1968)
[19] Ishii, K., Shiio, I.: J. Biochem.,63,661–669 (1968)
[20] Brox, L.W., Hampton, A.: Biochemistry,7,2589–2596 (1968)
[21] Heyde, E., Nagabhushanam, A., Vonarx, M., Morrison, J. F.: Biochim. Biophys. Acta,429,645–660 (1976)
[22] Heyde, E., Morrison, J.F.: Biochim. Biophys. Acta,429,661–671 (1976)
[23] Brox, L.W., Hampton, A.: Biochim. Biophys. Acta,206,215–223 (1970)
[24] Nagata, K., Mitsui, A., Tsushima, K.: Biochim. Biophys. Acta,177,680–682 (1969)
[25] Hampton, A., Brox, L.W., Bayer, M.: Biochemistry,8,2303–2311 (1969)
[26] Gilbert, H.J., Drabble, W.T.: Biochem. J.,191,533–541 (1980)
[27] Pourquie, J.: Biochim. Biophys. Acta,185,310–315 (1969)
[28] Holmes, E.W., Pehlke, D.M., Kelley, W.N.: Biochim. Biophys. Acta,364,209–217 (1974)
[29] Jackson, R.C., Weber, G.: Nature,256,331–333 (1975)
[30] Shelp, B.J., Atkin, C.A.: Plant Physiol.,72,1029–1034 (1983)
[31] Powell, G.F.: Biochemistry,12,1592–1595 (1973)
[32] Heyde, E., Morrison, J.F.: Biochim. Biophys. Acta,429,635–644 (1976)
[33] Atkins, C.A., Shelp, B.J., Storer, P.J.: Arch. Biochem. Biophys.,236,807–814 (1985)
[34] Okada, M., Shimura, K., Shiraki, H., Nakagawa, H.: J. Biochem.,94,1605–1613 (1983)
[35] Lowe, C.R., Hans, M., Spibey, N., Drabble, W.T.: Anal. Biochem.,104,23–28 (1980)
[36] Gilbert, H.J., Lowe, C.R., Drabble, W.T.: Biochem. J.,183,481–494 (1979)
[37] Jackson, R.C., Morris, H.P., Weber, G.: Biochem. J.,166,1–10 (1977)
[38] Krishnaiah, K.V.: Arch. Biochem. Biophys.,170,567–575 (1975)
[39] Yokosawa, H., Tobita, T., Yamada, T.: Biochim. Biophys. Acta,227,538–553 (1971)
[40] Turner, J.F., King, J.E.: Biochem. J.,79,147–151 (1961)
[41] Magasanik, B., Moyed, H.S., Gehring, L.B.: J. Biol. Chem.,226,339–350 (1957)
[42] Anderson, J.H., Sartorelli, A.C.: J. Biol. Chem.,243,4762–4768 (1968)
[43] Powell, G., Rajagopalan, K.V., Handler, P.: J. Biol. Chem.,244,4793–4797 (1969)

1 NOMENCLATURE

EC number
1.1.1.206

Systematic name
Tropine:NADP+ 3alpha-oxidoreductase

Recommended name
Tropine dehydrogenase

Synonymes

CAS Reg. No.
82532-89-6

2 REACTION AND SPECIFICITY

Catalysed reaction
Tropine + NADP+ →
→ tropinone + NADPH

Reaction type
Redox reaction

Natural substrates
Tropine + NADP+ [1]

Substrate spectrum
1 Tropine + NADP+ (r) [1]
2 Tropane-3alpha-ols + NADP+ (r) [1]
3 Scopine + NADP+ [1]
4 Nortropine + NADP+ [1]

Product spectrum
1 Tropinone + NADPH [1]
2 Corresponding tropanones + NADPH [1]
3 ?
4 ?

Inhibitor(s)
p-Hydroxymercuribenzoate [1]

Cofactor(s)/prosthetic group(s)/activating agents
NADP+ [1]; NADPH [1]

Metal compounds/salts

Turnover number (min⁻¹)

Specific activity (U/mg)

K_m-value (mM)
   0.77 (tropine) [1]; 0.83 (tropinone) [1]; 0.013 (NADP⁺) [1]; 0.023 (NADPH)
   [1]; 0.91 (scopine) [1]; 2.0 (nortropine) [1]; 3.03 (nortropinone)

pH-optimum
   9.5 (tropine + NADP⁺) [1]; 6.8 (tropinone + NADPH) [1]

pH-range

Temperature optimum (°C)
   60 (tropine + NADP⁺) [1]

Temperature range (°C)

---

3 ENZYME STRUCTURE

Molecular weight
   56000 (Datura stramonium, gel filtration) [1]

Subunits

Glycoprotein/Lipoprotein
   –

---

4 ISOLATION/PREPARATION

Source organism
   Datura stramonium [1]

Source tissue
   Root cultures [1]

Localisation in source

Purification
   Datura stramonium (partial) [1]

Crystallization
   –

Cloned
   –

Renaturated
   –

2

5 STABILITY

pH
  6.0–10.5 [1]

Temperature (°C)
  50 (unstable above) [1]

Oxidation

Organic solvent

General stability information

Storage
  –18°C, 50% glycerol, 3 weeks [1]

---

6 CROSSREFERENCES TO STRUCTURE DATABANKS

PIR/MIPS code

Brookhaven code

---

7 LITERATURE REFERENCES

[1] Koelen, K.J., Gross, G.G.: J. Med. Plant Res.,44,227–230 (1982)

1 NOMENCLATURE

EC number
1.1.1.207

Systematic name
(-)-Menthol:NADP+ oxidoreductase

Recommended name
(-)-Menthol dehydrogenase

Synonymes
Monoterpenoid dehydrogenase
More (not identical with EC 1.1.1.208)

CAS Reg. No.
81811-46-3

2 REACTION AND SPECIFICITY

Catalysed reaction
(-)-Menthone + NADPH →
→ (-)-menthol + NADP+

Reaction type
Redox reaction

Natural substrates
(-)-Menthone + NADPH [1, 2]

Substrate spectrum
1 (-)-Menthone + NADPH (ir) [1, 2]
2 d-Isomenthone + NADPH [1]
3 Cyclohexanones + NADPH [1]
4 Cyclohexenones + NADPH [1]

Product spectrum
1 (-)-Menthol + NADP+ [1, 2]
2 d-Neoisomenthol + NADP+ [1]
3 Corresponding cyclohexanols + NADP+ [1]
4 Corresponding cyclohexenols + NADP+ [1]

Inhibitor(s)
p-Hydroxymercuribenzoate [1]; Hg^{2+} [1]; N-Ethylmaleimide [1]; Iodoacetamide [1]

Cofactor(s)/prosthetic group(s)/activating agents
NADPH [1]

Metal compounds/salts

Turnover number (min^{-1})

Specific activity (U/mg)

K_m-value (mM)
0.015 (NADPH) [1]; 0.25 ((-)-menthone) [1]

pH-optimum
7.5 ((-)-menthone + NADPH) [1]

pH-range

Temperature optimum (°C)

Temperature range (°C)

---

3 ENZYME STRUCTURE

Molecular weight
35000 (Mentha piperita, gel filtration) [1]

Subunits

Glycoprotein/Lipoprotein
–

---

4 ISOLATION/PREPARATION

Source organism
Mentha piperita (peppermint) [1, 2]

Source tissue
Leaf (epidermis) [2]

Localisation in source

Purification
Mentha piperita [1]

Crystallization
–

Cloned

–

Renaturated

–

5 STABILITY

pH

Temperature (°C)

Oxidation

Organic solvent

General stability information

Storage

6 CROSSREFERENCES TO STRUCTURE DATABANKS

PIR/MIPS code

Brookhaven code

7 LITERATURE REFERENCES

[1] Kjonaas, R., Martinkus-Taylor, C., Croteau, R.: Plant Physiol.,69,1013–1017 (1982)
[2] Croteau, R., Winters, J.N.: Plant Physiol.,69,975–977 (1982)

3

1 NOMENCLATURE

EC number
1.1.1.208

Systematic name
(+)-Neomenthol:NADP$^+$ oxidoreductase

Recommended name
(+)-Neomenthol dehydrogenase

Synonymes
Monoterpenoid dehydrogenase
More (not identical with EC 1.1.1.207)

CAS Reg. No.
81811-47-4

---

2 REACTION AND SPECIFICITY

Catalysed reaction
(-)-Menthone + NADPH →
→ (+)-neomenthol + NADP$^+$

Reaction type
Redox reaction

Natural substrates
(-)-Menthone + NADPH [1, 2]

Substrate spectrum
1 (-)-Menthone + NADPH (ir) [1, 2]
2 d-Isomenthone + NADPH [1]
3 Cyclohexanones + NADPH [1]
4 Cyclohexenones + NADPH [1]

Product spectrum
1 (+)-Neomenthol + NADP$^+$ [1, 2]
2 d-Isomenthol + NADP$^+$ [1]
3 Corresponding cyclohexanols + NADP$^+$ [1]
4 Corresponding cyclohexenols + NADP$^+$ [1]

Inhibitor(s)
p-Hydroxymercuribenzoate [1]; Hg^{2+} [1]; N-Ethylmaleimide [1]; Iodoacetamide [1]

Cofactor(s)/prosthetic group(s)/activating agents
NADPH [1]

Metal compounds/salts

Turnover number (min^{-1})

Specific activity (U/mg)

K_m-value (mM)
0.022 (NADPH) [1]; 0.022 ((-)-menthone) [1]

pH-optimum
7.6 ((-)-menthone + NADPH) [1]

pH-range

Temperature optimum (°C)

Temperature range (°C)

---

3 ENZYME STRUCTURE

Molecular weight
35000 (Mentha piperita, gel filtration) [1]

Subunits

Glycoprotein/Lipoprotein
–

---

4 ISOLATION/PREPARATION

Source organism
Mentha piperita (peppermint) [1, 2]

Source tissue
Leaf (mesophyll) [2]

Localisation in source

Purification
Mentha piperita [1]

Crystallization
–

Cloned

–

Renaturated

–

5 STABILITY

pH

Temperature (°C)

Oxidation

Organic solvent

General stability information

Storage

6 CROSSREFERENCES TO STRUCTURE DATABANKS

PIR/MIPS code

Brookhaven code

7 LITERATURE REFERENCES

[1] Kjonaas, R., Martinkus-Taylor, C., Croteau, R.: Plant Physiol.,69,1013–1017 (1982)
[2] Croteau, R., Winters, J.N.: Plant Physiol.,69,975–977 (1982)

1 NOMENCLATURE

EC number
1.1.1.209

Systematic name
3(or 17)alpha-Hydroxysteroid:NAD(P)$^+$ oxidoreductase

Recommended name
3(or 17)alpha-Hydroxysteroid dehydrogenase

Synonymes
Dehydrogenase, 3(17)alpha-hydroxy steroid
3(17)alpha-Hydroxysteroid dehydrogenase
More (cf. EC 1.1.1.51)

CAS Reg. No.
83294-77-3

2 REACTION AND SPECIFICITY

Catalysed reaction
Androsterone + NAD(P)$^+$ →
→ 5alpha-androstane-3,17-dione + NAD(P)H

Reaction type
Redox reaction

Natural substrates

Substrate spectrum
1 Androsterone + NAD(P)$^+$ (i.e. 3alpha-hydroxy-5alpha-androstan-17-one) [1, 2]
2 17alpha-Estradiol + NAD(P)$^+$ (r [2]) [1, 2]
3 17alpha-Estradiol 3-glucuronide + NAD(P)$^+$ (r [2]) [1, 2]
4 Epitestosterone + NAD(P)$^+$ (r [2]) [1, 2]
5 5alpha-Androstan-3alpha,17beta-diol + NAD(P)$^+$ (r [2]) [1, 2]
6 Estrone 3-sulfate + NAD(P)H [2]

Product spectrum
1 5alpha-Androstan-3,17-dione + NAD(P)H [2]
2 Estrone + NAD(P)H
3 Estrone 3-glucuronide + NAD(P)H [2]
4 Androst-4-en-3,17-dione + NAD(P)H [2]
5 5alpha-Dihydrotestosterone + NAD(P)H (i.e. 17beta-hydroxy-5alpha-androstan-3-one) [2]
6 17alpha-Estradiol 3-sulfate + NAD(P)+

Inhibitor(s)
Androsterone (competitive to epitestosterone) [2]

Cofactor(s)/prosthetic group(s)/activating agents
NADP+ (preferred) [1, 2]; NADPH (preferred) [1, 2]; NAD+ (low activity) [1]; NADH (low activity) [1, 2]

Metal compounds/salts

Turnover number (min^{-1})

Specific activity (U/mg)

K$_m$-value (mM)
0.00053 (epitestosterone, liver isozyme IB) [2]; 0.00061 (epitestosterone, kidney isozyme II) [2]; 0.0012 (androsterone, liver isozyme IC) [2]; 0.0021–0.0024 (epitestosterone, liver isozyme IB, androsterone, kidney, isozyme IB) [2]; 0.00291 (epitestosterone, kidney isozyme IB) [2]; 0.004–0.0047 (androsterone, kidney isozyme II, liver isozyme IB) [2]

pH-optimum
6–6.5 [2]

pH-range

Temperature optimum (°C)

Temperature range (°C)

---

3 ENZYME STRUCTURE

Molecular weight
40000 (female rabbit, form IB, gel filtration) [1]
41000 (female rabbit, forms II and III, gel filtration) [1]

Subunits
Monomer (1 × 40000, female rabbit, form IB, 1 × 41000, female rabbit, forms II and III, SDS-PAGE) [1]

Glycoprotein/Lipoprotein
–

4 ISOLATION/PREPARATION

Source organism
Rabbit (female) [1, 2]

Source tissue
Kidney [1, 2]; Liver [2]

Localisation in source
Cytosol [1, 2]

Purification
Rabbit (female, 4 forms of enzyme) [1]

Crystallization
–

Cloned
–

Renaturated
–

5 STABILITY

pH

Temperature (°C)

Oxidation

Organic solvent

General stability information

Storage
–20°C [1]

6 CROSSREFERENCES TO STRUCTURE DATABANKS

PIR/MIPS code

Brookhaven code

7 LITERATURE REFERENCES

[1] Lau, P.C.K., Layne, D.S., Williamson, D.G.: J. Biol. Chem.,257,9444–9449 (1982)
[2] Lau, P.C.K., Layne, D.S., Williamson, D.G.: J. Biol. Chem.,257,9450–9456 (1982)

1 NOMENCLATURE

EC number
1.1.1.210

Systematic name
3beta(or 20alpha)-Hydroxysteroid:NADP⁺ oxidoreductase

Recommended name
3beta(or 20alpha)-Hydroxysteroid dehydrogenase

Synonymes
Progesterone reductase
Dehydrogenase, 3beta,20alpha-hydroxy steroid
3beta,20alpha-Hydroxysteroid oxidoreductase

CAS Reg. No.
82869-26-9

2 REACTION AND SPECIFICITY

Catalysed reaction
17beta-Hydroxy-5alpha-androstan-3-one + NADPH →
→ 5alpha-androstan-3beta,17beta-diol + NADP⁺ (3beta and 20alpha-catalytic activity share the same active site [1, 2])

Reaction type
Redox reaction

Natural substrates
Progesterone + NADPH (reaction in progesterone catabolism) [1, 2]

Substrate spectrum
1 17beta-Hydroxy-5alpha-androstan-3-one + NADPH (i.e. 5alpha-dihydro-testosterone [2]) [1, 2]
2 Progesterone + NADPH [1, 2]

Product spectrum
1 5alpha-Androstane-3beta,17beta-diol + NADP⁺ [1, 2]
2 4-Pregnene-20alpha-ol-3-one + NADP⁺ [1, 2]

Inhibitor(s)

17beta-Hydroxy-5alpha-androstan-3-one (competitive inhibitor to 20alpha-re-
duction of progesterone) [1, 2]; 19-Nortestosterone 17-bromoacetate (leads
to simultaneous, time-dependent loss of 3beta- and 20alpha-reductase activity, first order kinetic, either of both substrates protects) [1]; 16alpha-(Bro-
moacetoxy)-progesterone (competitive inhibitor of 3beta- and 20alpha-re-
ductase activity, either of both substrates protects) [2]

Cofactor(s)/prosthetic group(s)/activating agents

NADPH (optimal concentration: 0.5 mM) [1, 2]

Metal compounds/salts

Turnover number (min^{-1})

Specific activity (U/mg)

9.6 (20alpha-reductase) [2]

K_m-value (mM)

0.0025 (progesterone, 20alpha-reductase activity) [1]; 0.0094 (17beta-hy-
droxy-5alpha-androstan-3-one, 3beta-reductase activity) [1]; 0.0308 (progesterone) [2]; 0.074 (17beta-hydroxy-5alpha-androstan-3-one) [2]

pH-optimum

5.5–6.0 (in 0.1 M phosphate buffer) [1]

pH-range

Temperature optimum (°C)

37 (assay at) [1, 2]

Temperature range (°C)

3 ENZYME STRUCTURE

Molecular weight

35000 (sheep, gel filtration) [2]
55000 (bovine, gel filtration) [1]

Subunits

Monomer (1 × 35000, sheep, SDS-PAGE [2], 1 × 55000, bovine, SDS-PAGE
[1]) [1, 2]

Glycoprotein/Lipoprotein

–

4 ISOLATION/PREPARATION

Source organism

Bovine [1]; Sheep [2]

Source tissue
  Fetal erythrocytes [1, 2]

Localisation in source
  Soluble [1, 2]

Purification
  Bovine (affinity chromatography followed by Ca-phosphate gel adsorption)
  [1]; Sheep [2]

Crystallization
  –

Cloned
  –

Renaturated
  –

5 STABILITY

pH

Temperature (°C)

Oxidation

Organic solvent

General stability information
  Lyophilization leads to complete inactivation in a few days [1]

Storage
  –20°C, in 0.1 M phosphate buffer, pH 6.0, stable for months [1]; –20°C, purified enzyme indefinitely stable [1]; Frozen, as ammonium sulfate precipitate,
  stable [1]

6 CROSSREFERENCES TO STRUCTURE DATABANKS

PIR/MIPS code
  PIR2:JQ1850 (variola major virus)

Brookhaven code

7 LITERATURE REFERENCES

[1] Sharaf, M.A., Sweet, F.: Biochemistry,21,4615–4620 (1982)
[2] Chen, Q., Rosik, L.O., Nancarrow, C.D., Sweet, F.: Biochemistry,28,8856–8863 (1989)

1 NOMENCLATURE

EC number
1.1.1.211

Systematic name
Long-chain-(S)-3-hydroxyacyl-CoA:NAD$^+$ oxidoreductase

Recommended name
Long-chain-3-hydroxyacyl-CoA dehydrogenase

Synonymes
Dehydrogenase, long-chain 3-hydroxyacyl coenzyme A
beta-Hydroxyacyl-CoA dehydrogenase
3-Hydroxyacyl-CoA dehydrogenase
LCHAD [2]
More (cf. EC 1.1.1.35)

CAS Reg. No.
84177-52-6

2 REACTION AND SPECIFICITY

Catalysed reaction
(S)-3-Hydroxyacyl-CoA + NAD$^+$ →
→ 3-oxoacyl-CoA + NADH

Reaction type
Redox reaction

Natural substrates
More (multifunctional beta-oxidation enzyme) [2]

Substrate spectrum
1 (S)-3-Hydroxyacyl-CoA + NAD$^+$ (acts most rapidly on derivatives with chain-length 10 to 16, inactive with acetoacetyl-CoA [2], maximum activity with 3-ketohexadecanoyl-CoA [1]) [1, 2]
2 More (in addition to 3-hydroxyacyl-CoA dehydrogenase activity the enzyme posseses 2-enoyl-CoA hydratase and 3-ketoacyl-CoA thiolase activity which cannot be separated) [2]

Product spectrum
1 3-Oxoacyl-CoA + NADH
2 ?

Inhibitor(s)

Cofactor(s)/prosthetic group(s)/activating agents
    NAD+ [1, 2]

Metal compounds/salts

Turnover number (min⁻¹)

Specific activity (U/mg)

K_m-value (mM)
    16.3 (3-ketohexadecanoyl-CoA) [2]; 3.2–7.4 (3-hydroxyoctanoyl-CoA) [2]

pH-optimum

pH-range

Temperature optimum (°C)

Temperature range (°C)

3 ENZYME STRUCTURE

Molecular weight
    186000 (and higher, rat, gel filtration) [1]
    230000 (human, gel filtration) [2]

Subunits
    Tetramer (2 × 71000 + 2 × 47000, human, SDS-PAGE) [2]

Glycoprotein/Lipoprotein
    –

4 ISOLATION/PREPARATION

Source organism
    Rat [1]; Rabbit (neonatal) [1]; Human [2]

Source tissue
    Liver [1, 2]; Heart [1]; Kidney [1]; Brown adipose tissue [1]

Localisation in source
    Mitochondria (bound to inner membrane [1], bound [2]) [1, 2]

Purification
    Human [2]; Rat [1]

Crystallization
    –

Cloned

–

Renaturated

–

5 STABILITY

pH

Temperature (°C)

Oxidation

Organic solvent

General stability information

Storage

6 CROSSREFERENCES TO STRUCTURE DATABANKS

PIR/MIPS code

Brookhaven code

7 LITERATURE REFERENCES

[1] El-Fakhri, M., Middleton, B.: Biochim. Biophys. Acta,713,270–279 (1982)
[2] Carpenter, K., Pollitt, R.J., Middleton, B.: Biochem. Biophys. Res. Commun.,183, 443–448 (1992)

1 NOMENCLATURE

EC number
1.1.1.212

Systematic name
(3R)-3-Hydroxyacyl-[acyl-carrier-protein]:NAD+ oxidoreductase

Recommended name
3-Oxoacyl-[acyl-carrier-protein] reductase (NADH)

Synonymes
Reductase, 3-oxoacyl-[acyl carrier protein] (reduced nicotinamide adenine dinucleotide)

CAS Reg. No.
82047-86-7

2 REACTION AND SPECIFICITY

Catalysed reaction
3-Oxoacyl-[acyl-carrier-protein] + NADH →
→ (3R)-3-hydroxyacyl-[acyl-carrier-protein] + NAD+

Reaction type
Redox reaction

Natural substrates
More (part of the fatty acid synthase system in plants) [1]

Substrate spectrum
1 Acetoacetyl-[acyl-carrier-protein] + NADH [1]
2 Acetoacetyl-CoA + NADH [1]
3 Acetoacetyl-N-acetylcysteamine + NADH (low activity) [1]

Product spectrum
1 3-Hydroxybutanoyl-[acyl-carrier-protein] + NAD+
2 3-Hydroxybutanoyl-CoA + NAD+
3 3-Hydroxybutanoyl-N-acetylcysteamine + NAD+

Inhibitor(s)

Cofactor(s)/prosthetic group(s)/activating agents
NADH [1]

Metal compounds/salts

Turnover number (min⁻¹)

Specific activity (U/mg)

K_m-value (mM)
0.00273 (NADH) [1]; 0.00965 (acetoacetyl-[acyl-carrier-protein]) [1]; 0.155 (acetoacetyl-CoA) [1]; 6.2 (acetoacetyl-N-acetylcysteamine) [1]

pH-optimum
7.0 [1]

pH-range
6.2–8.5 (6.2: about 80% of activity maximum, 8.5: about 25% of activity maximum) [1]

Temperature optimum (°C)
30 (assay at) [1]

Temperature range (°C)

---

3 ENZYME STRUCTURE

Molecular weight
168000 (Persea americana, sucrose density gradient centrifugation) [1]

Subunits

Glycoprotein/Lipoprotein
–

---

4 ISOLATION/PREPARATION

Source organism
Persea americana [1]

Source tissue
Fruit [1]

Localisation in source
Plastids [1]

Purification
Persea americana [1]

Crystallization
–

Cloned
–

Renaturated
–

5 STABILITY

pH

Temperature (°C)

Oxidation

Organic solvent

General stability information

Storage
−20°C, 0.3 M potassium phosphate buffer, pH 7.0, 10% w/v glycerol,
2 weeks [1]

6 CROSSREFERENCES TO STRUCTURE DATABANKS

PIR/MIPS code

Brookhaven code

7 LITERATURE REFERENCES

[1] Caughey, I., Kekwick, R.G.O.: Eur. J. Biochem.,123,553–561 (1982)

1 NOMENCLATURE

EC number
1.1.1.213

Systematic name
3alpha-Hydroxysteroid:NAD(P)+ oxidoreductase (A-specific)

Recommended name
3alpha-Hydroxysteroid dehydrogenase (A-specific)

Synonymes
More (see also EC 1.1.1.50 for enzymes, catalyzing the same reaction, but which are B-specific with respect to hydrogen transfer to/from NAD(P)+/NAD(P)H. Enzymes without information of stereospecificity are summerized under EC 1.1.1.50, but may in fact belong to EC 1.1.1.213)

CAS Reg. No.

2 REACTION AND SPECIFICITY

Catalysed reaction
Androsterone + NAD(P)+ →
→ 5alpha-androstan-3,17-dione + NAD(P)H (mechanism [1, 2])

Reaction type
Redox reaction

Natural substrates
More (function in metabolism of androgens, glucocorticoids, biosynthesis of bile acids, function of dihydrodiol dehydrogenase and thus oxidation of trans-dihydrodiols of polycyclic aromatic hydrocarbons, involvement in regulation of inflammatory prostaglandins) [4]

Substrate spectrum
1 Androsterone + NAD(P)$^+$ (r [5]) [1–3, 5, 6]
2 9,10-Phenanthrenequinone + NAD(P)H [5]
3 4-Nitrobenzaldehyde + NAD(P)H [5]
4 4-Nitroacetophenone + NAD(P)H [5]
5 1-Acenaphthenol + NAD(P)$^+$ [5]
6 Benzenedihydrodiol + NAD(P)$^+$ [5]
7 Prostaglandins + NADP$^+$ (types A_1, A_2, B_1, B_2, D_2, E_1, E_2, F_1, F_{2alpha}, and 15-keto-prostaglandins) [3]
8 5beta-Androstan-3alpha,17beta-diol + NADP$^+$ (r) [6]
9 5alpha-Androstan-3alpha,17beta-diol + NADP$^+$ [6]
10 5alpha-Pregnan-3alpha,21-diol-20-one + NADP$^+$ (r) [6]
11 Lithocholic acid + NADP$^+$ [6]
12 Glycolithocholic acid + NADP$^+$ [6]
13 5beta-Pregnan-3alpha,21-diol-20-one + NADP$^+$ [6]
14 5beta-Pregnan-3alpha-ol-20-one + NADP$^+$ (r) [6]
15 Glycochenodeoxycholic acid + NADP$^+$ (i.e 3alpha,7alpha-dihydroxy-5beta-cholanoyl glycine) [6]
16 5beta-Dihydrocortisone + NADPH (i.e. 17alpha,21-dihydroxy-5beta-pregnan-3,11,20-trione) [6]

Product spectrum
1 5alpha-Androstan-3,17-dione + NADPH [1, 2, 5, 6]
2 ?
3 ?
4 ?
5 ?
6 ?
7 ?
8 5beta-Androstan-17beta-ol-3-one + NADPH
9 5alpha-Androstan-17beta-ol-3-one + NADPH [6]
10 5alpha-Pregnan-21-ol-3,20-dione + NADPH [6]
11 Dehydrolithocholic acid + NADPH (i.e. 3-oxo-5beta-cholan-24-oic acid) [6]
12 3-Oxo-5beta-cholanoyl glycine + NADPH
13 5beta-Pregnan-21-ol-3,20-dione + NADPH
14 5beta-Pregnan-3,20-dione + NADPH [6]
15 7alpha-Hydroxy-3-oxo-5beta-cholanoyl glycine + NADPH
16 3alpha,17alpha,21-Trihydroxy-5beta-pregnan-11,20-dione + NADP$^+$

Inhibitor(s)
Cibacron blue [1]; Hexestrol [1]; Indomethacin [3, 5]; Arachidonic acid [3]; Prostaglandin A_{2alpha} [3]; Pyrazole (10% at 0.4 mM) [5]; Meclofenamic acid [5]; Tolmetin [5]; Zomepirac [5]; Ibuprofen [5]; Oxyphenybutazone [5]; Betamethasone [5]; Dexamethasone [5]; 6alpha-Methylprednisolone [5]; Prednisolone [5]; Prednisone [5]; Cortisone [5]; Cortisol [5]; Prostaglandins [5]; More (not dicoumarol, disulfiram) [5]

Cofactor(s)/prosthetic group(s)/activating agents

NAD+ (A-specific [2]) [2, 5]; NADH (A-specific [2]) [2, 5]; NADP+ (A-specific [1]) [1, 5, 6]; NADPH (A-specific [1]) [1, 5, 6]; More (see also EC 1.1.1.50 for enzymes catalyzing the same reaction but with B-specificity concerning the cofactor. All enzymes without information of stereospecificity for cofactors are included in EC 1.1.1.50 but some of those may in fact belong to EC 1.1.1.213)

Metal compounds/salts

Turnover number (min⁻¹)

12 (glycochenodeoxycholic acid) [6]; 13 (lithocholic acid) [6]; 14 (5alpha-pregnan-3alpha,21-diol-20-one, glycolithocholic acid) [6]; 21 (5alpha-androstan-17beta-ol-3-one) [6]; 25 (5beta-pregnan-3,20-dione) [6]; 26 (5beta-dihydrocortisone) [6]; 27 (5beta-androstan-3,17-dione) [6]; 34 (5alpha-androstan-3,17-dione) [6]; 38 (5beta-pregnan-3alpha,21-diol-20-one) [6]; 44 (dehydrolithocholic acid) [6]; 46 (5beta-pregnan-3alpha-ol-20-one, 5alpha-pregnan-21-ol-3,20-one) [6]; 72 (5beta-androstan-3alpha,17beta-diol) [6]; 80 (5alpha-androstan-3alpha,17beta-diol) [6]

Specific activity (U/mg)

1.93 [2]; 2.0 [3]; 3.19 [5]; 0.71 [6]

K_m-value (mM)

0.0008 (5beta-pregnan-3,20-dione) [6]; 0.001–0.002 (5alpha-androstan-3,17-dione [2, 5], 9,10-phenanthrenequinone (+ NADPH, at pH 6.0) [5], 5beta-androstan-3,17-dione, dehydrolithocholic acid, 5alpha-androstan-17beta-ol-3-one, dihydrocortisone [6]) [2, 5, 6]; 0.0025–0.004 (NADH [2], lithocholic acid, glycolithocholic acid, glycochenodeoxycholic acid, 5alpha-androstan-3,17-dione, 5beta-pregnan-3alpha-ol-20-one [6]) [2, 6]; 0.0055 (5beta-pregnan-3alpha,21-diol-20-one) [6]; 0.012–0.016 (5beta-androstan-3alpha-ol-17-one, 5beta-androstan-3alpha,17beta-diol, 5alpha-androstan-3alpha,17beta-diol) [6]; 0.047 (androsterone) [2, 6]; 0.76 (NAD+) [2]; More (values for steroids, quinones, dihydrodiols with NADH and NADPH as cofactor and various pH-values) [5]

pH-optimum

10.2 [6]

pH-range

Temperature optimum (°C)

Temperature range (°C)

3 ENZYME STRUCTURE

Molecular weight
37029 (rat, sequence of cDNA) [4]
33000–34000 (rat, gel filtration, SDS-PAGE) [5]

Subunits
Monomer (1 × 37029, rat, sequence of cDNA [4], SDS-PAGE [5]) [4, 5]
? (x × 38000, hamster, SDS-PAGE) [6]

Glycoprotein/Lipoprotein
–

4 ISOLATION/PREPARATION

Source organism
Hamster [1, 6, 7]; Rat [2–5, 8]

Source tissue
Liver [1–8]

Localisation in source
Cytosol [3–6, 8]

Purification
Rat [5, 8]; Hamster [6]

Crystallization
–

Cloned
[4]

Renaturated
–

5 STABILITY

pH

Temperature (°C)
45 (inactivation at, protection by NADP+ [1], half-life: 17 min [6]) [1, 6]

Oxidation

Organic solvent

General stability information

Storage
–80°C, 20 mM potassium phosphate buffer, pH 7.0, 1 mM EDTA, 1 mM DTT, 30% glycerol [5]

6 CROSSREFERENCES TO STRUCTURE DATABANKS

PIR/MIPS code

Brookhaven code

7 LITERATURE REFERENCES

[1] Sawada, H., Hara, A., Ohmura, M., Nakayama, T., Deyashiki, Y.: J. Biochem.,109, 770–775 (1991)
[2] Askonas, L.J., Ricigliano, J.W., Penning, T.M.: Biochem. J.,278,835–841 (1991)
[3] Penning, T.M., Sharp, R.B: Biochem. Biophys. Res. Commun.,148,646–652 (1987)
[4] Pawlowski, J.E., Huizinga, M., Penning, T.M.: J. Biol. Chem.,266,8820–8825 (1991)
[5] Penning, T.M., Mukharji, I., Barrows, S., Talalay, P.: Biochem. J.,222,601–611 (1984)
[6] Ohmura, M., Hara, A., Nakagawa, M., Sawada, H.: Biochem.,266,583–589 (1990)
[7] Sawada, H., Hara, A., Nakagawa, M., Tsukada, F., Ohmura, M., Marsuura, K.: Int. J. Biochem.,21,367–375 (1989)
[8] Penning, T.M., Smithgall, T.E., Askona, L.J., Sharp, R.B.: Steroids,47,221–247 (1986)

Enzyme Handbook © Springer-Verlag Berlin Heidelberg 1995
Duplication, reproduction and storage in data banks are only
allowed with the prior permission of the publishers

1 NOMENCLATURE

EC number
1.1.1.214

Systematic name
(R)-Pantoyl-lactone:NADP+ oxidoreductase (B-specific)

Recommended name
2-Dehydropantoyl-lactone reductase (B-specific)

Synonymes
Reductase, 2-oxopantoyl lactone
2-Ketopantoyl lactone reductase
Ketopantoyl lactone reductase
More (cf. EC 1.1.1.168)

CAS Reg. No.
37211-75-9

2 REACTION AND SPECIFICITY

Catalysed reaction
2-Dehydropantoyl lactone + NADPH →
→ (R)-pantoyl lactone + NADP+

Reaction type
Redox reaction

Natural substrates
2-Dehydropantoyl lactone + NADPH [1]

Substrate spectrum
1 2-Dehydropantoyl lactone + NADPH [1]
2 More (keto-omega-methylpantoyl lactone is the only compound other than ketopantoyl lactone which is a substrate, less than 5% of the rate of keto-pantoyl lactone with: 2-keto-4-hydroxy-3-methylbutyric acid-gamma-lacto-ne, 2-keto-4-hydroxybutyric acid-gamma-lactone, ascorbic acid, dehydro-ascorbic acid, ketopantoic acid) [1]

Product spectrum
1 (R)-Pantoyl lactone + NADP+ [1]
2 ?

Inhibitor(s)

Cofactor(s)/prosthetic group(s)/activating agents
　NADPH (B-specific with respect to NADP$^+$) [1]; More (not NADH) [1]

Metal compounds/salts

Turnover number (min^{-1})

Specific activity (U/mg)

K_m-value (mM)

pH-optimum

pH-range

Temperature optimum (°C)

Temperature range (°C)

---

3 ENZYME STRUCTURE

Molecular weight

Subunits

Glycoprotein/Lipoprotein
　–

---

4 ISOLATION/PREPARATION

Source organism
　E. coli [1]

Source tissue

Localisation in source

Purification

Crystallization
　–

Cloned
　–

Renaturated
　–

5 STABILITY

pH

Temperature (°C)

Oxidation

Organic solvent

General stability information

Storage

6 CROSSREFERENCES TO STRUCTURE DATABANKS

PIR/MIPS code

Brookhaven code

7 LITERATURE REFERENCES

[1] Wilken, D.R., King, H.L., Dyar, R.E.: J. Biol. Chem.,250,2311–2314 (1975)

1 NOMENCLATURE

EC number
1.1.1.215

Systematic name
D-Gluconate:NADP+ oxidoreductase

Recommended name
Gluconate 2-dehydrogenase

Synonymes
2-Keto-D-gluconate reductase
Reductase, 2-ketogluconate
2-Ketogluconate reductase

CAS Reg. No.
68417-42-5

2 REACTION AND SPECIFICITY

Catalysed reaction
D-Gluconate + NADP+ →
→ 2-dehydro-D-gluconate + NADPH

Reaction type
Redox reaction

Natural substrates
2-Dehydro-D-gluconate + NADPH (2-keto-D-gluconate, part of non-phosphorylative pathway of carbohydrates predominantly in acetic acid bacteria [4], reverse reaction at the same rate by Gluconobacter enzyme [3]) [1, 3, 4]

Substrate spectrum

1 2-Dehydro-D-gluconate + NADPH (i.e. 2-keto-D-gluconate, r [1–3], best substrate for Acetobacter ascendens [2, 3] and Gluconobacter [3], strongly preferred reaction of Acetobacter enzyme, equal reaction rate of reduction of 2-ketogluconate and oxidation of gluconate with Gluconobacter enzyme [2]) [1–5]

2 L-Idonate + NADP+ (best substrate for Gluconobacter, poor substrate for Acetobacter) [1–3]

3 D-Galactonate + NADP+ (r [1–3], third best substrate for Gluconobacter [3], poor substrate for Acetobacter) [1–3]

4 5-Keto-D-gluconate + NADPH (poor substrate for Acetobacter) [1–3]

5 2-Keto-L-gulonate + NADPH (best substrate for Acetobacter rancens, third best substrate for Acetobacter ascendens and Gluconobacter [3]) [1–3]

6 D-Xylonate + NADPH (no substrate for Acetobacter [2]) [3]

7 Pyruvate + NADPH (reduction at 7% the rate of 2-ketogluconate reduction) [1]

8 Hydroxypyruvate + NADPH (reduction at 733% the rate of 2-keto-gluconate reduction) [1]

9 Glyoxylate + NADPH (reduction at 350% the rate of 2-ketogluconate reduction) [1]

10 Glyoxal + NADPH (reduction at 33% the rate of 2-ketogluconate reduction) [1]

11 Acetaldehyde + NADPH (poor substrate) [1]

12 More (no substrates are: 6-phospho-D-gluconate, D-mannonate, D-arabonate [3], D-xylonate, D-glucose, D-fructose, L-sorbose, 5-keto-D-fructose, D-sorbitol, glycerol [2]) [2, 3]

Product spectrum

1 Gluconate + NADP+ [1–5]

2 ?

3 2-Keto-D-galactonate + NADPH

4 ?

5 ?

6 ?

7 Lactate + NADP+

8 2,3-Dihydroxypropanoate + NADP+

9 Glycolate + NADP+

10 ?

11 Ethanol + NADP+

12 ?

Inhibitor(s)

Hg^{2+} (complete inhibition) [1, 3]; Sulfhydryl reagents (strong) [3]; Zn^{2+} [1]; Co^{2+} [1]; Cd^{2+} [1]; Oxamate (slight inhibition of reverse reaction) [1]; Tartronate (slight inhibition of reverse reaction) [1]; Oxalate [1]; Pyruvate (weak) [1]; More (monoiodoacetate does not inhibit) [1]

Cofactor(s)/prosthetic group(s)/activating agents
NADP+ [1–3]; NADPH [1–3]; More (no activity with NADH) [3]

Metal compounds/salts
More (no metal requirement) [1–3]

Turnover number (min⁻¹)

Specific activity (U/mg)
61.03 (Gluconobacter liquefaciens) [3]; 187.0 [2, 3]; 336.0 [1, 3]

Kₘ-value (mM)
0.0074 (NADPH) [2]; 0.01 (NADPH (+ 5-keto-D-gluconate)) [1]; 0.062 (NADPH (+ 2-keto-D-galactonate)) [1]; 0.065 (hydroxypyruvate) [1]; 0.083 (NADP+ (+ D-gluconate)) [1]; 0.11 (NADPH (+ 2-keto-L-gulonate)) [1]; 0.14 (NADPH (+ hydroxypyruvate)) [1]; 0.38 (glyoxylate) [1]; 0.71 (NADPH (+ glyoxylate)) [1]; 0.86 (5-keto-D-gluconate) [1]; 1.2 (NADP+) [2]; 2.4 (D-gluconate) [1]; 4.4 (gluconate) [2]; 5.3 (2-keto-D-gluconate) [2]; 16.0 (2-keto-D-galactonate) [1]; 91.0 (2-keto-L-gulonate) [1]

pH-optimum
6.0 (reduction of 2-ketogluconate) [2, 3]; 7.0 (reduction of 2-ketogluconate) [1]; 10.5 (oxidation of gluconate, Gluconobacter liquefaciens [3]) [2, 3]; 11.0 (oxidation of gluconate, Acetobacter ascendens) [3]; 12.0 (oxidation of gluconate, Acetobacter rancens [3]) [1, 3]

pH-range
4.5–8.5 (about half-maximal activity at pH 4.5 and 8.5, reduction of ketogluconate) [2]; 5.0–9.2 (about half-maximal activity at pH 5.0 and 9.2, reduction of ketogluconate) [1]; 10.0–12.0 (about half-maximal activity at pH 10 and 12, oxidation of gluconate) [2]; 10.8–12.5 (about half-maximal activity at pH 10.8 and 12.5, oxidation of gluconate) [1]

Temperature optimum (°C)
50 [1]; 55 [2]

Temperature range (°C)
35–65 (about half-maximal activity at 35°C and 65°C) [2]; 40–55 (about half-maximal activity at 40°C and 55°C) [1]

---

3 ENZYME STRUCTURE

Molecular weight
120000 (Acetobacter rancens, gel filtration [1, 3], Acetobacter ascendens, gel filtration [2, 3], Gluconobacter liquefaciens, gel filtration [3]) [1–3]

Subunits
Trimer (2 × 43000 + 1 × 34000, Gluconobacter liquefaciens, SDS-PAGE [3],
3 × 40000, Acetobacter ascendens, SDS gel electrophoresis [2]) [2, 3]
Octamer (8 × 15000, Acetobacter rancens, SDS-PAGE) [1]

Glycoprotein/Lipoprotein
–

---

4 ISOLATION/PREPARATION

Source organism
Acetobacter rancens [1, 3]; Acetobacter ascendens [2, 3]; Gluconobacter
liquefaciens [3]; Gluconobacter suboxydans [3]; Penicillium notatum (strain
Westling) [5]; Acetic acid bacteria (maximal formation at the end of expo-
nential growth phase) [4]; More (2-ketogluconate reductase activity is not
found in a number of oxidative or aerobic bacteria) [4]

Source tissue
Cell [1–5]

Localisation in source
Cytoplasm [1–3]

Purification
Acetobacter rancens (affinity chromatography on blue-dextran sepharose B)
[1, 3]; Acetobacter ascendens (affinity chromatography on blue-dextran sepharose B) [2, 3]; Gluconobacter liquefaciens [3]; Gluconobacter suboxydans (partial) [3]

Crystallization
(Acetobacter rancens [1, 3], Acetobacter ascendens [2, 3], Gluconobacter
liquefaciens [3]) [1–3]

Cloned
–

Renaturated
–

---

5 STABILITY

pH
6.0 (stable for a week in the cold, increasing inactivation below and above,
stable for a month or more with additional 2-mercaptoethanol [1], crude and
crystallized, stable in 0.01 M potassium phosphate buffer with the addition
of 2-mercaptoethanol at 5°C [3]) [1, 3]

Temperature (°C)

42 (stable for 10 min, oxidation activity) [2]; 53 (stable for 10 min, reduction activity) [2]; 55 (stable for 10 min [1], stable for 5 min, enhanced by the addition of substrate [3]) [1, 3]; 60 (rapid inactivation) [1]; 70 (stable for 10 min in the presence of either 2-ketogluconate or gluconate) [2]

Oxidation

Organic solvent

General stability information

2-Mercaptoethanol stabilizes [1]; Dialysis, stable to [1]; Dialysis leads to general loss of activity [5]; Gluconate enhances thermal and storage stability even of dilute enzyme solutions, 0.1 mg protein/l [2]; 2-Ketogluconate enhances thermal and storage stability even of dilute enzyme solutions, 0.1 mg protein/l [2]; 5-Ketogluconate does not stabilize [2]

Storage

−20°C, stable in the presence of sulfhydryl agents and ammonium sulfate [1]; −20°C, crude extract stable for up to 7 days under nitrogen [5]; 0°C, stable for at least a month in phosphate buffer [1]; 5°C, stable in 0.01 M potassium phosphate buffer with the addition of 2-mercaptoethanol [3]; Ammonium sulfate solution of crystalline enzyme stable for over 2 years [2]

---

6 CROSSREFERENCES TO STRUCTURE DATABANKS

PIR/MIPS code

Brookhaven code

---

7 LITERATURE REFERENCES

[1] Chiyonobu, T., Shinagawa, E., Adachi, O., Ameyama, M.: Agric. Biol. Chem.,40,175–184 (1976)
[2] Adachi, O., Chiyonobu, T., Shinagawa, E., Matsushita, K., Ameyama, M.: Agric. Biol. Chem.,42,2057–2062 (1978)
[3] Ameyama, M., Adachi, O.: Methods Enzymol.,89,203–210 (1982)
[4] Shinagawa, E., Chiyonobu, T., Matsushita, K., Adachi, O., Ameyama, M.: Agric. Biol. Chem.,42,1055–1057 (1978)
[5] Pitt, D., Mosley, M.J.: Antonie Leeuwenhoek,51,353–364 (1985)

1 NOMENCLATURE

EC number
1.1.1.216

Systematic name
2-trans,6-trans-Farnesol:NADP+ 1-oxidoreductase

Recommended name
Farnesol dehydrogenase

Synonymes
NADP-Farnesol dehydrogenase
Dehydrogenase, farnesol (nicotinamide adenine dinucleotide phosphate)

CAS Reg. No.
90804-55-0

2 REACTION AND SPECIFICITY

Catalysed reaction
2-trans,6-trans-Farnesol + NADP+ →
→ 2-trans,6-trans-farnesal + NADPH

Reaction type
Redox reaction

Natural substrates

Substrate spectrum
1 trans,trans-Farnesol + NADP+ (r, preferred substrate) [1]
2 cis,trans-Farnesol + NADP+ [1]
3 Geraniol + NADP+ [1]
4 Citronerol + NADP+ [1]
5 Nerol + NADP+ [1]
6 Decanol + NADP+ [1]

Product spectrum
1 trans,trans-Farnesal + NADPH [1]
2 ?
3 ?
4 ?
5 ?
6 ?

Enzyme Handbook © Springer-Verlag Berlin Heidelberg 1995
Duplication, reproduction and storage in data banks are only
allowed with the prior permission of the publishers

Inhibitor(s)
EDTA [1]; N-Ethylmaleimide [1]; Monoiodoacetamide [1]; $CuCl_2$ [1]; $ZnCl_2$ [1]; More (inhibitory effect of thiol reagents indicates that the SH-group of the enzyme participates in its activity) [1]

Cofactor(s)/prosthetic group(s)/activating agents
$NADP^+$ (NAD^+ cannot replace $NADP^+$) [1]

Metal compounds/salts

Turnover number (min^{-1})

Specific activity (U/mg)

K_m-value (mM)
0.0156 (trans,trans-farnesol) [1]; 0.729 (geraniol) [1]

pH-optimum
9.5 (oxidation of farnesol) [1]; 7.0–7.5 (reduction of farnesal) [1]

pH-range
6–9 (reduction of farnesal: 60% of maximal activity at pH 6.5 and at pH 8) [1]; 7.0–10.5 (oxidation of farnesol: 10% of maximal activity at pH 7.0, 30% of maximal activity at pH 10.2) [1]

Temperature optimum (°C)
25 (assay at) [1]

Temperature range (°C)

3 ENZYME STRUCTURE

Molecular weight
90000 (Ipomoea batatas, gel permeation chromatography) [1]

Subunits
Dimer (2 × 47000, Ipomoea batatas, SDS-PAGE) [1]

Glycoprotein/Lipoprotein
–

4 ISOLATION/PREPARATION

Source organism
Ipomoea batatas (sweet potato, enzyme is formed in response to infection by the black rot fungus Ceratocystis fimbriata) [1]

Source tissue
Roots [1]

Localisation in source

Purification
Ipomoea batatas [1]

Crystallization
–

Cloned
–

Renaturated
–

5 STABILITY

pH

Temperature (°C)

Oxidation

Organic solvent

General stability information

Storage

6 CROSSREFERENCES TO STRUCTURE DATABANKS

PIR/MIPS code

Brookhaven code

7 LITERATURE REFERENCES

[1] Inoue, H., Tsuji, H., Uritani, I.: Agric. Biol. Chem.,48,733–738 (1984)

1 NOMENCLATURE

EC number
1.1.1.217

Systematic name
Benzyl-(2R,3S)-2-methyl-3-hydroxybutanoate:NADP$^+$ 3-oxidoreductase

Recommended name
Benzyl-2-methyl-hydroxybutyrate dehydrogenase

Synonymes
Dehydrogenase, benzyl 2-methyl-3-hydroxybutyrate
Benzyl 2-methyl-3-hydroxybutyrate dehydrogenase [1]

CAS Reg. No.
99332-62-4

2 REACTION AND SPECIFICITY

Catalysed reaction
Benzyl(2R,3S)-2-methyl-3-hydroxybutanoate + NADP$^+$ →
→ benzyl 2-methyl-3-oxobutanoate + NADPH

Reaction type
Redox reaction

Natural substrates

Substrate spectrum
1 Benzyl 2-methyl-3-oxobutyrate + NADPH (no reduction of aldehydes or alcohols) [1]

Product spectrum
1 Benzyl(2R,3S)-2-methyl-3-hydroxybutanoate and benzyl
(2S,3S)-2-methyl-3-hydroxybutanoate + NADP$^+$ [1]

Inhibitor(s)
$CuSO_4$ [1]; $HgCl_2$ [1]; $NiCl_2$ [1]; $AgNO_3$ [1]; KCN [1]; NaF [1]; More (not inhibited by EDTA and thiol reagents e.g. cysteine) [1]

Cofactor(s)/prosthetic group(s)/activating agents
NADP$^+$ [1]; NAD$^+$ (56% of activity with NADP$^+$) [1]; FMN (42% of activity with NADP$^+$) [1]

Metal compounds/salts
$BaCl_2$ (stimulates) [1]

Turnover number (min^{-1})

Specific activity (U/mg)
   0.9 (benzyl 2-methyl-3-oxobutyrate) [1]

K_m-value (mM)
   0.42 (benzyl 2-methyl-3-oxobutyrate) [1]

pH-optimum
   7.0 (in 100 mM phosphate buffer) [1]

pH-range
   6–8 (90% of maximal activity at pH 6, 100% at pH 8) [1]

Temperature optimum (°C)
   45 [1]

Temperature range (°C)
   40–55 (relative activity: 90% at 40°C, 80% at 55°C) [1]

---

3 ENZYME STRUCTURE

Molecular weight
   32000 (Saccharomyces cerevisiae, gel filtration) [1]

Subunits
   Monomer (1 × 31000, Saccharomyces cerevisiae, SDS-PAGE) [1]

Glycoprotein/Lipoprotein
   –

---

4 ISOLATION/PREPARATION

Source organism
   Saccharomyces cerevisiae [1]

Source tissue
   Cell [1]

Localisation in source

Purification
   Saccharomyces cerevisiae [1]

Crystallization
   –

Cloned
   –

Renaturated
   –

5 STABILITY

pH
  6–9 [1]

Temperature (°C)
  20–50 (incubation for 10 min: 80% activity retained at 45°C, 10% at 50°C)
  [1]

Oxidation

Organic solvent

General stability information
  Purification procedure at 4°C [1]

Storage

6 CROSSREFERENCES TO STRUCTURE DATABANKS

PIR/MIPS code

Brookhaven code

7 LITERATURE REFERENCES

[1] Furuichi, A., Akita, H., Matsukura, H., Oishi, T., Horikoshi, K.: Agric. Biol. Chem.,49,2563–2570 (1985)

1 NOMENCLATURE

EC number
1.1.1.218

Systematic name
Morphine:NAD(P)+ 6-oxidoreductase

Recommended name
Morphine 6-dehydrogenase

Synonymes
Dehydrogenase, morphine 6-
Reductase, naloxone
Naloxone reductase

CAS Reg. No.
97002-71-6; 106640-77-1

2 REACTION AND SPECIFICITY

Catalysed reaction
Morphine + NAD(P)+ →
→ morphinone + NAD(P)H

Reaction type
Redox reaction

Natural substrates
Morphine + NADP+ (highly specific alkaloid dehydrogenase, oxidizing only
the C-6 hydroxy group of morphine and codeine [4], initial reaction in morphine catabolism [1–4], guinea pig enzyme acts with NAD(P)+ [1, 2], Pseudomonas enzyme: constitutive, acts only with NADP+ [3, 4]) [1–4]

Substrate spectrum

1 Morphine + NADP$^+$ (r) [1–4]
2 Nalorphine + NAD(P)$^+$ (best substrate, oxidation at 120% the rate of morphine oxidation) [1]
3 Codeine + NADP$^+$ (oxidation at 80% [1] or 120% [4] the rate of morphine oxidation, r [1, 4]) [1, 3, 4]
4 Normorphine + NAD(P)$^+$ (oxidation at 120% the rate of morphine oxidation) [1]
5 Ethylmorphine + NAD(P)$^+$ (oxidation at 51% the rate of morphine oxidation) [1]
6 Dihydrocodeine + NADP$^+$ (oxidation at 19% [1] or 7.1% [4] the rate of morphine oxidation) [1, 4]
7 trans-Decalindiol + NAD(P)$^+$ [1]
8 Naloxone + NAD(P)H (r) [2]
9 More (specific for oxidoreduction of the 6-hydroxyl group of codeine [4], morphine [2, 4] and morphine analogues having an unsaturated bond at C-7,8 and/or a C-14 hydroxyl group [2], acts only slowly on 7,8-saturated derivatives such as dihydromorphine, dihydrocodeine [1], poor substrates: indanol, acenaphthenol, tetralol, 2-cyclohexen-1-ol, geraniol [1], no substrates: thebaine [3], several prostaglandins: E_1, E_2, F_{1alpha}, F_{2alpha}, D_2, testosterone, estradiol, nortestosterone, 3-hydroxy-hexobarbital, 5beta- or 5alpha-androstan-17beta-ol-3-one, methanol, ethanol, 1-buten-3-ol, 1-penten-3-ol, cyclohexanol, allyl alcohol, cinnamyl alcohol, benzyl alcohol, p-nitrobenzyl alcohol, 3-phenyl-1-propanol, 4-phenyl-2-butanol, 4-chromanol [1]) [1–4]

Product spectrum

1 Morphinone + NADPH [1–4]
2 ?
3 Codeinone + NADPH [1–4]
4 ?
5 ?
6 ?
7 ?
8 6alpha-Naloxol + NAD(P)$^+$ [2]
9 ?

Inhibitor(s)

Phenylarsine oxide (dithiol modifier, strong competitive inhibitor to cofactor [1]) [1]; CdCl$_2$ (dithiol modifier, strong competitive inhibitor to cofactor) [1]; CuSO$_4$ [2, 4]; Lithocholic acid [2]; Quercitrin [2]; Prostaglandin E_1 [2]; Ketamine [2]; Naloxone (competitive to morphine) [2]; 1,10-Phenanthroline [4]; PCMB (partly reversible by DTT) [4]; 2,2'-Bipyridyl [4]; More (no inhibition by NEM, PCMB, iodoacetic acid [1], weak or no inhibition by 8-hydroxyquinoline, EDTA, iodoacetate, NEM [4]) [1, 4]

Cofactor(s)/prosthetic group(s)/activating agents
NADP+ (the Pseudomonas enzyme acts only with NADP+, not NAD+ [3, 4])
[1–4]; NAD+ (50% as effective as NADP+ [1], cannot replace NADP+ [3, 4])
[1, 2]; NADPH (2.7 times as effective as NADH) [2]; 2-Mercaptoethanol (activation, removal inactivates, reversible) [1, 2]; More (flavin, cytochrome or pyrroloquinoline quinone cannot replace NADP+) [4]

Metal compounds/salts

Turnover number (min^{-1})

Specific activity (U/mg)
0.53 [1]; 0.855 (naloxone) [2]; 0.897 (morphine) [2]; 2.67 [3]; 75.4 [4]

K_m-value (mM)
0.0022 (NADH) [2]; 0.0065 (NADPH) [2]; 0.015 (nalorphine (+ NAD+)) [1];
0.033 (codeine (+ NADP+)) [1]; 0.044–0.048 (codeine [4], normorphine (+ NAD+), nalorphine (+ NADP+) [1]) [1]; 0.079 (codeine (+ NAD+)) [1]; 0.12 (NAD+, morphine (+ NAD+)) [1]; 0.27 (naloxone (+ NADPH)) [2]; 0.35 (NADP+) [4]; 0.4–0.44 (ethylmorphine, NADP+ [1], naloxone (+ NADH [2])) [1, 2]; 0.46–0.49 (morphine [4], morphine (+ NADP+) [1]) [1, 4]; 0.68 (normorphine (+ NADP+)) [1]; 1.54 (ethylmorphine (+ NADP+), dihydromorphine (+ NAD+)) [1]; 2.11 (dihydromorphine (+ NADP+)) [1]; 2.91 (dihydrocodeine) [4]

pH-optimum
6.2 (with NADH as cofactor) [2]; 6.5 (oxidation) [4]; 6.8 (with NADPH as cofactor) [2]; 9.5 (reduction) [4]

pH-range
More (reduction at around neutral pH values) [1]; 6.0–8.5 (detectable activity) [4]; 7.0–10.5 (detectable activity) [4]

Temperature optimum (°C)
25 (assay at) [1, 2]; 30 (assay at) [3, 4]

Temperature range (°C)

---

3 ENZYME STRUCTURE

Molecular weight
29000 (guinea pig, gel filtration) [1]
32000 (Pseudomonas putida M10, gel filtration) [4]

Subunits
Monomer (1 × 31000, Pseudomonas putida M10, SDS-PAGE) [4]

Glycoprotein/Lipoprotein
–

4 ISOLATION/PREPARATION

Source organism
  Guinea pig (strain Hartley) [1, 2]; Pseudomonas putida M10 [3, 4]

Source tissue
  Liver [1, 2]; Cell [3, 4]

Localisation in source
  Cytosol [1–4]

Purification
  Guinea pig (dye-ligand-affinity chromatography) [1, 2]; Pseudomonas putida
  M10 (affinity chromatography) [3, 4]

Crystallization
  –

Cloned
  –

Renaturated
  –

5 STABILITY

pH
  6.0 (in 50 mM sodium phosphate buffer plus NAD(P)$^+$ at least 10 days stable at 2–4°C [1], at 4°C more stable than at pH 8.0 [2]) [1, 2]; 7.0 ($t_{1/2}$:
  6.5 min at 50°C in potassium phosphate buffer) [4]; 8.0 (in 50 mM sodium
  phosphate buffer plus NAD$^+$: at least 10 days stable at 2–4°C, plus NADP$^+$:
  50% loss of activity after 5 days at 2–4°C) [1]

Temperature (°C)
  50 ($t_{1/2}$: 6.5 min in potassium phosphate buffer, pH 7.0) [4]

Oxidation

Organic solvent

General stability information
  2-Mercaptoethanol stabilizes during purification and storage, removal inactivates, reversible [1, 2]; Dithiothreitol stabilizes [3]

Storage
  –80°C, stable for over 2 months [4]; 2–4°C, stable at least 10 days in 50 mM
  sodium phosphate buffer, pH 6.0 or 8.0 plus NAD$^+$ or NADP$^+$ [1]; 2–4°C,
  50% loss of activity in 5 days [1]; 4°C, more stable at pH 6.0 than at pH 8.0
  [2]

6 CROSSREFERENCES TO STRUCTURE DATABANKS

PIR/MIPS code

Brookhaven code

7 LITERATURE REFERENCES

[1] Yamano, S., Kageura, E., Ishida, T., Toki, S.: J. Biol. Chem.,260,5259–5264 (1985)

[2] Yamano, S., Nishida, F., Toki, S.: Biochem. Pharmacol.,35,4321–4326 (1986)

[3] Bruce, N.C., Wilmot, C.J., Jordan, K.N., Trebilcock, A.E., Stephens, L.D.G., Lowe, C.R.: Arch. Microbiol.,154,465–470 (1990)

[4] Bruce, N.C., Wilmot, C.J., Jordan, K.N., Trebilcock, A.E., Stephens, L.D.G., Lowe, C.R.: Biochem. J.,274,875–880 (1991)

1 NOMENCLATURE

EC number
1.1.1.219

Systematic name
cis-3,4-Leucopelargonidin:NADP+ 4-oxidoreductase

Recommended name
Dihydrokaempferol 4-reductase

Synonymes
Reductase, dihydromyricetin
NADPH-dihydromyricetin reductase
Reductase, dihydroquercetin
Dihydroflavanol 4-reductase

CAS Reg. No.
98668-58-7; 83682-99-9

2 REACTION AND SPECIFICITY

Catalysed reaction
cis-3,4-Leucopelargonidin + NADP+ →
→ (+)-dihydrokaempferol + NADPH

Reaction type
Redox reaction

Natural substrates
More (reaction in anthocyanidin biosynthesis in plants) [2]

Substrate spectrum
1 (+)-Dihydroquercetin + NADPH (best substrate [4]) [1–5]
2 (+)-Dihydrokaempferol + NADPH (30% as active as dihydroquercetin [4])
[2, 4]
3 (+)-Dihydromyrecitin + NADPH (i.e. 5'-hydroxydihydroquercetin) [1, 2]
4 More (no substrates: (+)-dihydromorin, i.e. 3,5,7,2',4'-pentahydroxyflava-
none, and pinobanksin, i.e. 3,5,7-trihydroxyflavanone) [4]

Product spectrum

1 cis-3,4-Leucocyanidin + NADP$^+$ (i.e.
  5,7,3',4'-tetrahydroxyflavan-3,4-cis-diol, 2,3-trans-configuration retained
  [5]) [1, 3–5]
2 cis-3,4-Pelargonidin + NADP$^+$ [2, 4]
3 cis-3,4-Leuco-delphinidin + NADP$^+$ (i.e. 5,7,4'-trihydroxyflavan-3,4-cis-diol)
  [1, 2]
4 ?

Inhibitor(s)

Cu^{2+} (strong inhibitor) [4]; Iodoacetate (to some extent) [4]; HEPES-buffer
(gives poor activity) [5]; More (no inhibition with PCMB, EDTA, Mg^{2+}, Co^{2+},
Fe^{2+}, Ca^{2+}, Mn^{2+}, Zn^{2+}) [4]

Cofactor(s)/prosthetic group(s)/activating agents

NADPH [1–5]; NADH (half as effective as NADPH [2], cannot replace
NADPH [3, 4]) [2]; More (neither FMN, FAD, 6,7-dimethyl-5,6,7,8-tetrahy-
dropterine [2], nor ascorbate [3] can replace NADPH, the reaction occurs
equally well anaerobically [3]) [2, 3]

Metal compounds/salts

Turnover number (min^{-1})

Specific activity (U/mg)

More [4]; 0.024 [1]

K$_m$-value (mM)

More (around 37 mM, (+)-dihydroquercetin, estimation) [5]; 0.048 ((+)-dihy-
droquercetin) [4]

pH-optimum

5.5–6.5 (broad optimum) [2]; 7.0 (broad optimum with Tris buffer, sharp optimum with citric acid-sodium phosphate buffer) [4]

pH-range

5.1–7.2 (about half-maximal activity at pH 5.1 and 7.2) [2]; 6.4–8.0 (no significant variations in activity) [5]

Temperature optimum (°C)

25 (assay at) [2]; 30 (assay at) [4]

Temperature range (°C)

3 ENZYME STRUCTURE

Molecular weight

133000 (Cryptomeria japonica, gel filtration) [4]

Subunits

Glycoprotein/Lipoprotein

–

4 ISOLATION/PREPARATION

Source organism
Gingko biloba [1]; Pseudotsuga menziesii (douglas fir) [1, 3, 5]; Matthiola in-
cana (gillyflower) [2]; Cryptomeria japonica (D.Don cv.'kumotoshi', japanese
cedar) [4]

Source tissue
Callus or cell suspension culture (from petioles (Gingko biloba), needles
(Pseudotsuga menziesii) [1, 5]) [1, 3–5]; Flower [2]

Localisation in source
Cytoplasm [2, 4]

Purification
Cryptomeria japonica [4]

Crystallization
–

Cloned
–

Renaturated
–

5 STABILITY

pH

Temperature (°C)

Oxidation

Organic solvent

General stability information

Storage
–70°C, crude extract stable for several weeks [2]; –20°C, 20% loss of activity within 2 days, 2-mercaptoethanol protects [4]; –20°C, 57% loss of activity
after 4 months, 2-mercaptoethanol protects [4]

6 CROSSREFERENCES TO STRUCTURE DATABANKS

PIR/MIPS code
PIR2:JQ1688 (Arabidopsis thaliana)

Brookhaven code

7 LITERATURE REFERENCES

[1] Stafford, H.A., Lester, H.H.: Plant Physiol.,78,791–794 (1985)
[2] Heller, W., Forkmann, G., Britsch, L., Grisebach, H.: Planta,165,284–287 (1985)
[3] Stafford, H.A., Lester, H.H.: Plant Physiol.,70,695–698 (1982)
[4] Ishikura, N., Murakami, H., Fujii, Y.: Plant Cell Physiol.,29,795–799 (1988)
[5] Stafford, H.A., Lester, H.H.: Plant Physiol.,76,184–186 (1984)

1 NOMENCLATURE

EC number
1.1.1.220

Systematic name
6-Lactoyl-5,6,7,8-tetrahydropterin:NADP⁺ 2'-oxidoreductase

Recommended name
6-Pyruvoyltetrahydropterin 2'-reductase

Synonymes
Reductase, 6-pyruvoyltetrahydropterin
6PPH4(2'-oxo) reductase
6-Pyruvoyl tetrahydropterin (2'-oxo)reductase
6-Pyruvoyl-tetrahydropterin 2'-reductase
Pyruvoyl-tetrahydropterin reductase
More (not identical with EC 1.1.1.153)

CAS Reg. No.
97089-79-7

2 REACTION AND SPECIFICITY

Catalysed reaction
6-Pyruvoyltetrahydropterin + NADPH →
→ 6-lactoyl-5,6,7,8-tetrahydropterin + NADP⁺

Reaction type
Redox reaction

Natural substrates
More (when sepiapterin reductase activity is limiting, a large proportion of 6-[dihydroxypropyl-(L-erythro)-5,6,7,8-tetrahydropterin]synthesis proceeds through the 6-lactoyl intermediate [2], implicated in biosynthesis of tetrahydrobiopterin, may play an additional role in reduction of aldehydes derived from biogenic amine neurotransmitters and corticosteroid hormones as well as in pathogenesis of diabetic complications [3]) [2, 3]

Substrate spectrum
1 6-Pyruvoyltetrahydropterin + NADPH (only catalyzes reduction of C-2'-oxo group, inactive towards the C-1'-oxo group [2]) [1–4]
2 4-Nitrobenzaldehyde + NADPH [2]
3 Phenanthrenequinone + NADPH [2]
4 Menadione + NADPH [2]

Enzyme Handbook © Springer-Verlag Berlin Heidelberg 1995
Duplication, reproduction and storage in data banks are only
allowed with the prior permission of the publishers

Product spectrum
1 6-Lactoyl-5,6,7,8-tetrahydropterin + NADP+
2 ?
3 ?
4 ?

Inhibitor(s)

Cofactor(s)/prosthetic group(s)/activating agents
NADPH (absolutely dependent on) [1, 2]

Metal compounds/salts

Turnover number (min^{-1})

Specific activity (U/mg)
49.3 [2]

K_m-value (mM)

pH-optimum
7.4 (assay at) [2]

pH-range

Temperature optimum (°C)
37 (assay at) [2]

Temperature range (°C)

---

3 ENZYME STRUCTURE

Molecular weight
37000 (rat, gel filtration) [2]

Subunits
Monomer (1 × 36500, rat, SDS-PAGE) [2]

Glycoprotein/Lipoprotein
−

---

4 ISOLATION/PREPARATION

Source organism
Rat [1, 2]; Human (enzyme is identical to aldose reductase) [3, 4]

Source tissue
Brain [1, 2]; Liver [3, 4]

Localisation in source

Purification
Rat [2]; Human [4]

Crystallization
–

Cloned
–

Renaturated
–

5 STABILITY

pH

Temperature (°C)

Oxidation

Organic solvent

General stability information

Storage

6 CROSSREFERENCES TO STRUCTURE DATABANKS

PIR/MIPS code

Brookhaven code

7 LITERATURE REFERENCES

[1] Milstien, S., Kaufman, S.: Biochem. Biophys. Res. Commun.,128,1099–1107 (1985)
[2] Milstien, S., Kaufman, S.: J. Biol. Chem.,264,8066–8073 (1989)
[3] Steinerstauch, P., Wermuth, B., Leimbacher, W., Curtius, H.-C.: Biochem. Biophys. Res. Commun.,164,1130–1136 (1989)
[4] Curtius, H.-C., Steinerstauch, P., Leimbacher, W., Redweik, U., Takikawa, S., Ghishla, S. in "Unconjugated Pterins and Related Biogenic Amines" (Curtius, H.-C., Blau, N., Levine, R.A., eds.) ,89–98, Walter de Gruyter, Berlin, New York (1987)

1 NOMENCLATURE

EC number
1.1.1.221

Systematic name
Vomifoliol:NAD⁺ 4'-oxidoreductase

Recommended name
Vomifoliol 4'-dehydrogenase

Synonymes

CAS Reg. No.
94949-18-5

2 REACTION AND SPECIFICITY

Catalysed reaction
(+,-)-6-Hydroxy-3-oxo-alpha-ionol + NAD⁺ →
→ (+,-)-6-hydroxy-3-oxo-alpha-ionone + NADH

Reaction type
Redox reaction

Natural substrates
Vomifoliol + NAD⁺ (involved in the metabolism of abscisic acid in Corynebacterium) [1]

Substrate spectrum
1 Vomifoliol + NAD⁺ [1]

Product spectrum
1 Dehydrovomifoliol + NADH [1]

Inhibitor(s)

Cofactor(s)/prosthetic group(s)/activating agents
NAD⁺ [1]

Metal compounds/salts

Turnover number (min⁻¹)

Specific activity (U/mg)

K_m-value (mM)

pH-optimum

pH-range

Temperature optimum (°C)

Temperature range (°C)

3 ENZYME STRUCTURE

Molecular weight

Subunits

Glycoprotein/Lipoprotein

–

4 ISOLATION/PREPARATION

Source organism
   Corynebacterium sp. [1]

Source tissue

Localisation in source

Purification

Crystallization

–

Cloned

–

Renaturated

–

5 STABILITY

pH

Temperature (°C)

Oxidation

Organic solvent

General stability information

Storage

6 CROSSREFERENCES TO STRUCTURE DATABANKS

PIR/MIPS code

Brookhaven code

7 LITERATURE REFERENCES

[1] Hasegawa, S., Poling, S.M., Maier, V.P., Bennett, R.D.: Phytochemistry,23,2769–2771 (1984)

1 NOMENCLATURE

EC number
1.1.1.222

Systematic name
(R)-3-(4-Hydroxyphenyl)lactate:NAD(P)$^+$ 2-oxidoreductase

Recommended name
(R)-4-Hydroxyphenyllactate dehydrogenase

Synonymes
D-Hydrogenase, D-aryllactate
(R)-Aromatic lactate dehydrogenase

CAS Reg. No.
101754-02-3

2 REACTION AND SPECIFICITY

Catalysed reaction
(R)-3-(4-Hydroxyphenyl)lactate + NAD(P)$^+$ →
→ 3-(4-hydroxyphenyl)pyruvate + NAD(P)H

Reaction type
Redox reaction

Natural substrates
3-(4-Hydroxyphenyl)lactate + NAD(P)$^+$ (reaction in aromatic amino acid catabolism) [1]

Substrate spectrum
1 3-(4-Hydroxyphenyl)pyruvate + NAD(P)H (i.e. p-hydroxyphenylpyruvate,
 best substrate, r, the initial rate of oxidation is 100-fold smaller than that of
 reduction) [1]
2 Phenylpyruvate + NAD(P)H (r, reduction at 27% the rate of p-hydroxyphenylpyruvate reduction [1]) [1, 2]
3 Indolepyruvate + NAD(P)H (r, reduction at 30% the rate of p-hydroxyphenylpyruvate reduction [1]) [1, 2]
4 Pyruvate + NAD(P)H (poor substrate) [1]
5 2-Ketobutyrate + NAD(P)H (poor substrate) [1]
6 2-Ketoisopentanoate + NAD(P)H (poor substrate) [1]
7 More (best substrate for oxidation is p-hydroxyphenyllactate, D-phenyllactate is oxidized at 75% and indole-lactate at 52% the rate of p-hydroxyphenyllactate oxidation) [1]

Product spectrum
1 (R)-3-(4-Hydroxyphenyl)lactate + NAD(P)$^+$ (i.e. p-hydroxyphenyllactate) [1]
2 D-Phenyllactate + NAD(P)$^+$ [1]
3 Indolelactate + NAD(P)$^+$ [1, 2]
4 Lactate + NAD(P)$^+$
5 2-Hydroxybutyrate + NAD(P)$^+$
6 2-Hydroxyisopentanoate + NAD(P)$^+$
7 ?

Inhibitor(s)
Divalent metal ions (i.e., Cu^{2+}, Hg^{2+}, Co^{2+}, Fe^{2+}, Zn^{2+}) [1]; EDTA (partly reversible by Mn^{2+}) [1]; 1,10-Phenanthroline [1]; More (no inhibitors are L-enantiomers of substrates, thio-glycolate, 2-mercaptoethanol, 2,3-dimer-captoethanol, 1,4-dithioerythrit, L-cysteine, GSH) [1]

Cofactor(s)/prosthetic group(s)/activating agents
NADH [1, 2]; NADPH (50% decrease of reaction velocity if NADPH substitutes NADH) [1]; NAD$^+$ [1]; NADP$^+$ [1]

Metal compounds/salts
Mn^{2+} (requirement, K_m: 0.01 mM [1]) [1, 2]

Turnover number (min^{-1})

Specific activity (U/mg)
70.2 [1]

K_m-value (mM)
0.0077 (NADH) [1]; 0.0086 (NADPH) [1]; 0.021 (NADP$^+$) [1]; 0.044 (p-hydroxyphenylpyruvate, phenylpyruvate) [1]; 0.067 (indolepyruvate) [1]; 0.076 (p-hydroxyphenyllactate) [1]; 0.083 (D-phenyllactate) [1]; 0.093 (NAD$^+$) [1]; 0.094 (indolelactate) [1]

pH-optimum
6.5 (reduction) [1]; 9.5 (oxidation) [1]

pH-range
4.2–9.0 (about half-maximal activity at pH 4.2 and 9.0, reduction) [1]; 7.6–11.3 (about half-maximal activity at pH 7.6 and 11.3, oxidation) [1]

Temperature optimum (°C)
50 [1]

Temperature range (°C)
32–68 (about half-maximal activity at 32°C and 68°C) [1]

3 ENZYME STRUCTURE

Molecular weight
250000 (Candida maltosa, gel filtration) [1]
280000 (Candida maltosa, glycerol density gradient centrifugation) [1]

Subunits
Tetramer (4 × 68000, Candida maltosa, SDS-PAGE) [1]

Glycoprotein/Lipoprotein
–

4 ISOLATION/PREPARATION

Source organism
Candida maltosa [1]; Leishmania donovani donovani [2]

Source tissue
Cell [1]; Promastigotes (most active from mid-log growth-phase) [2]

Localisation in source
Cytoplasm [1]

Purification
Candida maltosa (affinity chromatography on 5'AMP-Sepharose 4B) [1]

Crystallization
–

Cloned
–

Renaturated
–

5 STABILITY

pH

Temperature (°C)
30 (and below, stable) [1]; 42 ($t_{1/2}$: 30 min) [1]; 50 (complete inactivation after 30 min) [1]

Oxidation

Organic solvent

General stability information
Glycerol stabilizes at high concentrations, 30–50% v/v, during purification and storage [1]

Storage

−25°C, purified preparation stable in potassium phosphate buffer, pH 6.5, containing 50% v/v glycerol [1]

6 CROSSREFERENCES TO STRUCTURE DATABANKS

PIR/MIPS code

Brookhaven code

7 LITERATURE REFERENCES

[1] Bode, R., Lippoldt, A., Birnbaum, D.: Biochem. Physiol. Pflanz.,181,189–198 (1986)
[2] Leelayoova, S., Marbury, D., Rainey, P.M., MacKenzie, N.E., Hall, J.E.: J. Protozool., 39,350–358 (1992)

1 NOMENCLATURE

EC number
1.1.1.223

Systematic name
(-)-trans-Isopiperitenol:NAD⁺ oxidoreductase

Recommended name
Isopiperitenol dehydrogenase

Synonymes
Dehydrogenase, isopiperitenol

CAS Reg. No.
96595-05-0

---

2 REACTION AND SPECIFICITY

Catalysed reaction
(-)-trans-Isopiperitenol + NAD⁺ →
→ (-)-isopiperitenone + NADH

Reaction type
Redox reaction

Natural substrates
More (involved in the biosynthesis of menthol and related monoterpenes in peppermint (Mentha piperita) leaves) [1]

Substrate spectrum
1 (-)-trans-Isopiperitenol + NAD⁺ (preferred substrate) [1]
2 (+)-trans-Isopiperitenol + NAD⁺ (95% activity of that of (-)-trans-isopiperitenol) [1]
3 (+)-Isomenthol + NAD⁺ (activity less than 2% of that of (-)-trans-isopiperitenol) [1]
4 (+)-trans-Pulegol + NAD⁺ (19% activity of that of (-)-trans-isopiperitenol) [1]

Product spectrum
1 (-)-Isopiperitenone + NADH [1]
2 (+)-Isopiperitenone + NADH [1]
3 ?
4 ?

Inhibitor(s)
$HgCl_2$ (strong) [1]; p-Hydroxymercuribenzoate (strong) [1]; N-Ethylmaleimide (weak) [1]; Iodoacetamide (weak) [1]; More (1,10-phenanthroline has no effect) [1]

Cofactor(s)/prosthetic group(s)/activating agents
NAD^+ [1]; $NADP^+$ (less than 10% of the activity with NAD) [1]; More (FAD completely ineffective) [1]

Metal compounds/salts

Turnover number (min^{-1})

Specific activity (U/mg)

K_m-value (mM)
0.017 ((+)- or (-)-trans-isopiperitenol) [1]; 0.08 ((+)- or (-)-cis-isopiperitenol) [1]

pH-optimum
8.5–10.5 [1]

pH-range
7.5–11 (half-maximal activity at pH 7.5 and 11) [1]

Temperature optimum (°C)

Temperature range (°C)

3 ENZYME STRUCTURE

Molecular weight
66000 (Mentha piperita, gel filtration) [1]

Subunits

Glycoprotein/Lipoprotein
–

4 ISOLATION/PREPARATION

Source organism
Mentha piperita (peppermint) [1]

Source tissue
Leaf [1]

Localisation in source

Purification
  Mentha piperita [1]

Crystallization
  –

Cloned
  –

Renaturated
  –

---

5 STABILITY

pH

Temperature (°C)

Oxidation
  DTT at 1 mM stabilizes [1]

Organic solvent

General stability information
  Bovine serum albumin stabilizes [1]

Storage

---

6 CROSSREFERENCES TO STRUCTURE DATABANKS

PIR/MIPS code

Brookhaven code

---

7 LITERATURE REFERENCES

[1] Kjonaas R.B., Venkatachalam K.V., Croteau R.: Arch. Biochem. Biophys.,238,49–60 (1985)

---

1 NOMENCLATURE

EC number
1.1.1.224

Systematic name
D-Mannitol-1-phosphate:NADP+ 6-oxidoreductase

Recommended name
Mannose-6-phosphate 6-reductase

Synonymes
NADPH-dependent mannose 6-phosphate reductase
Mannose-6-phosphate reductase
Reductase, 6-phosphomannose
NADP-dependent mannose-6-P:mannitol-1-P oxidoreductase [1]
NADPH-dependent M6P reductase [2]
NADPH-mannose-6-P reductase [1]

CAS Reg. No.
88747-79-9

2 REACTION AND SPECIFICITY

Catalysed reaction
D-Mannitol 1-phosphate + NADP+ →
→ D-mannose 6-phosphate + NADPH

Reaction type
Redox reaction

Natural substrates
Mannose 6-phosphate + NADPH (involved in the biosynthesis of mannitol in celery (Apium graveolens) leaves) [1]

Substrate spectrum
1 Mannose 6-phosphate + NADPH (r, no activity with fructose 6-phosphate, glucose 6-phosphate, mannose 1-phosphate, mannose [2]) [1, 2]

Product spectrum
1 Mannitol 1-phosphate + NADP+ [1, 2]

Inhibitor(s)

Cofactor(s)/prosthetic group(s)/activating agents
NADPH (NADH cannot replace NADPH in reducing mannose 6-phosphate [2]) [1]; NADP+ (NAD+ partially (8% activity) substitutes for NADP+ in oxidizing mannitol 1-phosphate) [2]

Metal compounds/salts

Turnover number (min^{-1})

Specific activity (U/mg)
16.4 (mannose, celery protoplast extract) [1]; 35.9 (mannose, whole celery leaf extract) [1]

K_m-value (mM)
15.8 (mannose 6-phosphate) [2]; 45 (mannitol 1-phosphate) [2]

pH-optimum
7.5 (mannose 6-phosphate) [2]; 8.5 (mannitol 1-phosphate) [2]

pH-range
6.0–9.0 (inactivated at pH below 6) [2]

Temperature optimum (°C)
25 (assay at) [1]

Temperature range (°C)

3 ENZYME STRUCTURE

Molecular weight
58000 (Apium graveolens, gel filtration) [2]

Subunits
Dimer (2 × 35000, Apium graveolens, SDS-PAGE) [2]

Glycoprotein/Lipoprotein
–

4 ISOLATION/PREPARATION

Source organism
Apium graveolens (celery) [1, 2]; Ligustrum vulgare (privet) [2]

Source tissue
Protoplasts of leaves [1, 2]

Localisation in source
Cytoplasma [1, 2]

Purification
Apium graveolens [1, 2]

Crystallization
–

Cloned
–

Renaturated
–

5 STABILITY

pH
6.0–9.0 [2]

Temperature (°C)

Oxidation
DTT stabilizes [2]

Organic solvent

General stability information
DTT stabilizes [2]; Degassed buffers stabilize [2]

Storage
4°C, lyophilized active fractions stable for several months [2]; –21°C, 50% activity retained after 1 month [2]

6 CROSSREFERENCES TO STRUCTURE DATABANKS

PIR/MIPS code

Brookhaven code

7 LITERATURE REFERENCES

[1] Rumpho, M..E., Edwards, G.D., Loescher, W.H.: Plant Physiol.,73,869–873 (1983)
[2] Loescher, W.H., Tyson, R.H., Everard, J.D., Redgewell, J., Bieleski, R.L.: Plant Physiol.,98,1396–1402 (1992)

Enzyme Handbook © Springer-Verlag Berlin Heidelberg 1995
Duplication, reproduction and storage in data banks are only
allowed with the prior permission of the publishers

1 NOMENCLATURE

EC number
1.1.1.225

Systematic name
Chlordecone-alcohol:NADP+ 2-oxidoreductase

Recommended name
Chlordecone reductase

Synonymes
Reductase, chlordecone
CDR [2]

CAS Reg. No.
102484-73-1

2 REACTION AND SPECIFICITY

Catalysed reaction
Chlordecone + NADPH →
→ chlordecone alcohol + NADP+

Reaction type
Redox reaction

Natural substrates
Chlordecone + NADPH (bioreduction of the organochlorine pesticide to chlordecone alcohol in liver) [1]

Substrate spectrum
1 Chlordecone + NADPH (chlordecone an organochlorine pesticide is 1,1a,3,3b,4,5,5a,5b,6-decachlorooctahydro-1,3,4-metheno-2H-cyclo-buta[cd]pentalen-2-one) [1, 3]
2 More (not: pyridine 4-aldehyde, 4-benzoylpyridine, 9,10-phenanthrenedio-ne) [3]

Product spectrum
1 Chlordecone alcohol + NADP+
2 ?

Enzyme Handbook © Springer-Verlag Berlin Heidelberg 1995
Duplication, reproduction and storage in data banks are only
allowed with the prior permission of the publishers

Inhibitor(s)
Quercetin (not [1]) [3]; p-Chloromercuribenzoate [1, 3]; Iodomethacin [1, 3];
Ethacrynic acid [1, 3]; Hg^{2+} (slight) [3]; Cu^{2+} (slight) [3]; More (insensitive to
phenobarbital) [1, 3]

Cofactor(s)/prosthetic group(s)/activating agents
NADPH [1, 3]; More (no activity with NADH) [3]

Metal compounds/salts

Turnover number (min^{-1})

Specific activity (U/mg)
0.008 [3]

K_m-value (mM)
0.0026 (chlordecone) [1]; 0.003 (chlordecone) [3]; 0.005 (NADPH) [1]

pH-optimum
6.5 [1]; 7.5 (2 optima: pH 7.5 and 9.0) [3]; 9.0 (2 optima: pH 7.5 and 9.0) [3]

pH-range

Temperature optimum (°C)

Temperature range (°C)

---

3 ENZYME STRUCTURE

Molecular weight

Subunits

Glycoprotein/Lipoprotein
–

---

4 ISOLATION/PREPARATION

Source organism
Human [1–3]; Mongolian gerbil [1, 3]; Rabbit [1, 3]; More (undetectable in:
rat [1, 3], mouse [1, 3], guinea pig [1], hamster [3]) [1, 3]

Source tissue
Liver [1, 3]; Intestine [1]; Kidney [1]; Lung [1]; Heart [1]; Spleen [1]; Brain
[1]; More (found in all tissues, highest activity in liver, intestine and kidney)
[1]

Localisation in source
Cytoplasm [1, 3]

Purification
  Human [3]

Crystallization
  –

Cloned
  [2]

Renaturated
  –

---

5 STABILITY

pH

Temperature (°C)

Oxidation

Organic solvent

General stability information

Storage
  –70°C, 50% glycerol [3]

---

6 CROSSREFERENCES TO STRUCTURE DATABANKS

PIR/MIPS code
  PIR2:A34263 (human)

Brookhaven code

---

7 LITERATURE REFERENCES

[1] Molowa, D.T., Wrighton, S.A., Blanke, R.V., Guzelian, P.S.: J. Toxicol. Environ. Health,17,375–384 (1986)
[2] Winters, C.J., Molowa, D.T., Guzelian, P.S.: Biochemistry,29,1080–1087 (1990)
[3] Molowa, D.T., Shayne, A.G., Guzelian, P.S.: J. Biol. Chem.,261,12624–12627 (1986)

1 NOMENCLATURE

EC number
1.1.1.226

Systematic name
trans-4-Hydroxycyclohexanecarboxylate:NAD$^+$ 4-oxidoreductase

Recommended name
4-Hydroxycyclohexanecarboxylate dehydrogenase

Synonymes
Dehydrogenase, trans-4-hydroxycyclohexanecarboxylate

CAS Reg. No.
67272-36-0

2 REACTION AND SPECIFICITY

Catalysed reaction
trans-4-Hydroxycylohexanecarboxylate + NAD$^+$ →
→ 4-oxocyclohexanecarboxylate + NADH

Reaction type
Redox reaction

Natural substrates

Substrate spectrum
1 trans-4-Hydroxycyclohexanecarboxylate + NAD$^+$ (r, highly specific for trans-isomer) [1]

Product spectrum
1 4-Oxocyclohexanecarboxylate + NADH [1]

Inhibitor(s)
EDTA (weak) [1]; N-Bromosuccinimide (strong) [1]; More (no inhibition by various sulfhydryl reagents and by iodoacetic acid, iodoacetamide, N-ethyl-maleimide, p-hydroxymercuribenzoic acid) [1]

Cofactor(s)/prosthetic group(s)/activating agents
NADH (very specific, no activity with NADPH) [1]; NAD$^+$ [1]

Metal compounds/salts

Turnover number (min⁻¹)

Specific activity (U/mg)
   0.1041 [1]

K_m-value (mM)
   0.23 (NAD⁺) [1]; 0.28 (NADH) [1]; 0.5 (4-oxocyclohexanecarboxylate) [1];
   0.51 (trans-4-hydroxycyclohexanecarboxylate) [1]

pH-optimum
   6.8 (reduction of 4-oxoacid) [1]; 7.2 (reduction of 4-oxoacid) [2]; 8.0 (oxidation of trans-4-hydroxyacid) [1]

pH-range
   6–8 [1]

Temperature optimum (°C)

Temperature range (°C)

---

3 ENZYME STRUCTURE

Molecular weight
   53600 (Corynebacterium cyclohexanicum, gel filtration) [1]

Subunits
   Dimer (2 × 27600, Corynebacterium cyclohexanicum, SDS-PAGE) [1]

Glycoprotein/Lipoprotein
   –

---

4 ISOLATION/PREPARATION

Source organism
   Corynebacterium cyclohexanicum [1]; Alcaligenes sp. [2]

Source tissue
   Cell [1, 2]

Localisation in source

Purification
   Corynebacterium cyclohexanicum [1]; Alcaligenes sp. (partial) [2]

Crystallization
   –

Cloned
   –

Renaturated
   –

5 STABILITY

pH

Temperature (°C)

Oxidation

Organic solvent

General stability information

Storage
-80°C [1]

6 CROSSREFERENCES TO STRUCTURE DATABANKS

PIR/MIPS code

Brookhaven code

7 LITERATURE REFERENCES

[1] Obata, H., Uebayasi, M., Kaneda, T.: Eur. J. Biochem.,174,451–458 (1988)
[2] Taylor, D. G., Trudgill, P.: J. Bacteriol.,5,401–411 (1978)

1 NOMENCLATURE

EC number
1.1.1.227

Systematic name
(-)-Borneol:NAD⁺ oxidoreductase

Recommended name
(-)-Borneol dehydrogenase

Synonymes
Dehydrogenase, (-)-borneol

CAS Reg. No.
111940-48-8

---

2 REACTION AND SPECIFICITY

Catalysed reaction
(-)-Borneol + NAD⁺ →
→ (-)-camphor + NADH

Reaction type
Redox reaction

Natural substrates
Borneol + NAD⁺ (reaction in camphor biosynthesis) [1]

Substrate spectrum
1 (-)-Borneol + NAD⁺ [1]

Product spectrum
1 (-)-Camphor + NADH [1]

Inhibitor(s)

Cofactor(s)/prosthetic group(s)/activating agents
NAD⁺ [1]; NADP⁺ (65%-85% as effective as NAD⁺) [1]; More (flavin nucleotides are ineffective) [1]

Metal compounds/salts

Turnover number (min^{-1})

Specific activity (U/mg)

K$_m$-value (mM)

pH-optimum

pH-range

Temperature optimum (°C)
    30 (assay at) [1]

Temperature range (°C)

---

3 ENZYME STRUCTURE

Molecular weight

Subunits

Glycoprotein/Lipoprotein
    –

---

4 ISOLATION/PREPARATION

Source organism
    Tanacetum vulgare (tansy) [1]

Source tissue
    Leaf epidermis (oil glands) [1]

Localisation in source
    Soluble [1]

Purification

Crystallization
    –

Cloned
    –

Renaturated
    –

5 STABILITY

pH

Temperature (°C)

Oxidation

Organic solvent

General stability information
Lyophilization, stable to [1]

Storage
−40°C, lyophilized, under argon, stable for several months without deterioration [1]

6 CROSSREFERENCES TO STRUCTURE DATABANKS

PIR/MIPS code

Brookhaven code

7 LITERATURE REFERENCES

[1] Dehal, S.S., Croteau, R.: Arch. Biochem. Biophys.,258,287–291 (1987)

1 NOMENCLATURE

EC number
1.1.1.228

Systematic name
(+)-cis-Sabinol:NAD$^+$ oxidoreductase

Recommended name
(+)-Sabinol dehydrogenase

Synonymes
Dehydrogenase, (+)-cis-sabinol

CAS Reg. No.
111940-50-2

2 REACTION AND SPECIFICITY

Catalysed reaction
(+)-cis-Sabinol + NAD$^+$ →
→ (+)-sabinone + NADH

Reaction type
Redox reaction

Natural substrates
(+)-cis-Sabinol + NAD$^+$ (pathway in the biosynthesis of (-)-3-isothujone
(sage) or (+)-3-thujone (tansy)) [1]

Substrate spectrum
1 (+)-cis-Sabinol + NAD$^+$ [1]

Product spectrum
1 (+)-Sabinone + NADH [1]

Inhibitor(s)

Cofactor(s)/prosthetic group(s)/activating agents
NAD$^+$ [1]; NADP$^+$ (65%-85% as effective as NAD$^+$) [1]; More (flavin nucleotides are ineffective) [1]

Metal compounds/salts

Turnover number (min⁻¹)

Specific activity (U/mg)

K_m-value (mM)

pH-optimum

pH-range

Temperature optimum (°C)
    30 (assay at) [1]

Temperature range (°C)

---

3 ENZYME STRUCTURE

Molecular weight

Subunits

Glycoprotein/Lipoprotein
    –

---

4 ISOLATION/PREPARATION

Source organism
    Salvia officinalis (sage) [1]; Tanacetum vulgare (tansy) [1]

Source tissue
    Leaf epidermis (oil glands) [1]

Localisation in source
    Soluble [1]

Purification

Crystallization
    –

Cloned
    –

Renaturated
    –

5 STABILITY

pH

Temperature (°C)

Oxidation

Organic solvent

General stability information
Lyophilization, stable to [1]

Storage
−40°C, lyophilized, under argon, stable for several months without deterioration [1]

6 CROSSREFERENCES TO STRUCTURE DATABANKS

PIR/MIPS code

Brookhaven code

7 LITERATURE REFERENCES

[1] Dehal, S.S., Croteau, R.: Arch. Biochem. Biophys.,258,287–291 (1987)

1 NOMENCLATURE

EC number
1.1.1.229

Systematic name
Diethyl-(2R,3R)-2-methyl-3-hydroxysuccinate:NADP+ 3-oxidoreductase

Recommended name
Diethyl 2-methyl-3-oxosuccinate reductase

Synonymes
Reductase, diethyl 2-methyl-3-oxosuccinate

CAS Reg. No.
110369-21-6

2 REACTION AND SPECIFICITY

Catalysed reaction
Diethyl (2R,3R)-2-methyl-3-hydroxysuccinate + NADP+ →
→ diethyl 2-methyl-3-oxosuccinate + NADPH

Reaction type
Redox reaction

Natural substrates
Diethyl 2-methyl-3-oxosuccinate + NADPH [1] (asymmetric reduction in
yeast, the reduction is strongly preferred over the weak oxidation activity)
[1]

Substrate spectrum

1 Diethyl 2-methyl-3-oxosuccinate + NADPH (r, the reduction is strongly preferred, the catalyzed reaction is an asymmetric reduction, best substrate)
  [1]
2 Dimethyl 2-methyl-3-oxosuccinate + NADPH (poor substrate) [1]
3 More (no aldehydes or several synthetic substrates are reduced, such as:
  acet-, form-, isobutyr-, n-valer-, n-hexyl-, n-hept-, p-hydroxy-benz-, p-dime-
  thylamino benz-, benzaldehyde, furfural, benzyl 2-methyl-3-oxobutyrate,
  ethyl 2-chloro-3-oxobutyrate, methyl 3-phenyl-2-oxobutyrate, ethyl
  2-methyl-3-oxobutyrate, methyl 3-(2'-furyl)-2-methyl-3-oxopropionate, none
  of the following alcohols or other substrates is oxidized: methanol, ethanol, n-propyl-, iso-propyl-, n-butyl-, isobutyl-, n-amyl-, isoamyl-, benzyl alcohol, glycerin, shikimic acid, 2-deoxy-D-glucose, glucose 6-phosphate, in
  the presence of NAD^+ or $NADP^+$) [1]

Product spectrum

1 Diethyl 2-methyl-3-hydroxysuccinate + $NADP^+$ (alpha-methyl-beta-hydroxy
  ester, the enzyme catalyzes an asymmetric reduction, the product is a
  mixture of the (2R,3R)-syn- and (2S,3R)-anti-beta-hydroxyester) [1]
2 ?
3 ?

Inhibitor(s)

1,10-Phenanthroline [1]; Monoiodoacetate [1]; High salt concentration (i.e.
NaCl, KCl) [1]; Ag^+ (strong) [1]; Cu^{2+} (strong) [1]; Fe^{3+} (strong) [1]; Zn^{2+} [1];
Cysteine [1]; KCN [1]

Cofactor(s)/prosthetic group(s)/activating agents

NADPH [1]; $NADP^+$ (weak activity) [1]

Metal compounds/salts

Turnover number (min^{-1})

Specific activity (U/mg)

K_m-value (mM)

1.25 (diethyl 2-methyl-3-oxosuccinate) [1]; 111.0 (diethyl 2-methyl-3-hydroxy-
succinate) [1]

pH-optimum

6.0 [1]

pH-range

5.3–7.3 (about half-maximal activity at pH 5.3 and 7.3) [1]

Temperature optimum (°C)

50 [1]

Temperature range (°C)

3 ENZYME STRUCTURE

Molecular weight
61000 (Saccharomyces fermentati, gel filtration) [1]

Subunits
Monomer (1×63000, Saccharomyces fermentati, SDS-PAGE) [1]

Glycoprotein/Lipoprotein
–

---

4 ISOLATION/PREPARATION

Source organism
Saccharomyces fermentati [1]

Source tissue
Cell [1]

Localisation in source
Cytoplasm [1]

Purification
Saccharomyces fermentati [1]

Crystallization
–

Cloned
–

Renaturated
–

---

5 STABILITY

pH
6.0–9.0 (most stable) [1]

Temperature (°C)
30 (and below, stable) [1]; 60 (and above, readily inactivated) [1]

Oxidation

Organic solvent

General stability information

Storage
4°C, stable [1]

---

6 CROSSREFERENCES TO STRUCTURE DATABANKS

PIR/MIPS code

Brookhaven code

7 LITERATURE REFERENCES

[1] Furuichi, A., Akita, H., Matsukura, H., Oishi, T., Horikoshi, K.: Agric. Biol. Chem.,51,293–299 (1987)

1 NOMENCLATURE

EC number
1.1.1.230

Systematic name
3alpha-Hydroxyglycyrrhetinate:NADP⁺ 3-oxidoreductase

Recommended name
3alpha-Hydroxyglycyrrhetinate dehydrogenase

Synonymes
Dehydrogenase, 3alpha-hydroxyglycyrrhetinate
More (not identical with EC 1.1.1.50)

CAS Reg. No.
114308-07-5

---

2 REACTION AND SPECIFICITY

Catalysed reaction
3alpha-Hydroxyglycyrrhetinate + NADP⁺ →
→ 3-oxoglycyrrhetinate + NADPH

Reaction type
Redox reaction

Natural substrates

Substrate spectrum
1 3alpha-Hydroxyglycyrrhetic acid + NADP⁺ (r, oxidation of 3alpha-hydroxy group faster than reduction of 3-keto group) [1]
2 3alpha-Hydroxy-18alpha-glycyrrhetic acid + NADP⁺ (r) [1]
3 More (not: 3beta-hydroxyglycyrrhetic acid, steroids or bile acids containing 3alpha-hydroxy or 3-keto group) [1]

Product spectrum
1 3-Ketoglycyrrhetic acid + NADPH [1]
2 3-Keto-18alpha-glycyrrhetic acid + NADPH [1]
3 ?

Inhibitor(s)

Cofactor(s)/prosthetic group(s)/activating agents
NADP⁺ [1]; NADPH [1]

Metal compounds/salts

Turnover number (min⁻¹)

Specific activity (U/mg)

K_m-value (mM)
  0.002 (3alpha-hydroxyglycyrrhetic acid) [1]; 0.001 (3-ketoglycyrrhetic acid)
  [1]; 0.00027 (NADP⁺) [1]; 0.024 (NADPH) [1]

pH-optimum
  6.0 (reduction of 3-keto group) [1]; 8.0 (oxidation of 3alpha-hydroxy group)
  [1]

pH-range

Temperature optimum (°C)

Temperature range (°C)

---

3 ENZYME STRUCTURE

Molecular weight
  53000 (Clostridium innocuum, gel filtration) [1]

Subunits
  Dimer (2 × 30000, Clostridium innocuum, SDS-PAGE) [1]

Glycoprotein/Lipoprotein
  –

---

4 ISOLATION/PREPARATION

Source organism
  Clostridium innocuum [1]

Source tissue

Localisation in source
  Soluble part of cell [1]

Purification
  Clostridium innocuum [1]

Crystallization
  –

Cloned
  –

Renaturated
  –

5 STABILITY

pH

Temperature (°C)

Oxidation

Organic solvent

General stability information

Storage

---

6 CROSSREFERENCES TO STRUCTURE DATABANKS

PIR/MIPS code

Brookhaven code

---

7 LITERATURE REFERENCES

[1] Akao, T., Akao, T., Hattori, M., Namba, T., Kobashi, K.: J. Biochem.,103,504–507 (1988)

1 NOMENCLATURE

EC number

1.1.1.231

Systematic name

(5Z,13E)-(15S)-6,9alpha-Epoxy-11alpha,15-dihydroxyprosta-5,13-dienoate:
NADP⁺ 15-oxidoreductase

Recommended name

15-Hydroxyprostaglandin-I dehydrogenase (NADP⁺)

Synonymes

Dehydrogenase, prostacyclin

PG I_2 dehydrogenase [1]

Prostacyclin dehydrogenase

NADP-linked 15-hydroxyprostaglandin (prostacyclin) dehydrogenase [2]

NADP⁺-dependent PGI₂-specific 15-hydroxyprostaglandin dehydrogenase
[1]

CAS Reg. No.

79468-49-8

---

2 REACTION AND SPECIFICITY

Catalysed reaction

(5Z,13E)-(15S)-6,9alpha-Epoxy-11alpha,15-dihydroxyprosta-5,13-dienoate +
NADP⁺ →

→ (5Z,13E)-6,9alpha-epoxy-11alpha-hydroxy-15-oxoprosta-5,13-dienoate +
NADPH (random mechanism or ordered mechanism with NADP⁺ binding
first [2])

Reaction type

Redox reaction

Natural substrates

(5Z,13E)-(15S)-6,9alpha-Epoxy-11alpha,15-dihydroxyprosta-5,13-dienoate +
NADP⁺ [2]

Substrate spectrum

1 (5Z,13E)-(15S)-6,9alpha-Epoxy-11alpha,15-dihydroxyprosta-5,13-dienoate + NADP⁺ (i.e. prostaglandin I_2, prostacyclin) [1]

2 Prostaglandin B_1 + NADP⁺ (at 8% the reaction rate of prostaglandin I_2) [1]

3 Prostaglandin E_2 + NADP⁺ (at 4% the reaction rate of prostaglandin I_2) [1]

4 Prostaglandin F_{2alpha} + NADP⁺ (at 2.2% the reaction rate of prostaglandin I_2) [1]

5 6-Keto-prostaglandin F_{1alpha} + NADP⁺ (at 1.8% the reaction rate of prostaglandin I_2) [1]

6 4-Nitrobenzaldehyde + NADP⁺ (more rapid reaction than with prostacyclin) [2]

7 Phenanthrene quinone + NADP⁺ (more rapid reaction than with prostacyclin) [2]

8 Menadione + NADP⁺ (more rapid reaction than with prostacyclin) [2]

9 More (enzyme catalyzes reduction of 9-keto group in prostaglandin E_2 at less than 1% the rate of prostaglandin I_2 oxidation) [1]

Product spectrum

1 (5Z,13E)-6,9alpha-Epoxy-11alpha-hydroxy-15-oxoprosta-5,13-dienoate + NADPH (i.e. 15-keto prostaglandin I_2 [1], the product is not stable and not available for analysis of the reverse reaction [2]) [1, 2]

2 ?

3 ?

4 ?

5 ?

6 ?

7 ?

8 ?

9 ?

Inhibitor(s)
NADPH [2]; Furosemide [2]

Cofactor(s)/prosthetic group(s)/activating agents
NADP⁺ [1, 2]; More (no activity with NAD⁺ [1], not a flavoprotein [2]) [1, 2]

Metal compounds/salts

Turnover number (min⁻¹)

Specific activity (U/mg)
More [2]; 19.9 [1]

K_m-value (mM)
0.0005 (phenanthrene quinone) [2]; 0.00538 (NADP⁺) [2]; 0.0134 (NADP⁺) [1]; 0.0667 (4-nitrobenzaldehyde) [2]; 0.250 (menadione) [2]; 0.278 (prostaglandin I_2) [1]; 0.364 (prostaglandin I_2) [2]

pH-optimum
   10.3 (assay at) [1, 2]

pH-range

Temperature optimum (°C)
   25 (assay at) [1, 2]

Temperature range (°C)

---

3 ENZYME STRUCTURE

Molecular weight
   56000 (rabbit, SDS-PAGE of amidinated enzyme) [2]
   62000 (rabbit, gel filtration) [1]

Subunits
   Dimer (2 × 28000, rabbit, SDS-PAGE) [2]

Glycoprotein/Lipoprotein
   No glycoprotein [2]

---

4 ISOLATION/PREPARATION

Source organism
   Rabbit [1, 2]

Source tissue
   Kidney [1, 2]

Localisation in source

Purification
   Rabbit [1, 2]

Crystallization
   –

Cloned
   –

Renaturated
   –

5 STABILITY

pH

Temperature (°C)

Oxidation

Organic solvent

General stability information
Freezing and thawing: repeatedly, slow loss of activity [1]

Storage
Stable for months when frozen [1]

6 CROSSREFERENCES TO STRUCTURE DATABANKS

PIR/MIPS code

Brookhaven code

7 LITERATURE REFERENCES

[1] Korff, J.M., Jarabak, J.: Methods Enzymol.,86,152–155 (1982) (Review)
[2] Korff, J.M., Jarabak, J.: J. Biol. Chem.,257,2177–2181 (1982)

1 NOMENCLATURE

EC number
1.1.1.232

Systematic name
(15S)-15-Hydroxy-5,8,11-cis-13-trans-eicosatetraenoate:NAD(P)$^+$ 15-oxidore-
ductase

Recommended name
15-Hydroxyeicosatetraenoate dehydrogenase

Synonymes
Dehydrogenase, 15-hydroxyeicosatetraenoate

CAS Reg. No.
117910-46-0

2 REACTION AND SPECIFICITY

Catalysed reaction
(15S)-15-Hydroxy-5,8,11-cis-13-trans-eicosatetraenoate + NAD(P)$^+$ →
→ 15-oxo-5,8,11-cis-13-trans-eicosatetraenoate + NAD(P)H

Reaction type
Redox reaction

Natural substrates

Substrate spectrum
1 (15S)-15-Hydroxy-5,8,11-cis-13-trans-eicosatetraenoate + NAD(P)$^+$ [1]

Product spectrum
1 15-Oxo-5,8,11-cis-13-trans-eicosatetraenoate + NAD(P)H

Inhibitor(s)

Cofactor(s)/prosthetic group(s)/activating agents
NAD$^+$ (no great difference of specificity between NAD$^+$ and NADP$^+$) [1];
NADP$^+$ (no great difference of specificity between NAD$^+$ and NADP$^+$) [1]

Metal compounds/salts

Turnover number (min^{-1})

Specific activity (U/mg)

K_m-value (mM)
   0.0083 ((15S)-15-hydroxy-5,8,11-cis-13-trans-eicosatetraenoic acid) [1]

pH-optimum
   9.8 [1]

pH-range
   7.6–11.2 (about 30% of activity maximum at pH 7.6 and 11.2) [1]

Temperature optimum (°C)
   37 (assay at) [1]

Temperature range (°C)

---

3 ENZYME STRUCTURE

Molecular weight

Subunits

Glycoprotein/Lipoprotein
   –

---

4 ISOLATION/PREPARATION

Source organism
   Mouse [1]

Source tissue
   Liver [1]

Localisation in source
   Microsomes (tightly bound to membranes) [1]

Purification

Crystallization
   –

Cloned
   –

Renatured
   –

5 STABILITY

pH

Temperature (°C)

Oxidation

Organic solvent

General stability information

Storage

6 CROSSREFERENCES TO STRUCTURE DATABANKS

PIR/MIPS code

Brookhaven code

7 LITERATURE REFERENCES

[1] Sok, D.-E., Kang, J.B., Shin, H.D.: Biochem. Biophys. Res. Commun.,156,524–529 (1988)

1 NOMENCLATURE

EC number
1.1.1.233

Systematic name
N-Acyl-D-mannosamine:NAD+ 1-oxidoreductase

Recommended name
N-Acylmannosamine 1-dehydrogenase

Synonymes
Dehydrogenase, N-acylmannosamine
N-Acetyl-D-mannosamine dehydrogenase
N-Acyl-D-mannosamine dehydrogenase
N-Acylmannosamine dehydrogenase

CAS Reg. No.
117698-08-5

2 REACTION AND SPECIFICITY

Catalysed reaction
N-Acyl-D-mannosamine + NAD+ →
→ N-acyl-D-mannosaminolactone + NADH

Reaction type
Redox reaction

Natural substrates
N-Acetyl-D-mannosamine + NAD+ (reaction in N-acetyl-D-mannosamine catabolism) [1]

Substrate spectrum
1 N-Acetyl-D-mannosamine + NAD+ (ir, highly specific) [1]
2 N-Glycolyl-D-mannosamine + NAD+ (oxidation at 62% the rate of N-acetyl-D-mannosamine oxidation) [1]
3 More (no substrates are N-acetylglucosamine, N-acetylgalactosamine, amino sugars, neutral hexoses and pentoses) [1]

Product spectrum
1 N-Acetyl-D-mannosaminolactone + NADH (N-acetyl-D-mannosaminic acid is identified as product, the lactone is an immediate product) [1]
2 ?
3 ?

Inhibitor(s)

Hg^{2+} (strong) [1]; Sodium dodecylsulfate (strong) [1]; Cd^{2+} (weak) [1]; Ni^{2+} (weak) [1]; Zn^{2+} (weak) [1]; Cu^{2+} (weak) [1]; More (no inhibition with metal-chelating or SH-group-blocking reagents, e.g. EDTA, 2,2'-bipyridyl, 8-hydroxyquinoline, PCMB) [1]

Cofactor(s)/prosthetic group(s)/activating agents

NAD^+ (specific for) [1]; More (no cofactor: $NADP^+$, ferricyanide, 2,6-dichloro-phenolindophenol, phenazine methosulfate) [1]

Metal compounds/salts

More (no metal requirement) [1]

Turnover number (min^{-1})

Specific activity (U/mg)

340 [1]; 545 (E. coli JM109 (pNAM307)) [2]

K_m-value (mM)

0.41 (NAD^+, value unchanged by cloning and expression in E. coli JM109 [2]) [1, 2]; 1.0 (N-acetyl-D-mannosamine, value unchanged by cloning and expression in E. coli JM109 [2]) [1, 2]; 13.3 (N-glycolyl-D-mannosamine) [1]

pH-optimum

8.0–9.5 [1]; 8.5 (E. coli JM109 (pNAM307)) [2]

pH-range

7.2–10.5 (about half-maximal activity at pH 7.2 and 10.5) [1]

Temperature optimum (°C)

30 (assay at) [1]

Temperature range (°C)

---

3 ENZYME STRUCTURE

Molecular weight

120000 (Flavobacterium sp., gel filtration) [1]

Subunits

Tetramer (4 × 29000, Flavobacterium sp., SDS-PAGE) [1]

Glycoprotein/Lipoprotein

–

4 ISOLATION/PREPARATION

Source organism
Flavobacterium sp. strain 141–8 [1, 2]

Source tissue

Localisation in source

Purification
Flavobacterium sp. (affinity chromatography, preparative PAGE [1], cloned
in E. coli JM109 strain pNAM307 [2]) [1, 2]

Crystallization
–

Cloned
(Flavobacterium sp., cloned and expressed under control of a lac-promoter
in E. coli JM109, the E. coli transformants JM109(pNAM307) and
JM109(pNAM308) show up to 200-fold higher activity than Flavobacterium
sp.) [1]

Renaturated
–

5 STABILITY

pH
5.0 (inactivation at 25°C after 16 h, gradually reversed to about 80% of original activity by shifting the pH back to 8.2 and keeping it at room temperature for 36 h) [1]; 7.0 (and below, more than 70% loss of activity) [1]; 8.5–9.5
(stable for 10 min at 45°C) [1]

Temperature (°C)
25 (inactivation at pH 5.0 after 16 h, gradually reversed to about 80% of original activity by shifting the pH back to 8.2 and keeping it at room temperature for 36 h) [1]; 42 (and below, stable for at least 10 min) [1]; 50 ($t_{1/2}$:
10 min) [1]; 60 (complete inactivation after 10 min) [1]

Oxidation

Organic solvent

General stability information
Repeated freezing leads to gradual inactivation [1]

Storage
–20°C, stable for at least several months [1]

6 CROSSREFERENCES TO STRUCTURE DATABANKS

PIR/MIPS code
  PIR2:A43744 (Flavobacterium sp. (strain 141–8))

Brookhaven code

7 LITERATURE REFERENCES

[1] Horiuchi, T., Kurokawa, T.: J. Biochem.,104,466–471 (1988)
[2] Yamamoto-Otake, H., Koyama, Y., Horiuchi, T., Nakano, E.: Appl. Environ. Microbiol.,
    57,1418–1422 (1991)

1 NOMENCLATURE

EC number
1.1.1.234

Systematic name
(2S)-Flavan-4-ol:NADP+ 4-oxidoreductase

Recommended name
Flavanone 4-reductase

Synonymes
Reductase, flavanone 4-

CAS Reg. No.
115232-53-6

2 REACTION AND SPECIFICITY

Catalysed reaction
(2S)-Flavan-4-ol + NADP+ →
→ (2S)-flavanone + NADPH

Reaction type
Redox reaction

Natural substrates
(2S)-Flavanone + NADPH (involved in the biosynthesis of 3-deoxyanthocya-
nidines, such as apigeninidin or luteolinidin from (2S)-flavanones naringenin
or eriodictyol) [1]

Substrate spectrum
1 (2S)-Naringenin + NADPH [1, 2]
2 (2S)-Eriodictyol + NADPH [1, 2]
3 5,7,3',4',5'-Pentahydroxyflavanone + NADPH [1]
4 More (as dihydroxyflavonols, such as dihydrokaempferol, dihydroquercetin and dihydromyricetin are substrates for the Sinningia enzyme preparation, too, the enzyme might be identical with dihydroflavonol-4-reductase)
[1]

Product spectrum
1 Apiforol + NADP+ [1, 2]
2 Luteoforol + NADP+ [1, 2]
3 5,7,3',4',5'-Pentahydroxyflavan-4-ol + NADP+ [1]
4 ?

Inhibitor(s)

Naringenin (above 0.1 mM) [1]; p-Chloromercuribenzoate (0.5 mM) [1]; Glycerol (10% v/v, complete irreversible inactivation) [1]; Sucrose (20% w/v, complete irreversible inactivation) [1]; More (no inhibition by KCN, EDTA, diethyl dithiocarbamate, phenanthroline or N-ethylmaleimide, 2 mM each) [1]

Cofactor(s)/prosthetic group(s)/activating agents

NADPH (requirement) [1, 2]; NADH (77% as active as NADPH) [1]; More (no cofactors are FAD, FMN or tetrahydropterin, with or without NADPH) [1]

Metal compounds/salts

Turnover number (min^{-1})

Specific activity (U/mg)

0.0009 [1]

K_m-value (mM)

0.021–0.023 (naringenin) [1]

pH-optimum

5.5 [2]; 6.0 [1]

pH-range

4.7–6.9 (about half-maximal activity at pH 4.7 and 6.9) [1]

Temperature optimum (°C)

Temperature range (°C)

---

3 ENZYME STRUCTURE

Molecular weight

Subunits

Glycoprotein/Lipoprotein

–

---

4 ISOLATION/PREPARATION

Source organism

Sinningia cardinalis (syn. Rechsteineria cardinalis, strain " Feuerschein") [1]; Columnea hybrida (strain "Heklua") [2]

Source tissue

Flower [1, 2]

Localisation in source
Soluble [2]

Purification
More (freezing fresh flowers in liquid nitrogen and storing them at –80°C enhances extractable activity compared to fresh flowers) [1]

Crystallization
–

Cloned
–

Renaturated
–

5 STABILITY

pH

Temperature (°C)
95 (inactivation after 5 min) [1]

Oxidation

Organic solvent
Glycerol, 10% v/v, inactivates irreversibly [1]

General stability information
Sucrose, 20% w/v, inactivates irreversibly [1]

Storage

6 CROSSREFERENCES TO STRUCTURE DATABANKS

PIR/MIPS code

Brookhaven code

7 LITERATURE REFERENCES

[1] Stich, K., Forkmann, G.: Biochemistry,27,785–789 (1988)
[2] Stich, K.: Z. Naturforsch.,43c,311–314 (1987)

1 NOMENCLATURE

EC number
1.1.1.235

Systematic name
Coformycin:NADP+ 8-oxidoreductase

Recommended name
8-Oxocoformycin reductase

Synonymes
Reductase, 8-ketodeoxycoformycin

CAS Reg. No.
114995-16-3

2 REACTION AND SPECIFICITY

Catalysed reaction
8-Oxocoformycin + NADPH →
→ coformycin + NADP+ (possible mechanism [1])

Reaction type
Redox reaction

Natural substrates
8-Ketodeoxycoformycin + NADPH (production of the antibiotic 2'-deoxyco-
formycin) [1]
8-Ketocoformycin + NADPH (production of the antibiotic coformycin) [1]

Substrate spectrum
1 8-Ketodeoxycoformycin + NADPH [1]
2 8-Ketocoformycin + NADPH [1]
3 More (stereospecifically reduces the 8-keto group of both ketonucleoside
substrates to a hydroxyl group with the R configuration at C-8) [1]

Product spectrum
1 2'-Deoxycoformycin + NADP+ (r) [1]
2 Coformycin + NADP+ [1]
3 ?

Enzyme Handbook © Springer-Verlag Berlin Heidelberg 1995
Duplication, reproduction and storage in data banks are only
allowed with the prior permission of the publishers

Inhibitor(s)
    $NADP^+$ [1]

Cofactor(s)/prosthetic group(s)/activating agents
    NADPH (B-specific with respect to NADPH, 25-fold higher activity than with
    NADH) [1]; NADH (NADPH is 25-fold higher active than NADH) [1]

Metal compounds/salts

Turnover number (min^{-1})

Specific activity (U/mg)

K_m-value (mM)
    0.150 (8-ketocoformycin) [1]; 0.250 (8-ketodeoxycoformycin) [1]

pH-optimum
    8.0 [1]

pH-range

Temperature optimum (°C)
    25 [1]

Temperature range (°C)

---

3 ENZYME STRUCTURE

Molecular weight

Subunits

Glycoprotein/Lipoprotein
    –

---

4 ISOLATION/PREPARATION

Source organism
    Streptomyces antibioticus [1]

Source tissue
    Cell [1]

Localisation in source

Purification
    Streptomyces antibioticus (partial) [1]

Crystallization
    –

Cloned

–

Renaturated

–

5 STABILITY

pH

Temperature (°C)

Oxidation

Organic solvent

General stability information

Storage

6 CROSSREFERENCES TO STRUCTURE DATABANKS

PIR/MIPS code

Brookhaven code

7 LITERATURE REFERENCES

[1] Hanvey, J.C., Hawkins, E.S., Baker, D.C., Suhadolnik, R.J.: Biochemistry,27, 5790–5795 (1988)

1 NOMENCLATURE

EC number
1.1.1.236

Systematic name
Pseudotropine:NADP$^+$ 3-oxidoreductase

Recommended name
Tropinone reductase

Synonymes
Reductase, tropinone (psi-tropine-forming)
Tropinone reductase II
Pseudotropine forming tropinone reductase [1]

CAS Reg. No.
136111-61-0

2 REACTION AND SPECIFICITY

Catalysed reaction
Pseudotropine + NADP$^+$ →
→ tropinone + NADPH

Reaction type
Redox reaction

Natural substrates
Tropinone + NADPH (product is a precursor of tropan-3beta-ol–esters) [4]

Substrate spectrum
1 Tropinone + NADPH (ir [2, 5], stereospecific reduction [1], B-specific oxidoreductase: transfers pro-S hydrogen at C$_4$ of NADPH to tropinone [2]) [1–5]
2 N-Propyl-4-piperidone + NADPH (reduced at 300% [1], 388% [2] the rate of tropinone reduction) [1, 2]
3 N-Methyl-4-piperidone + NADPH (reduced at 230% [1], 381% [2] the rate of tropinone reduction) [1, 2]
4 Tetrahydrothiopyran–4-one + NADPH (reduced at 25% [1], 288% [2] the rate of tropinone reduction) [1, 2]
5 4-Piperidone + NADPH (reduced at 70% [1], 182% [2] the rate of tropinone reduction) [1, 2]
6 6-Hydroxytropinone + NADPH (reduced at 25% the rate of tropinone reduction [1]) [1, 2]

7 Nortropinone + NADPH (reduced at 10% the rate of tropinone reduction [1]) [1, 2]
8 3-Methylcyclohexanone + NADPH [2]
9 4-Methylcyclohexanone + NADPH (reduced at 31% the rate of tropinone reduction [1]) [1, 2]
10 4-Ethylcyclohexanone + NADPH [2]
11 More (no substrates are N-methyl-2-piperidone, N-methyl-2-pyrrolidone, 2-pyrrolidone, 3-quinuclidinone [1, 2], N-methylpiperidine, N-methylpiperazine, N-methylmorpholine, 4-chlor-1-methylpiperidine [1], 2-piperidone, hygrine, 4-pelletierine, N-propyl-4-piperidone [2], 2-carbomethoxy-3-tropinone [2, 5], 8-thiabicyclo[3.2.1]octane-3-one [2, 4], 8-thiabicyclo[3.2.1]octane-3-alpha-ol, 8-thiabicyclo[3.2.1]octane-3-beta-ol [4]) [1, 2, 4, 5]

Product spectrum
1 Pseudotropine + NADP+ (i.e. 3-beta-hydroxytropine or psi-tropine) [1–5]
2 ?
3 ?
4 ?
5 ?
6 ?
7 ?
8 ?
9 ?
10 ?
11 ?

Inhibitor(s)
Cu^{2+} (strong, 1mM) [2]; Hg^{2+} (strong, 1mM) [2]; PCMB (strong) [2]; Diethyl dicarbonate (moderate) [2]; Metal ions (e.g. Fe^{3+}, Fe^{2+}, Mg^{2+}, Ca^{2+}, Mn^{2+}, Co^{2+}, Ni^{2+}, Li$^+$: slight to negligible) [2]; Iodoacetate (slight) [2]; N-Ethylmaleimide (slight) [2]; More (no inhibition by EDTA, 8-thiabicyclo[3.2.1]octane-3-one [2, 4], 8-thiabicyclo[3.2.1]octane-3-alpha-ol, 8-thiabicyclo[3.2.1]octane-3-beta-ol [4]) [2, 4]

Cofactor(s)/prosthetic group(s)/activating agents
NADPH (requirement, specific for [3], B-specific [2]) [1–5]; NADH (activation, 5.4% as effective as NADPH [1], cannot replace NADPH [3]) [1, 5]

Metal compounds/salts
More (no metal ion requirement) [2]

Turnover number (min^{-1})

Specific activity (U/mg)
0.128 [3]; 1.05 [1]; 23.22 (Hyoscyamus niger) [2]

K_m-value (mM)
0.0061 (NADPH, Hyoscyamus niger) [2]; 0.017–0.021 (NADPH) [1, 3];
0.034–0.035 (tropinone, Hyoscyamus niger) [1, 2]; 0.057 (N-methyl-4-piperi-
done) [1]; 0.098 (tropinone) [4]; 0.220 (tropinone) [3]; 0.251–0.299 (NADP⁺)
[2], N-propyl-4-piperidone, Hyoscyamus niger) [1, 2]; 0.534 (4-ethylcyclohe-
xanone, Hyoscyamus niger) [2]; 0.687 (pseudotropine, Hyoscyamus niger)
[2]; 0.77 (N-methyl-4-piperidone, Hyoscyamus niger) [2]; 7.58–7.60 (3-me-
thylcyclohexanone, NADH, Hyoscyamus niger) [2]; 21.6 (4-piperidone,
Hyoscyamus niger) [2]

pH-optimum
5.3–6.5 (Hyoscyamus niger) [2]; 5.8–6.3 [1]; 6.3 (broad) [3]; 6.4 (broad) [5]

pH-range
4.5–8.5 (about 75% of maximal activity at pH 4.5 and 8.5) [3]

Temperature optimum (°C)
30 (assay at) [1, 3]; 37 (assay at) [2]

Temperature range (°C)

---

3 ENZYME STRUCTURE

Molecular weight
69000 (Hyoscyamus niger, gel filtration, minor peak) [2]
84000 (Hyoscyamus niger, gel filtration) [1]
103000 (Hyoscyamus niger, gel filtration, major peak) [2]

Subunits
Dimer (2 × 39000, Hyoscyamus niger, SDS-PAGE) [1]
Homotrimer or homotetramer (3 (4) × 29000, Hyoscyamus niger, SDS-PAGE)
[2]

Glycoprotein/Lipoprotein
–

---

4 ISOLATION/PREPARATION

Source organism
Hyoscyamus niger [1, 2]; Datura innoxia [5]; Datura stramonium [3, 4];
Hyoscyamus muticus [2]; Hyoscyamus pusillus [2]; Hyoscyamus canarien-
sis [2]; Physochlaina orientalis [2]; Atropa acuminata [2]; Atropa belladonna
[2]; Physalis philadelphica [2]; Physalis edulis [2]; Physalis alkekengi [2];
Calystegia sepium [2]

Source tissue
Root (cell culture (hairy root cells transformed with Agrobacterium rhizogenes [3]) [1–3], branch root (Hyoscyamus niger, Datura stramonium, Atropa
belladonna), main root (Datura stramonium) [2], highest activity in branched
and cultured roots) [1–3]; Stem (Atropa belladonna, not Hyoscyamus niger)
[2]; Leaf (Atropa belladonna, not Datura stramonium and Hyoscyamus
niger) [2]; Flower (Atropa belladonna, not Datura stramonium and Hyoscyamus niger) [2]; More (not in fruit of Datura stramonium and Hyoscyamus
niger) [2]

Localisation in source

Purification
Hyoscyamus niger [1, 2]; Datura stramonium (hydrophobic interaction chromatography, affinity chromatography) [3]

Crystallization
–

Cloned
–

Renaturated
–

---

5 STABILITY

pH

Temperature (°C)

Oxidation

Organic solvent

General stability information
Glycerol, 23% v/v, stabilizes [1, 2]; NADPH stabilizes [2]; NADH stabilizes
[2]

Storage
–20°C, at least 3 months stable [1, 3]; 4°C, 4 days stable [1]

---

6 CROSSREFERENCES TO STRUCTURE DATABANKS

PIR/MIPS code
PIR2:A48674 (I jimsonweed); PIR2:B48674 (II jimsonweed)

Brookhaven code

7 LITERATURE REFERENCES

[1] Dräger, B., Hashimoto, T., Yamada, Y.: Agric. Biol. Chem.,52,2663–2667 (1988)
[2] Hashimoto, T., Nakajima, K., Ongena, G., Yamada, Y.: Plant Physiol.,100,836–845 (1992)
[3] Portsteffen, A., Dräger, B., Nahrstedt, A.: Phytochemistry,31,1135–1138 (1992)
[4] Dräger, B., Portsteffen, A., Schaal, A., McCabe, P.H., Peerless, A.C.J., Robins, R.J.: Planta,188,581–586 (1992)
[5] Couladis, M.M., Friesen, J.B., Landgrebe, M.E., Leete, E.: Phytochemistry,30, 801–805 (1991)

1 NOMENCLATURE

EC number
1.1.1.237

Systematic name
4-Hydroxyphenyllactate:NAD$^+$ oxidoreductase

Recommended name
Hydroxyphenylpyruvate reductase

Synonymes
Reductase, hydroxyphenylpyruvate
HPRP [2]

CAS Reg. No.
117590-77-9

2 REACTION AND SPECIFICITY

Catalysed reaction
3-(4-Hydroxyphenyl)lactate + NAD$^+$ →
→ 3-(4-hydroxyphenyl)pyruvate + NADH

Reaction type
Redox reaction

Natural substrates
3-(4-Hydroxyphenyl)lactate + NAD$^+$ [1]
3-(3,4-Dihydroxyphenyl)lactate + NAD$^+$ [1]
More (involved with EC 2.3.1.140 in the biosynthesis of rosmarinic acid) [1]

Substrate spectrum
1 3-(4-Hydroxyphenyl)lactate + NAD$^+$ (r [2]) [1, 2]
2 3-(3,4-Dihydroxyphenyl)lactate + NAD$^+$ (r [2]) [1, 2]
3 3-Methoxy-4-hydroxyphenylpyruvate + NADH [2]

Product spectrum
1 3-(4-Hydroxyphenyl)pyruvate + NADH
2 3,4-Dihydroxyphenylpyruvate + NADH
3 ?

Inhibitor(s)
Rosmarinic acid [2]; Pyruvate [2]; p-Coumaroyl-CoA [2]

Cofactor(s)/prosthetic group(s)/activating agents
Dithiothreitol (stimulates [1], no effect [2]) [1]; Ascorbic acid (stimulation [1], no effect [2]) [1]; NADH [1, 2]; NADPH (lower activity than with NADH [1]) [1, 2]

Metal compounds/salts

Turnover number (min⁻¹)

Specific activity (U/mg)
0.010 (p-hydroxyphenylpyruvate) [2]; 0.095 (NADPH) [2]; 0.130 (3,4-dihydroxyphenylpyruvate) [2]; 0.190 (NADH) [2]; 0.250 (3-methoxy-4-hydroxyphenylpyruvate) [2]

K_m-value (mM)

pH-optimum
6.5–7.0 [1]

pH-range

Temperature optimum (°C)
37 [2]

Temperature range (°C)

---

3 ENZYME STRUCTURE

Molecular weight

Subunits

Glycoprotein/Lipoprotein
–

---

4 ISOLATION/PREPARATION

Source organism
Coleus blumei [1, 2]

Source tissue
Cell culture [1, 2]

Localisation in source
Soluble [1, 2]

Purification

Crystallization

–

Cloned

–

Renaturated

–

---

5 STABILITY

pH

Temperature (°C)

Oxidation

Organic solvent

General stability information

Storage
–20°C, 30% loss of activity after 1 month, 85% loss of activity after 10 weeks
[2]

---

6 CROSSREFERENCES TO STRUCTURE DATABANKS

PIR/MIPS code

Brookhaven code

---

7 LITERATURE REFERENCES

[1] Petersen, M., Alfermann, A.W.: Z. Naturforsch.,43c,501–504 (1988)
[2] Häusler, E., Petersen, M., Alfermann, A.W.: Z. Naturforsch.,46c,371–376 (1991)

1 NOMENCLATURE

EC number
1.1.1.238

Systematic name
12beta-Hydroxysteroid:NADP⁺ 12-oxidoreductase

Recommended name
12beta-Hydroxysteroid dehydrogenase

Synonymes
Dehydrogenase, 12beta-hydroxy steroid (nicotinamide adenine dinucleotide phosphate)

CAS Reg. No.
118390-62-8

2 REACTION AND SPECIFICITY

Catalysed reaction
3alpha,7alpha,12beta-Trihydroxy-5beta-cholanate + NADP⁺ →
→ 3alpha,7alpha-dihydroxy-12-oxo-5beta-cholanate + NADPH

Reaction type
Redox reaction

Natural substrates

Substrate spectrum
1 3alpha-Hydroxy-12-oxo-5beta-cholanoate + NADPH (r, reverse reaction 10% the rate of forward reaction) [1]
2 3,12-Dioxo-5beta-cholanoate + NADPH [1]
3 7,12-Dioxo-5beta-cholanoate + NADPH [1]
4 3alpha,7alpha-Dihydroxy-12-oxo-5beta-cholanoate + NADPH [1]
5 3alpha-Hydroxy-7,12-dioxo-5beta-cholanoate + NADPH [1]
6 7alpha-Hydroxy-3,12-dioxo-5beta-cholanoate + NADPH [1]
7 3,7,12-Trioxo-5beta-cholanoate + NADPH [1]
8 3,7,12-Trioxotaureate + NADPH [1]

Product spectrum
 1 3alpha,12beta-Dihydroxy-5beta-cholanoate + NADP$^+$ [1]
 2 12beta-Hydroxy-3-oxo-5beta-cholanoate + NADP$^+$
 3 12beta-Hydroxy-7-oxo-5beta-cholanoate + NADP$^+$
 4 3alpha,7alpha,12beta-Trihydroxy-5beta-cholanoate + NADP$^+$
 5 3alpha,12beta-Dihydroxy-7-oxo-5beta-cholanoate + NADP$^+$
 6 7alpha,12beta-Dihydroxy-3-oxo-5beta-cholanoate + NADP$^+$
 7 12beta-Hydroxy-3,7-dioxo-5beta-cholanoate + NADP$^+$
 8 12beta-Hydroxy-3,7-dioxotaureate + NADP$^+$

Inhibitor(s)

Cofactor(s)/prosthetic group(s)/activating agents
 NADPH (specific for, not replaceable by NADH) [1]; NADP$^+$ [1]

Metal compounds/salts

Turnover number (min^{-1})

Specific activity (U/mg)
 2.1–6.4 [1]

K_m-value (mM)
 0.0837 (7,12-dioxo-5beta-cholanoate) [1]; 0.0854 (NADPH) [1]; 0.0895
 (3,12-dioxo-5beta-cholanoate) [1]; 0.138
 (7alpha-hydroxy-3,12-dioxo-5beta-cholanoate) [1]; 0.141
 (3alpha-hydroxy-12-oxo-5beta-cholanoate) [1]; 0.142 (3,7,12-trioxotaureate)
 [1]; 0.144 (3alpha-hydroxy-7,12-dioxo-5beta-cholanoate) [1]; 0.171
 (3,7,12-trioxo-5beta-cholanoate) [1]; 0.274
 (3alpha,7alpha-dihydroxy-12-oxo-5beta-cholanoate) [1]

pH-optimum
 7.8 (oxidation of 12beta-hydroxy bile acids) [1]; 10.0 (reduction of 12-oxo
 bile acids) [1]

pH-range

Temperature optimum (°C)

Temperature range (°C)

---

3 ENZYME STRUCTURE

Molecular weight
 126000 (Clostridium paraputrificum, gel filtration) [1]

Subunits

Glycoprotein/Lipoprotein
 –

4 ISOLATION/PREPARATION

Source organism
Clostridium paraputrificum [1]; Clostridium tertium [1]; Clostridium difficile [1]

Source tissue

Localisation in source

Purification

Crystallization
–

Cloned
–

Renaturated
–

5 STABILITY

pH
8.5 (best value for stability) [1]

Temperature (°C)
4 (at least 6 h) [1]; 37 (45 min, complete inactivation) [1]; 50 (15 min, complete inactivation) [1]

Oxidation

Organic solvent

General stability information

Storage

6 CROSSREFERENCES TO STRUCTURE DATABANKS

PIR/MIPS code

Brookhaven code

7 LITERATURE REFERENCES

[1] Edenharder, R., Pfützner, A.: Biochim. Biophys. Acta,962,362–370 (1988)

1 NOMENCLATURE

EC number
1.1.1.239

Systematic name
3alpha(or 17beta)-Hydroxysteroid:NAD+ oxidoreductase

Recommended name
3alpha(17beta)-Hydroxysteroid dehydrogenase (NAD+)

Synonymes
Dehydrogenase, 3alpha,17beta-hydroxy steroid
3alpha(17beta)-HSD [2]
More (different from EC 1.1.1.50 and EC 1.1.1.213)

CAS Reg. No.
126469-82-7

2 REACTION AND SPECIFICITY

Catalysed reaction
Testosterone + NAD+ →
→ androst-4-ene-3,17-dione + NADH (random mechanism [2])

Reaction type
Redox reaction

Natural substrates

Substrate spectrum
1 Testosterone + NAD+ [1]
2 5beta-Pregnan-3alpha-ol-20-one + NAD+ [1]
3 5beta-Androstan-3alpha,17beta-diol + NAD+ [1]
4 5alpha-Androstan-3alpha,17beta-diol + NAD+ [1]
5 5alpha-Pregnan-3alpha,21-diol-20-one + NAD+ [1]
6 Lithocholic acid + NAD+ [1]
7 Glycolithocholic acid + NAD+ [1]
8 5beta-Androstan-17beta-ol-3-one + NAD+ [1]
9 5beta-Pregnan-3alpha,21-diol-20-one + NAD+ [1]
10 Glycochenodeoxycholic acid + NAD+ [1]

Product spectrum

1 Androst-4-ene-3,17-dione + NADH [1]
2 5beta-Pregnan-3,20-dione + NADH [1]
3 5beta-Androstan-3alpha-ol-17-one + NADH [1]
4 5alpha-Androstan-3alpha-ol-17-one + NADH
5 5alpha-Pregnan-21-ol-3,20-dione + NADH
6 5beta-Cholan-3-one-24-oic acid + NADH
7 3-Oxo-5beta-cholanoyl glycine + NADH
8 5beta-Androstan-3,17-dione + NADH
9 5beta-Pregnan-21-ol-3,20-dione + NADH
10 7alpha-Hydroxy-3-oxo-5-beta-cholanoyl glycine + NADH

Inhibitor(s)

5beta-Androstan-3beta-ol-17-one [1]; 5beta-Pregnan-3beta-ol-20-one [1]; 5alpha-Androstan-3beta-ol-17-one [1]; 5alpha-Pregnan-3beta-ol-20-one [1]; 21-Hydroxypregn-4-ene-3,20-dione [1]; Hexoestrol [1, 2]; Stilboestrol [1]; Dienstrol [1]; Medroxyprogesterone acetate [1]; Indomethacin [1]; Dexamethasone [1]; Cibacron blue [2]; 5beta-Androstan-3,17-dione (product inhibition, forward reaction) [2]; 5beta-Dihydrotestosterone (product inhibition, reverse reaction) [2]

Cofactor(s)/prosthetic group(s)/activating agents

NAD⁺ (preferred) [1]; NADH (preferred [1], A-specific, i.e. transfer of pro-R-hydrogen from NADH to substrate [2]) [1, 2]; NADP⁺ (15% of NAD⁺-activity) [1]; NADPH (27% of NADH-activity) [1]

Metal compounds/salts

Turnover number (min⁻¹)

1.2 (glycochenodeoxycholic acid) [1]; 3.0 (lithocholic acid) [1]; 9.5 (5beta-pregnan-3,20-dione) [1]; 10 (5beta-androstan-3,17-dione) [1]; 14 (5beta-pregnan-3alpha-ol-20-one) [1]; 15 (glycolithocholic acid) [1]; 16 (5alpha-pregnan-3alpha,21-diol-20-one) [1]; 22 (5beta-androstan-17beta-ol-3-one, 5beta-pregnan-3alpha,21-diol-20-one) [1]; 25 (testosterone) [1]; 27 (5alpha-androstan-3alpha,17beta-diol) [1]; 51 (5beta-androstan-3alpha,17beta-diol) [1]

Specific activity (U/mg)

4.2 [1]

K_m-value (mM)

0.0005 (lithocholic acid) [1]; 0.0022 (testosterone) [1]; 0.003 (glycolithocholic acid) [1]; 0.0031 (5alpha-androstan-3alpha,17beta-diol) [1]; 0.0038 (5beta-pregnan-3,20-dione) [1]; 0.0046 (5beta-pregnan-3alpha-ol-20-one) [1]; 0.0048 (5beta-androstan-3alpha,17beta-diol) [1]; 0.0049 (5beta-androstan-17beta-ol-3-one) [1]; 0.0062 (5beta-pregnan-3alpha,17beta-diol-20-one) [1]; 0.0066 (5beta-androstan-3,17-dione) [1]; 0.0075 (glycochenodeoxycholic acid) [1]; 0.02 (NADH) [1]; 0.27 (NAD⁺) [1]

pH-optimum
10.4 (oxidation of 5beta-androstan-3alpha,17beta-diol) [1]

pH-range

Temperature optimum (°C)

Temperature range (°C)

---

3 ENZYME STRUCTURE

Molecular weight

Subunits

Glycoprotein/Lipoprotein
–

---

4 ISOLATION/PREPARATION

Source organism
Hamster [1, 2]

Source tissue
Liver [1, 2]

Localisation in source
Cytosol [1, 2]

Purification
Hamster [1]

Crystallization
–

Cloned
–

Renaturated
–

---

5 STABILITY

pH

Temperature (°C)
45 (half-life 2.3 min) [1]

Oxidation

Organic solvent

General stability information

Storage
4°C, 10 mM Tris/HCl buffer, pH 8.0, 5 mM mercaptoethanol, 0.5 mM EDTA
[1]

---

6 CROSSREFERENCES TO STRUCTURE DATABANKS

PIR/MIPS code

Brookhaven code

---

7 LITERATURE REFERENCES

[1] Ohmura, M., Hara, A., Nakagawa, M., Sawada, H.: Biochem. J.,266,583–589 (1990)
[2] Sawada, H., Hara, A., Ohmura, M., Nakayama, T., Deyashi, Y.: J. Biochem.,109, 770–779 (1991)

1 NOMENCLATURE

EC number
1.1.1.240

Systematic name
N-Acetyl-D-hexosamine:NAD⁺ 1-oxidoreductase

Recommended name
N-Acetylhexosamine 1-dehydrogenase

Synonymes
Dehydrogenase, N-acetylhexosamine
N-Acetyl-D-hexosamine dehydrogenase

CAS Reg. No.
122785-18-6

2 REACTION AND SPECIFICITY

Catalysed reaction
N-Acetyl-D-glucosamine + NAD⁺ →
→ N-acetyl-D-glucosaminate + NADH

Reaction type
Redox reaction

Natural substrates
N-Acetyl-D-glucosamine + NAD⁺ [1]

Substrate spectrum
1 N-Acetyl-D-glucosamine + NAD⁺ [1]
2 N-Acetylgalactosamine + NAD⁺ [1]
3 N-Acetylmannosamine + NAD⁺ (slowly) [1]

Product spectrum
1 N-Acetyl-D-glucosaminate + NADH
2 ?
3 ?

Inhibitor(s)
Hg^{2+} [1]; SDS [1]; NaN_3 (weak) [1]

Cofactor(s)/prosthetic group(s)/activating agents
NAD⁺ (strictly specific for NAD⁺ as hydrogen acceptor) [1]

Metal compounds/salts

Turnover number (min⁻¹)
39300 (N-acetyl-D-glucosamine) [1]; 55920 (N-acetyl-D-galactosamine) [1]

Specific activity (U/mg)
239 [1]

K_m-value (mM)
0.24 (NAD⁺ (+ N-acetyl-D-glucosamine)) [1]; 0.8 (N-acetyl-D-galactosamine) [1]; 5.3 (N-acetyl-D-glucosamine) [1]

pH-optimum
10.0 [1]

pH-range

Temperature optimum (°C)

Temperature range (°C)

---

3 ENZYME STRUCTURE

Molecular weight
124000 (Pseudomonas sp., gel filtration) [1]

Subunits
Tetramer (4 × 30000, Pseudomonas sp., SDS-PAGE) [1]

Glycoprotein/Lipoprotein
–

---

4 ISOLATION/PREPARATION

Source organism
Pseudomonas sp. No 53 [1]

Source tissue

Localisation in source

Purification
Pseudomonas sp. No 53 [1]

Crystallization
–

Cloned
–

Renaturated
–

5 STABILITY

pH
   8.0–10.5 (highest stability) [1]

Temperature (°C)
   45 (10 min, unstable below pH 5 and above pH 12) [1]; 55 (10 min, stable) [1]; 65 (10 min, 50% loss of activity) [1]; 70 (10 min, complete loss of activity) [1]

Oxidation

Organic solvent

General stability information
   NAD$^+$, 3 mM, stabilizes against heat inactivation [1]

Storage
   –20°C, several weeks [1]

---

6 CROSSREFERENCES TO STRUCTURE DATABANKS

PIR/MIPS code

Brookhaven code

---

7 LITERATURE REFERENCES

[1] Horiuchi, T., Kurokawa, T.: Agric. Biol. Chem.,53,1919–1925 (1989)

1 NOMENCLATURE

EC number

1.1.1.241

Systematic name

6-endo-Hydroxycineole:NAD+ 6-oxidoreductase

Recommended name

6-endo-Hydroxycineole dehydrogenase

Synonymes

Dehydrogenase, 6-endo-hydroxycineole

CAS Reg. No.

122933-68-0

2 REACTION AND SPECIFICITY

Catalysed reaction

6-endo-Hydroxycineole + NAD+ →

→ 6-oxocineole + NADH

Reaction type

Redox reaction

Natural substrates

Substrate spectrum

1 6-endo-Hydroxycineole + NAD(P)+ (r) [1]

Product spectrum

1 6-Oxocineole + NAD(P)H [1]

Inhibitor(s)

Cofactor(s)/prosthetic group(s)/activating agents

NAD+ [1]; NADH [1]; NADP+ [1]; NADPH (10% of activity with NADH) [1]

Metal compounds/salts

Turnover number (min^{-1})

Specific activity (U/mg)
0.62 (6-endo-hydroxycineole) [1]; 1.08 (6-oxocineole) [1]

K_m-value (mM)

pH-optimum
7.5 (reduction of 6-oxocineole) [1]; 10.5 (oxidation of 6-endo-hydroxycineole) [1]

pH-range

Temperature optimum (°C)
30 (assay at) [1]

Temperature range (°C)

---

3 ENZYME STRUCTURE

Molecular weight

Subunits

Glycoprotein/Lipoprotein
–

---

4 ISOLATION/PREPARATION

Source organism
Rhodococcus sp. [1]

Source tissue
Cell [1]

Localisation in source

Purification
Rhodococcus sp. [1]

Crystallization
–

Cloned
–

Renaturated
–

5 STABILITY

pH

Temperature (°C)

Oxidation

Organic solvent

General stability information

Storage
-20°C, after dialysis against 1 mM KH_2PO_4/Na_2HPO_4 buffer, pH 7.1 [1]

6 CROSSREFERENCES TO STRUCTURE DATABANKS

PIR/MIPS code

Brookhaven code

7 LITERATURE REFERENCES

[1] Williams, D.R., Trudgill, P.W., Taylor, D.G.: J. Gen. Microbiol.,135,1957–1967 (1989)

1 NOMENCLATURE

EC number
1.1.1.242

Systematic name
Dihydrozeatin:NADP$^+$ oxidoreductase

Recommended name
Zeatin reductase

Synonymes
Reductase, zeatin

CAS Reg. No.
123644-82-6

2 REACTION AND SPECIFICITY

Catalysed reaction
Dihydrozeatin + NADP$^+$ →
→ zeatin + NADPH

Reaction type
Redox reaction

Natural substrates
trans-Zeatin + NADPH (regulation of cytokinin levels in plant tissues, enzyme activity is located within two isoenzymes) [1]

Substrate spectrum
1 trans-Zeatin + NADPH [1]

Product spectrum
1 Dihydrozeatin + NADP$^+$ [1]

Inhibitor(s)

Cofactor(s)/prosthetic group(s)/activating agents
NADPH (NADH cannot substitute for NADPH) [1]

Metal compounds/salts

Turnover number (min^{-1})

Specific activity (U/mg)

K$_m$-value (mM)
0.07 (trans-zeatin, high molecular weight isoenzyme) [1]; 0.1–0.23
(trans-zeatin, low molecular weight isoenzyme) [1]

pH-optimum
7.5–8.0 [1]

pH-range

Temperature optimum (°C)
27 (assay at) [1]

Temperature range (°C)

---

3 ENZYME STRUCTURE

Molecular weight
55000 (Phaeseolus vulgaris, high molecular weight isoenzyme, gel filtration)
[1]
25000 (Phaeseolus vulgaris, low molecular weight isoenzyme, gel filtration)
[1]

Subunits

Glycoprotein/Lipoprotein
–

---

4 ISOLATION/PREPARATION

Source organism
Phaeseolus vulgaris [1]

Source tissue
Immature embryos [1]

Localisation in source

Purification
Phaeseolus vulgaris (partial) [1]

Crystallization
–

Cloned
–

Renaturated
–

5 STABILITY

pH

Temperature (°C)

Oxidation

Organic solvent

General stability information
Glycerol, 20%, stabilizes [1]

Storage
−20°C, in 0.05 M phosphate buffer containing 5 mM DTT, 0.5 mM EDTA,
5 mM NADPH, 0.5 M KCl and 20% glycerol: loss of about 10% of activity
after 3 days [1]

6 CROSSREFERENCES TO STRUCTURE DATABANKS

PIR/MIPS code

Brookhaven code

7 LITERATURE REFERENCES

[1] Martin, R.C., Mok, M.C., Shaw, G., Mok, D.W.: Plant Physiol.,90,1630–1635 (1989)

1 NOMENCLATURE

EC number
1.1.1.243

Systematic name
(-)-trans-Carveol:NADP+ oxidoreductase

Recommended name
Carveol dehydrogenase

Synonymes
Dehydrogenase, (-)-trans-carveol

CAS Reg. No.
122653-66-1

2 REACTION AND SPECIFICITY

Catalysed reaction
(-)-trans-Carveol + NADP+ →
→ (-)-carvone + NADPH

Reaction type
Redox reaction

Natural substrates
(-)-trans-Carveol + NADP+ (monoterpene synthesis in mints) [1]

Substrate spectrum
1 (-)-trans-Carveol + NADP+ [1]

Product spectrum
1 (-)-Carvone + NADPH [1]

Inhibitor(s)

Cofactor(s)/prosthetic group(s)/activating agents

NADP+ [1]

Metal compounds/salts

Turnover number (min^{-1})

Specific activity (U/mg)

K$_m$-value (mM)

pH-optimum

pH-range

Temperature optimum (°C)
    30 (assay at) [1]

Temperature range (°C)

---

3 ENZYME STRUCTURE

Molecular weight

Subunits

Glycoprotein/Lipoprotein
    –

---

4 ISOLATION/PREPARATION

Source organism
    Mentha spicata (spearmint) [1]

Source tissue
    Leaf [1]; Glandular trichomes [1]

Localisation in source

Purification
    Mentha spicata (partial) [1]

Crystallization
    –

Cloned
    –

Renaturated
    –

5 STABILITY

pH

Temperature (°C)

Oxidation
  1 mM DTT stabilizes [1]

Organic solvent

General stability information

Storage

6 CROSSREFERENCES TO STRUCTURE DATABANKS

PIR/MIPS code

Brookhaven code

7 LITERATURE REFERENCES

[1] Gershenzon, J., Maffei, M., Croteau, R.: Plant Physiol.,89,1351–1357 (1989)

1 NOMENCLATURE

EC number
1.1.1.244

Systematic name
Methanol:NAD+ oxidoreductase

Recommended name
Methanol dehydrogenase

Synonymes
Dehydrogenase, methanol

CAS Reg. No.
74506-37-9

2 REACTION AND SPECIFICITY

Catalysed reaction
Methanol + NAD+ →
→ formaldehyde + NADH

Reaction type
Redox reaction

Natural substrates
Methanol + NAD+ (involved in initial methanol oxidation) [1–4]

Substrate spectrum
1 Methanol + NAD+ (r, the reverse reaction is the much more stable activity, oxidation at 33% [1], 44% [2] the rate of ethanol oxidation, NADP+ cannot replace NAD+) [1–4]
2 Ethanol + NAD+ (best substrate) [1, 2]
3 n-Propanol + NAD+ (oxidation at 71% the rate of ethanol oxidation) [1, 2]
4 n-Butanol + NAD+ (oxidation at 87% the rate of ethanol oxidation) [1, 2]
5 More (no substrates are 2-propanol, 2,3-butandiol, mannitol and glycerol [4], in the presence of activator protein and Mg^{2+} the enzyme possesses a high affinity active site for alcohol and NAD+, not for formaldehyde) [4]

Product spectrum
1 Formaldehyde + NADH [1–4]
2 ?
3 ?
4 ?
5 ?

Inhibitor(s)
1,10-Phenanthroline (reduction of aldehyde) [3]; EDTA (reduction of aldehyde) [3]; N-Ethylmaleimide (weak inhibition of aldehyde reduction) [3]; Iodoacetate (weak inhibition of aldehyde reduction) [3]

Cofactor(s)/prosthetic group(s)/activating agents
NAD^+ (requirement, specific for) [1–4]; Activator protein (activation in presence of Mg^{2+}, in vitro: 3 mol soluble activator protein per mol enzyme, in vivo estimated as 1 mol per 17.5 mol enzyme, MW 50000 with two subunits of MW 27000 each, in presence of activator protein and Mg^{2+} enzyme possesses a high affinity active site for alcohol and NAD^+, up to 40-fold increase of activity, primarily of V_{max}, slight decrease of K_m-value for methanol, enzyme/activator-interaction is dilution-sensitive, no stimulation of formaldehyde reduction) [4]; More (no pyrroloquinoline quinone enzyme) [1]

Metal compounds/salts
Zn^{2+} (requirement, 1 mol per mol subunit, atomic absorption spectroscopy) [3]; Mg^{2+} (requirement, 1–2 mol per mol subunit, atomic absorption spectroscopy, activation by activator protein depends on Mg^{2+}. Zn^{2+}, Mn^{2+}, Ca^{2+} cannot replace Mg^{2+} [4]) [3, 4]

Turnover number (min⁻¹)

Turnover number (min^{-1})

Specific activity (U/mg)
19.6 (formaldehyde reduction) [1, 2]

K_m-value (mM)
More (3.8 mM and 166 mM (methanol), in cell-free extracts biphasic kinetics for methanol, ethanol or NAD^+, due to two active sites, not for formaldehyde [1], in presence of the activator protein biphasic kinetics with all alcoholic substrates and NAD^+ [4]) [1]; 2.0 (formaldehyde) [1]

pH-optimum
6.7 (formaldehyde reduction) [1, 2]; 9.5 (methanol oxidation) [1, 4]

pH-range

Temperature optimum (°C)
50 [1, 2]; 57–59 [4]

Temperature range (°C)

---

3 ENZYME STRUCTURE

Molecular weight
363000 (Bacillus sp., ultracentrifugation sedimentation equilibrium, during the centrifugation time of 80 h, dissociation occurs) [2]
430000 (Bacillus sp., calculated from electron microscopic analysis and MW of subunit) [2]

Subunits
Decamer (10 × 43000, Bacillus sp., SDS-PAGE [1, 2], the decameric structure is detected by electron microscopic analysis [3]) [1–3]

Glycoprotein/Lipoprotein
–

---

4 ISOLATION/PREPARATION

Source organism
Bacillus sp. (methylotrophic and thermotolerant strains C1 [1–4], PB1, AR2, TS1, TS2, TS4, S1, TF [1]) [1–4]

Source tissue
Cell [1–4]

Localisation in source
Cytoplasm [1–4]

Purification
Bacillus sp. (strain C1, dehydrogenase activity, not reductase activity is lost during purification [1], purification of activator protein [4]) [1–4]

Crystallization
–

Cloned
–

Renaturated
–

---

5 STABILITY

pH

Temperature (°C)
50 ($t_{1/2}$: above 2 min) [4]; 60 ($t_{1/2}$: 6 min) [4]; 70 ($t_{1/2}$: 2 min) [4]

Oxidation

Organic solvent

General stability information
Dilution inactivates methanol dehydrogenase, not reductase [1, 2]; Sucrose, 20% w/v, stabilizes formaldehyde reductase not methanol dehydrogenase activity [1, 2]; Metal ions, such as Fe^{2+} and Zn^{2+}, do not stabilize methanol dehydrogenase activity [1, 2]; Glycerol does not stabilize methanol dehydrogenase activity [1, 2]; Dithiothreitol stabilizes reductase, not methanol dehydrogenase activity [1, 2]; Storage leads to dissociation [3]

Storage
$-80°C - 4°C$, in the presence of DTT at least 4 days stable [1, 2]; $4°C - 20°C$, the enzyme dissociates upon storage [3]; $4°C - 20°C$, complete dissoziation after 48 h [4]

---

6 CROSSREFERENCES TO STRUCTURE DATABANKS

PIR/MIPS code

Brookhaven code

---

7 LITERATURE REFERENCES

[1] Arfman, N., Watling, E.M., Clement, W., Van Oosterwijk, R.J., De Vries, G.E., Harder, W., Attwood, M.M., Dijkhuizen, L.: Arch. Microbiol.,152,280–288 (1989)
[2] Arfman, N., Dijkhuizen, L.: Methods Enzymol.,188,223–226 (1990)
[3] Vonck, J., Arfman, N., De Vries, G.E., Van Beeumen, J., Van Bruggen, E.F.J., Dijkhuizen, L.: J. Biol. Chem.,266,3949–3954 (1991)
[4] Arfman, N., Van Beeumen, J., De Vries, G.E., Harder, W., Dijkhuizen, L.: J. Biol. Chem.,266,3955–3960 (1991)

1 NOMENCLATURE

EC number

1.1.1.245

Systematic name

Cyclohexanol:NAD$^+$ oxidoreductase

Recommended name

Cyclohexanol dehydrogenase

Synonymes

Dehydrogenase, cyclohexanol

CAS Reg. No.

63951-98-4

---

2 REACTION AND SPECIFICITY

Catalysed reaction

Cyclohexanol + NAD$^+$ →
→ cyclohexanone + NADH

Reaction type

Redox reaction

Natural substrates

Cyclohexanol + NAD$^+$ (reaction in cyclohexane metabolism [1], reaction in anaerobic cyclohexanol catabolism [2, 3], inducible [2–4]) [1–4]

Substrate spectrum

1 Cyclohexanol + NAD$^+$ (r [2], preferred substrate [1], specific for alicyclic alcohols) [1–4]
2 Cyclopentanol + NAD$^+$ (r, the reverse reaction proceeds at 4% of cyclohexanone reduction [2], oxidation at 232% the rate of cyclohexanol oxidation) [2, 4]
3 Cyclooctanol + NAD$^+$ (oxidation at 500% the rate of cyclohexanol oxidation) [4]
4 2-Cyclohexenol + NAD$^+$ (r, the reverse reaction proceeds at 1% the rate of cyclohexanone reduction, oxidation at 142% the rate of cyclohexanol oxidation) [2]
5 2-Methylcyclohexanol + NAD$^+$ (oxidation at 51% the rate of cyclohexanol oxidation) [4]
6 3-Methylcyclohexanol + NAD$^+$ (oxidation at 74% the rate of cyclohexanol oxidation) [4]

7 4-Methylcyclohexanol + NAD+ (oxidation at 66% the rate of cyclohexanol oxidation) [4]
8 2,3-Dimethylcyclohexanol + NAD+ (oxidation at 78% the rate of cyclohexanol oxidation) [4]
9 2,6-Dimethylcyclohexanol + NAD+ (oxidation at 14% the rate of cyclohexanol oxidation) [4]
10 Cyclohexane-1,2-diol + NAD+ (r, the reverse reaction proceeds at 3% the rate of cyclohexanone reduction [2], oxidation of the trans isomer at 15% [2], 43% [4] the rate of cyclohexanol oxidation) [1, 2, 4]
11 Cyclohexane-1,3-diol + NAD+ (r (in 100 mM MES-buffer, pH 6.0), the reverse reaction proceeds at 3% the rate of cyclohexanone reduction [2], oxidation at 14% [2], 96% [4] the rate of cyclohexanol oxidation) [1, 2, 4]
12 Cyclohexane-1,4-diol + NAD+ (r, the reverse reaction proceeds at 240% the rate of cyclohexanone reduction, both hydroxyl groups are reduced with 2 NADH [2], oxidation at 36% [2], 91% [4] the rate of cyclohexanol oxidation) [1, 2, 4]
13 1,2-Cycloheptanediol + NAD+ (r, the reverse reaction proceeds at 1% the rate of cyclohexanone reduction, oxidation at 25% the rate of cyclohexanol oxidation) [2]
14 2,3-Butandiol + NAD+ (oxidation at 4% the rate of cyclohexanol oxidation) [4]
15 1-Butanol + NAD+ (poor substrate) [4]
16 1-Hexanol + NAD+ (oxidation at 7% the rate of cyclohexanol oxidation) [2]
17 3-Hexanol + NAD+ (oxidation at 196% the rate of cyclohexanol oxidation) [4]
18 1-Heptanol + NAD+ (oxidation at 32% the rate of cyclohexanol oxidation [2], poor substrate [4]) [2, 4]
19 4-Heptanol + NAD+ (oxidation at 125% the rate of cyclohexanol oxidation) [4]
20 1-Octanol + NAD+ (poor substrate) [4]
21 1-Decanol + NAD+ (poor substrate) [4]
22 1-Dodecanol + NAD+ (poor substrate) [4]
23 2-Propanol + NAD+ (oxidation at 19% the rate of cyclohexanol oxidation) [4]
24 2-Pentanol + NAD+ (oxidation at 43% the rate of cyclohexanol oxidation) [4]
25 3-Cyclohexyl-1-propanol + NAD+ (poor substrate) [4]
26 2-Methyl-1-pentanol + NAD+ (poor substrate) [4]
27 4-Methyl-1-pentanol + NAD+ (poor substrate) [4]
28 Benzyl alcohol + NAD+ (poor substrate) [4]
29 2-Hydroxycyclohexanone + NADH (dimer, both hydroxyl groups are reduced with 2 NADH, the reduction proceeds at 50% the rate of cyclohexanone reduction) [2]

30 More (no substrates for oxidation with NAD⁺ are cyclohexane-1,2-dione
[1], 2-hydroxycyclohexanone (dimer [2]) [1, 2], 1,3,5-trihydroxycyclohe-
xane (cis,cis), ethanol, 1-propanol, 1-butanol, 1-pentanol, menthol [2],
phenol, 1,3-propanediol, ethyleneglycol, propyleneglycol [4], no substrates for reduction with NADH are epsilon-caprolactone, 2-cyclopente-
none, 1,3-cyclohexanedione (in Tris-buffer, pH 7.8), 1,3-cyclopentanedio-
ne, 5-oxocaproate [2]) [1, 2, 4]

Product spectrum
  1 Cyclohexanone + NADH [1–4]
  2 Cyclopentanone + NADH
  3 Cyclooctanone + NADH
  4 2-Cyclohexenone + NADH
  5 2-Methylcyclohexanone + NADH
  6 3-Methylcyclohexanone + NADH
  7 4-Methylcyclohexanone + NADH
  8 2,3-Dimethylcyclohexanone + NADH
  9 2,6-Dimethylcyclohexanone + NADH
10 Cyclohexan-1,2-dione + NADH [1, 2]
11 Cyclohexan-1,3-dione + NADH [1, 2]
12 Cyclohexan-1,4-dione + NADH [1, 2]
13 ?
14 ?
15 Butanal + NADH
16 Hexanal + NADH
17 3-Hexanone + NADH
18 Heptanal + NADH
19 4-Heptanone + NADH
20 Octanal + NADH
21 Decanal + NADH
22 Dodecanal + NADH
23 2-Propanone + NADH
24 2-Pentanone + NADH
25 3-Cyclohexyl-1-propanal + NADH
26 2-Methylpentanal + NADH
27 4-Methylpentanal + NADH
28 Benzylaldehyde + NADH
29 ?
30 ?

Inhibitor(s)
PCMB (strong) [4]; More (no inhibition by NaN₃, sodium arsenite, potassium
cyanide, 2,2'-bipyridyl, 4,5-dihydroxy-1,3-benzenedisulfonic acid, EDTA,
1,10-phenanthroline, iodoacetate) [4]

Enzyme Handbook © Springer-Verlag Berlin Heidelberg 1995
Duplication, reproduction and storage in data banks are only
allowed with the prior permission of the publishers

Cofactor(s)/prosthetic group(s)/activating agents
NAD+ [1–4]; NADP+ (6% [1], less than 5% [2] as effective as NAD+) [1, 2];
NADH (reverse reaction) [2]; More (NADP+, FAD, FMN, cytochrome c,
2,6-dichlorophenolindophenol, phenazine methosulfate, 2,6-dichlorophe-
nolindophenol + phenazine methosulfate or ferricyanide do not act as acceptors) [4]

Metal compounds/salts

Turnover number (min⁻¹)

Specific activity (U/mg)
0.28 (crude) [3]; 0.31 (aerobically grown cells) [2]; 0.80 (oxidation, anaerobically grown cells) [2]; 0.87 [1]; 1.9 (reduction, anaerobically grown cells)
[2]; 143.5 [4]

K_m-value (mM)
0.0013 (cyclohexanol) [1]; 0.024 (NAD+) [4]; 0.037 (cyclohexanol) [4]; 0.04
(NAD+) [2]; 0.05 (cyclohexanol) [2]; 0.41 (cyclohexan-1,2-diol) [1]; 0.50 (cy-
clohexan-1,3-diol) [1]; 0.59 (cyclohexan-1,4-diol) [1]

pH-optimum
7.0 (reduction of ketone) [4]; 10.1–10.5 (oxidation of alcohol) [1]; 10.5 (oxidation of alcohol) [4]

pH-range

Temperature optimum (°C)
30 (assay at) [2]

Temperature range (°C)

3 ENZYME STRUCTURE

Molecular weight
145000 (Nocardia sp., gel filtration) [4]

Subunits

Glycoprotein/Lipoprotein
–

4 ISOLATION/PREPARATION

Source organism
Xanthobacter sp. [1]; Pseudomonas sp. (strain K601) [2]; Acinetobacter sp.
(strain NCIB 9871) [3]; Nocardia sp. [4]

Source tissue
Cell [1–4]

Localisation in source
Cytoplasm [4]

Purification
Nocardia sp. (affinity chromatography) [4]

Crystallization
–

Cloned
–

Renaturated
–

---

5 STABILITY

pH

Temperature (°C)
45 ($t_{1/2}$: 1 min) [4]; 55 (85% loss of activity after 1 min) [4]; 60 (95% loss of activity after 1 min) [4]

Oxidation

Organic solvent

General stability information
Ion-exchange chromatography leads to significant loss of activity [4]; Gel filtration leads to significant loss of activity [4]; PAGE leads to protein dissociation [4]

Storage

---

6 CROSSREFERENCES TO STRUCTURE DATABANKS

PIR/MIPS code

Brookhaven code

---

7 LITERATURE REFERENCES

[1] Trower, M.K., Buckland, R.M., Higgins, R., Griffin, M.: Appl. Environ. Microbiol.,49, 1282–1289 (1985)
[2] Dangel, W. Czech, A., Fuchs, G.: Arch. Microbiol.,152,273–279 (1989)
[3] Donaghue, N.A., Trudgill, P.W.: Eur. J. Biochem.,60,1–7 (1975)
[4] Stirling, L.A., Perry, J.J.: Curr. Microbiol.,4,37–40 (1980)

Enzyme Handbook © Springer-Verlag Berlin Heidelberg 1995
Duplication, reproduction and storage in data banks are only
allowed with the prior permission of the publishers

1 NOMENCLATURE

EC number
1.1.1.246

Systematic name
Medicarpin:NADP+ 2'-oxidoreductase

Recommended name
Pterocarpin synthase

Synonymes
Synthase, pterocarpan
Pterocarpan synthase

CAS Reg. No.
118477-70-6

2 REACTION AND SPECIFICITY

Catalysed reaction
Vestitone + NADPH →
→ medicarpin + NADP+

Reaction type
Redox reaction

Natural substrates
Vestitone + NADPH (catalyzes the final step in biosynthesis of the pterocarpin phytoalexins medicarpin and maackiain) [1, 2]

Substrate spectrum
1 Vestitone + NADPH [1, 2]

Product spectrum
1 Medicarpin + NADP+ [1, 2]

Inhibitor(s)

Cofactor(s)/prosthetic group(s)/activating agents
NADPH [1]; NADH (10% of the activity with NADPH) [1]

Metal compounds/salts

Turnover number (min^{-1})

Specific activity (U/mg)
  More [1]

K$_m$-value (mM)
  0.017 (vestitone) [1]; 0.04 (NADPH) [1]

pH-optimum
  6 [1]

pH-range

Temperature optimum (°C)
  30 [1]

Temperature range (°C)

---

3 ENZYME STRUCTURE

Molecular weight

Subunits

Glycoprotein/Lipoprotein
  –

---

4 ISOLATION/PREPARATION

Source organism
  Cicer arietinum (chickpea) [1, 2]

Source tissue
  Cell suspension culture [1, 2]

Localisation in source
  Soluble [1]

Purification
  Cicer arietinum (partial) [1]

Crystallization
  –

Cloned
  –

Renaturated
  –

5 STABILITY

pH

Temperature (°C)

Oxidation

Organic solvent

General stability information

Storage

6 CROSSREFERENCES TO STRUCTURE DATABANKS

PIR/MIPS code

Brookhaven code

7 LITERATURE REFERENCES

[1] Bleß, W., Barz, W.: FEBS Lett.,235,47–50 (1988)
[2] Daniel, S., Tiemann, K., Wittkampf, U., Bless, W., Hinderer, W., Barz, W.: Planta,182, 270–278 (1990)

1 NOMENCLATURE

EC number
1.1.2.2

Systematic name
D-Mannitol:ferricytochrome-c 2-oxidoreductase

Recommended name
Mannitol dehydrogenase (cytochrome)

Synonymes
Polyol dehydrogenase

CAS Reg. No.
37250-78-5

2 REACTION AND SPECIFICITY

Catalysed reaction
D-Mannitol + ferricytochrome c →
→ D-fructose + ferrocytochrome c

Reaction type
Redox reaction

Natural substrates
D-Mannitol + ferricytochrome [1, 2]

Substrate spectrum
1 D-Mannitol + ferricytochrome (other hydrogen acceptors: phenol blue,
  ferricyanide, phenolindo-2,6-dichlorophenol, o-chlorophenolindo-2,6-di-
  chlorophenol) [1, 2]
2 Acyclic D-erythro polyols (e.g. erythritiol, ribitol, D-arabinitol, D-sorbitol,
  overview [1, 2]) + ferricytochrome [1, 2]

Product spectrum
1 D-Fructose + ferrocytochrome [1, 2]
2 Acyclic D-erythro ketoses + ferrocytochrome [1, 2]

Inhibitor(s)
　HCN [2]

Cofactor(s)/prosthetic group(s)/activating agents
　Ferricytochrome [1, 2]

Metal compounds/salts
　Ca^{2+} (restores activity) [1, 2]; Mg^{2+} (restores activity) [1, 2]

Turnover number (min^{-1})

Specific activity (U/mg)

K_m-value (mM)
　31–34 (D-mannitol) [1, 2]; 20 (sorbitol) [1]; 44 (ribitol) [1]; 17 (D-arabitol) [1]

pH-optimum
　5.4 [2]

pH-range

Temperature optimum (°C)

Temperature range (°C)

3 ENZYME STRUCTURE

Molecular weight

Subunits

Glycoprotein/Lipoprotein
　–

4 ISOLATION/PREPARATION

Source organism
　Acetobacter suboxydans [1, 2]

Source tissue

Localisation in source

Purification

Crystallization
　–

Cloned
　–

Renaturated
　–

5 STABILITY

pH

Temperature (°C)

Oxidation

Organic solvent

General stability information

Storage
  Several weeks, 0°C [1]

6 CROSSREFERENCES TO STRUCTURE DATABANKS

PIR/MIPS code

Brookhaven code

7 LITERATURE REFERENCES

[1] Edson, N.L., Shaw, D.R.D.: Methods Enzymol.,9,147–149 (1966)
[2] Arcus, A.C., Edson, N.L.: Biochem. J.,64,385–394 (1956)

1 NOMENCLATURE

EC number
1.1.2.3

Systematic name
(S)-Lactate:ferricytochrome-c 2-oxidoreductase

Recommended name
L-Lactate dehydrogenase (cytochrome)

Synonymes
Lactic acid dehydrogenase
Cytochrome b_2 (flavin-free derivative of flavocytochrome b_2) [3]
Flavocytochrome b_2
L-Lactate cytochrome c reductase [3]
L(+)-Lactate:cytochrome c oxidoreductase [7]
Dehydrogenase, lactate (cytochrome)
L-Lactate cytochrome c oxidoreductase
Lactate dehydrogenase (cytochrome)
Lactic cytochrome c reductase

CAS Reg. No.
9078-32-4

2 REACTION AND SPECIFICITY

Catalysed reaction
(S)-Lactate + 2 ferricytochrome c →
→ pyruvate + 2 ferrocytochrome c (mechanism [2])

Reaction type
Redox reaction

Natural substrates
More (enzyme can feed electrons to the respiratory chain at the level of cytochrome c and provide energy through the third site of oxidative phosphorylation) [7]

Substrate spectrum

1 (S)-Lactate + ferricytochrome c (D-isomer not oxidized [9]) [1–16]
2 2-Hydroxybutyrate + ferricytochrome c [1]
3 Glycolate + ferricytochrome c (at 5% the rate of lactate) [1]
4 More (other electron acceptors used: ferricyanide [1, 7, 9], 2,6-dichloro-
phenolindophenol [1, 7, 9], 1,2-naphthoquinone 4-sulfate [1], methylene
blue [7], 1,2-naphthoquinone 4-sulfonate [7], not: O_2 [9], NAD+ [9],
NADP+ [9]. No activity with L-malate [9]) [1, 7, 9]

Product spectrum

1 Pyruvate + ferrocytochrome c
2 2-Oxobutyrate + ferrocytochrome c
3 Glyoxylate + ferrocytochrome c
4 ?

Inhibitor(s)

Oxalate [1, 9]; Glycerate [1]; D-Malate [1]; L-Malate [1]; Mandelate [1];
Phenyl pyruvate [1]; Fatty acids [1]; Ethane nitronate [11]; Pyruvate [15];
More (unlike Saccharomyces cerevisiae enzyme activity of Hansenula cytochrome b_2 is inhibited in presence of excess substrate [7], inhibition by salts
is competitive and completely overcome by cytochrome c [1]) [1, 7]

Cofactor(s)/prosthetic group(s)/activating agents

FMN (flavohemoprotein, FMN:protoheme ratio is 1.1 [1], 1 FMN per enzyme
molecule of MW 230000 [4]) [1, 4]; Heme (1 heme per enzyme molecule of
MW 230000 [4], chemical characterization of the heme-binding core [16])
[4, 16]

Metal compounds/salts

More (no significant amount of nonheme iron, copper or molybdenum) [1];
Magnesium (1 gatom per 80000 MW protein, may be associated with
deoxyribonucleotide component) [1]

Turnover number (min⁻¹)

More (7000–8000 per equivalent of flavin [1]) [1–3]

Specific activity (U/mg)

More [9]

Kₘ-value (mM)

0.4 (L-lactate, Saccharomyces cerevisiae) [3]; 1.3 (L-lactate, Hansenula anomala) [3]; 3.85 (L-lactate (+ dichlorophenolindophenol)) [9]; More [2, 15]

pH-optimum

7.2–8.4 (ferricyanide) [9]; 8 (proteolytically cleaved enzyme with altered properties) [3]

pH-range

6.0–8.4 (6.0: about 50% of activity maximum, 7.2–8.4: activity maximum) [9]

Temperature optimum (°C)
20–25 (assay at) [8]; 30 (assay at) [8]

Temperature range (°C)

3 ENZYME STRUCTURE

Molecular weight
220000 (Hansenula anomala) [5]
220000–240000 (Saccharomyces cerevisiae, gel filtration, X-ray diffraction studies) [4]
235000 (Saccharomyces cerevisiae, X-ray diffraction studies) [7]
More (primary structure) [10]

Subunits
Tetramer (4 × 57500, Saccharomyces cerevisiae, SDS-PAGE [4, 6],
4 × 60000, Hansenula anomala [5]) [4–6]

Glycoprotein/Lipoprotein
–

4 ISOLATION/PREPARATION

Source organism
Rhizopus oryzae [9]; Yeast [1, 16]; Saccharomyces cerevisiae [3, 4, 6, 7, 10, 11, 13, 14]; Hansenula anomala [3, 5, 8, 12, 15]

Source tissue
Cell [4, 8]

Localisation in source
Mitochendria (soluble component of the mitochondrial intermembrane space [14]) [7, 14]

Purification
Saccharomyces cerevisiae [3, 4, 10]; Hansenula anomala [3, 5, 8]; Rhizopus oryzae [9]

Crystallization
(proteolytically cleaved enzyme with altered properties [3]) [1, 3, 4, 8]

Cloned
[12, 13]

Renaturated
–

5 STABILITY

pH

Temperature (°C)

Oxidation
  Photoinactivation [8]

Organic solvent

General stability information
  Low ionic strength: Saccharomyces cerevisiae enzyme is soluble but unstable [4]; Low ionic strength: Hansenula anomala enzyme exists as soluble inactive monomer [5]; Phenylmethylsulfonyl fluoride stabilizes during purification [8]

Storage
  4°C, as a suspension in 70% saturated ammonium sulfate solution under N_2, about 20% loss of activity after 2 months [4]; More [8]

6 CROSSREFERENCES TO STRUCTURE DATABANKS

PIR/MIPS code
  PIR2:S06600 (precursor yeast (Pichia anomala)); PIR1:CBBY2 (precursor yeast (Saccharomyces cerevisiae))

Brookhaven code
  1FCB (Yeast (Saccharomyces cerevisiae)); 1LTD (Yeast (Saccharomyces cerevisiae) recombinant form)

7 LITERATURE REFERENCES

[1] Nygaard, A.P. in " The Enzymes",2nd Ed. (Boyer, P.D., Lardy, H., Myrbäck, K., eds.)
     7,557–565 (1963) (Review)
[2] Lederer, F. in "Flavins and Flavoproteins1990" Proceedings of the Tenth Internatio-
     nal Symposium Como, Italy, July15–20 (Curti, B., Ronchi, S., Zanetti, G., eds.),
     773–782, Walter de Gruyter, Berlin, New York (1991) (Review)
[3] Labeyrie, F., Baudras, A., Lederer, F.: Methods Enzymol.,53,238–256 (1978)
     (Review)
[4] Jacq, C., Lederer, F.: Eur. J. Biochem.,41,311–320 (1974)
[5] Baudras, A., Spyridakis, A.: Biochimie,53,943ff. (1971)
[6] Jacq, C., Lederer, F.: Eur. J. Biochem.,25,41–48 (1972)
[7] Hatefi, Y., Stiggall, D.L. in "The Enzymes",3rd Ed. (Boyer,P.D., ed.) 13,175–297
     (1976) (Review)
[8] Blazy, B., Bardet, M., Baudras, A.: Anal. Biochem.,88,624–633 (1978)
[9] Pritchard, G.G.: Biochim. Biophys. Acta,250,25–34 (1971)
[10] Ghrir, R., Becam, A.-M., Lederer, F.: Eur. J. Biochem.,139,59–74 (1984)
[11] Genet, R., Lederer, F.: Biochem. J.,266,301–304 (1990)
[12] Black, M.T., Gunn, F.J., Chapman, S.K., Reid, G.A.: Biochem. J.,263,973–976
     (1989)
[13] Brunt, C.E., Cox, M.C., Thurgood, A.G.P., Moore, G.R., Reid, G.A., Chapman, S.K.:
     Biochem. J.,283,87–90 (1992)
[14] Daum, G., Böhni, P.C., Schatz, G.: J. Biol. Chem.,257,13028–13033 (1982)
[15] Tegoni, M., Janot, J.-M., Labeyrie, F.: Eur. J. Biochem.,190,329–342 (1990)
[16] Guiard, B., Groudinsky, O., Lederer, F.: Eur. J. Biochem.,34,241–247 (1973)

1 NOMENCLATURE

EC number
1.1.2.4

Systematic name
(R)-Lactate:ferricytochrome-c 2-oxidoreductase

Recommended name
D-Lactate dehydrogenase (cytochrome)

Synonymes
Lactic acid dehydrogenase
Dehydrogenase, D-lactate (cytochrome)
Cytochrome-dependent D-(-)-lactate dehydrogenase
D-Lactate-cytochrome c reductase [1]
D(-)-Lactic cytochrome c reductase [2]

CAS Reg. No.
37250-79-6

2 REACTION AND SPECIFICITY

Catalysed reaction
(R)-Lactate + 2 ferricytochrome c →
→ pyruvate + 2 ferrocytochrome c

Reaction type
Redox reaction

Natural substrates

Substrate spectrum
1 (R)-Lactate + 2 ferricytochrome c (ir [3, 5], specific for D-isomer [1]) [1–7]
2 DL-alpha-Hydroxybutyrate + cytochrome c [1–4]
3 DL-alpha-Hydroxyvalerate + cytochrome c (low activity) [1, 2]
4 DL-Glycerate + ferricyanide (at 10% the rate of D-lactate oxidation) [2]
5 More (enzyme from Peptostreptococcus elsdenii reduces ferricyanide dyes, O_2, cytochrome c [4], other acceptors: phenazine methosulfate [1–3], not: NAD^+ [4], $NADP^+$ [4], ferricyanide [1, 2], 2,6-dichlorophenolindophenol [1, 2], 2,3',6-trichlorophenolindophenol [1, 2], methylene blue [1, 2], suicide inactivation with alpha-hydroxy acid 2-hydroxy-3-butynoate as substrate [7]) [1–4, 7]

Product spectrum
1 Pyruvate + ferrocytochrome c
2 ?
3 ?
4 ?
5 ?

Inhibitor(s)
Fatty acids [1]; Glycerate [2]; Oxalate [2, 3, 5]; p-Chloromercuriphenylsulfonate [3]; EDTA [3, 5]; 1,10-Phenanthroline [3, 4]; p-Mercuriphenylsulfonate [5]; 2-Hydroxy-3-butynoate (suicide inactivation) [7]; Oxalacetate [1, 2]; Pyruvate [1]; alpha-Ketoglutarate [1]; Thiol reagents [1]; Acetic anhydride [1]; Mg^{2+} [1]; Ca^{2+} [1]; Monovalent cations [1]; More (not: atebrin [1], H_2O_2 [3, 5], L-lactate [6]) [1, 3, 5, 6]

Cofactor(s)/prosthetic group(s)/activating agents
FAD (flavoprotein [1–5], 1 mol FAD per 250000 MW protein [1], 1 mol FAD per 50000 MW protein [3, 5], possible presence of 2 mol flavin per enzyme molecule [3], 2 mol per mol of enzyme [5, 6]) [1–5, 6]

Metal compounds/salts
Phosphate (reaction velocity is linearly dependent on phosphate ion concentration, no requirement) [4]; Diphosphate (partial stimulation) [4]; Thiamine diphosphate (partial stimulation) [4]; Zn^{2+} (1 gatom per 22000–27000 MW protein [3, 5], Zn^{2+} moiety is very tightly bound [3], 4–6 gatom per mol of enzyme [5], 2 gatom per mol of enzyme [6]) [3, 5, 6]

Turnover number (min⁻¹)
More (15000 mol lactate per min and per mol of reducible flavin [2], 90000 mol lactate per min and per mol of flavin [5]) [2, 5]

Specific activity (U/mg)
More [3]

K_m-value (mM)
0.0054 (cytochrome c) [3]; 0.05 (D-lactate) [2]; 0.23 (ferricyanide) [4]; 0.285 (D-lactate) [3]; 1.34 (DL-alpha-hydroxybutyrate) [2]; 1.4 (D-alpha-hydroxybutyrate) [3]; 2.5 (DL-alpha-hydroxyvalerate) [2]; 4.45 (phenazine methosulfate) [3]; 50 (D-lactate) [4]; More [5, 6]

pH-optimum
5.5–6.0 (cytochrome assay, phosphate-acetate buffer) [3]; 7.0 (phosphate buffer, phenazine-2,6-dichlorophenolindophenol assay) [3]; 7.9 [4]; 8.0 (Tris buffer, cytochrome assay) [3]

pH-range

Temperature optimum (°C)
24 (assay at) [4]; 30 (assay at) [4]

Temperature range (°C)

---

3 ENZYME STRUCTURE

Molecular weight
100000 (Saccharomyces cerevisiae, measurement of sedimentation constant) [3, 5]
105000 (Megasphaera elsdenii, sedimentation equilibrium analysis) [6]

Subunits
Dimer (2 × 55000, Megasphaera elsdenii, SDS-PAGE) [6]

Glycoprotein/Lipoprotein
–

---

4 ISOLATION/PREPARATION

Source organism
Saccharomyces cerevisiae [2, 3, 5]; Megasphaera elsdenii [6, 7]; Yeast [1];
Peptostreptococcus elsdenii [4]

Source tissue
Cell [2, 3]

Localisation in source
More (associated with respiratory particles) [2, 3]; Mitochondria [5]

Purification
Peptostreptococcus elsdenii [4]; Saccharomyces cerevisiae [2, 3, 5]; Megasphaera elsdenii [6]

Crystallization
–

Cloned
–

Renaturated
–

5 STABILITY

pH
3.0 (30 s, stable) [3]; 6.5 (maximal stability at moderate ionic strength, 0.02–0.1 M phosphate) [3]; 7.5 (30 s, inactivation) [3]

Temperature (°C)

Oxidation

Organic solvent

General stability information
Dithiothreitol required for stability [6]; Very stable in the cold [3]

Storage

---

6 CROSSREFERENCES TO STRUCTURE DATABANKS

PIR/MIPS code
PIR2:S34813 (yeast (Saccharomyces cerevisiae))

Brookhaven code

---

7 LITERATURE REFERENCES

[1] Nygaard, A.P. in "The Enzymes",2nd Ed. (Boyer, P.D., Lardy, H., Myrbäck, K., eds.) 7,557–565 (1963) (Review)
[2] Nygaard, A.P.: J. Biol. Chem.,236,920–925 (1961)
[3] Gregolin, C., Singer, T.P.: Biochim. Biophys. Acta,67,201–218 (1963)
[4] Brockman, H.L., Wood, W.A.: Methods Enzymol.,41 PtB,309–312 (1975) (Review)
[5] Hatefi, Y., Stiggall, D.L. in "The Enzymes",3rd Ed. (Boyer, P.D., ed.) 13,175–297 (1976) (Review)
[6] Olson, S.T., Massey, V.: Biochemistry,18,4714–4723 (1979)
[7] Olson, S.T., Massey, V., Ghisla, S., Whitfield, C.D.: Biochemistry,18,4724–4731 (1979)

1 NOMENCLATURE

EC number
1.1.2.5

Systematic name
(R)-Lactate:ferricytochrome-c-553 2-oxidoreductase

Recommended name
D-Lactate dehydrogenase (cytochrome c-553)

Synonymes

CAS Reg. No.

2 REACTION AND SPECIFICITY

Catalysed reaction
D-Lactate + 2 ferricytochrome c-553 →
→ pyruvate + 2 ferrocytochrome c-553

Reaction type
Redox reaction

Natural substrates
D-Lactate + ferricytochrome c-553 [1]

Substrate spectrum
1 D-Lactate + ferricytochrome c-553 (artificial electron acceptors: 1-me-
thoxyphenazinium methyl sulfate, ferricyanide, tetrazolium dyes, methylene blue, 2,6-dichlorophenolindophenol, not: FAD, FMN, cytochrome c_3,
high molecular weight cytochrome, eukaryotic cytochrome c (yeast and
horse), O_2, NAD^+, $NADP^+$) [1]
2 DL-2-Hydroxybutyrate + ferricytochrome c-553 [1]

Product spectrum
1 Pyruvate + ferrocytochrome c-553 [1]
2 2-Ketobutyrate + ferrocytochrome c-553 [1]

Inhibitor(s)
1,10-Phenanthroline [1]; Cyanide [1]

Cofactor(s)/prosthetic group(s)/activating agents
Ferricytochrome c-553 [1]; EDTA (slight stimulation) [1]

Metal compounds/salts

Turnover number (min^{-1})

Specific activity (U/mg)

K$_m$-value (mM)
  0.8 (D-lactate) [1]; 2.4 (DL-2-hydroxybutyrate) [1]

pH-optimum
  7.5–9.0 (D-lactate + nitrotetrazolium blue) [1]

pH-range
  5.0 (not active below, D-lactate + nitrotetrazolium blue) [1]

Temperature optimum (°C)

Temperature range (°C)

---

3 ENZYME STRUCTURE

Molecular weight

Subunits

Glycoprotein/Lipoprotein
  –

---

4 ISOLATION/PREPARATION

Source organism
  Desulfovibrio vulgaris [1]

Source tissue
  Cell [1]

Localisation in source

Purification
  Desulfovibrio vulgaris (partially) [1]

Crystallization
  –

Cloned
  –

Renaturated
  –

5 STABILITY

pH

Temperature (°C)

Oxidation

Organic solvent

General stability information

Storage
-20°C, 2 days, 20% loss of activity [1]; 4°C, 0.1 M Tris-HCl, pH 7.3, 2 days,
20% loss of activity [1]

6 CROSSREFERENCES TO STRUCTURE DATABANKS

PIR/MIPS code

Brookhaven code

7 LITERATURE REFERENCES

[1] Ogata, M., Arihara, K., Yagi, T.: J. Biochem.,89,1423–1431 (1981)

1 NOMENCLATURE

EC number
1.1.3.3

Systematic name
(S)-Malate:oxygen oxidoreductase

Recommended name
Malate oxidase

Synonymes
Oxidase, malate
FAD-dependent malate oxidase
Malic oxidase
Malic dehydrogenase II

CAS Reg. No.
9028-73-3

---

2 REACTION AND SPECIFICITY

Catalysed reaction
(S)-Malate + O_2 →
→ oxaloacetate + H_2O_2

Reaction type
Redox reaction

Natural substrates

Substrate spectrum
1 (S)-Malate + O_2 (other electron acceptors: $K_3Fe(CN)_6$ [1, 2, 4, 5], vitamin K_1 [1], dichlorophenolindophenol [1, 3, 5], thiazolyl blue tetrazolium [1], phenazine methosulfate [5], not cytochrome c [1]) [1–5]
2 More (no substrates: succinate, fumarate, aspartate, oxaloacetate, 2-oxobutyrate, citrate, isocitrate) [1]

Product spectrum
1 Oxaloacetate + H_2O_2 [1]
2 ?

Inhibitor(s)
ATP [1]; ADP [1]; NAD^+ [1]; AMP [1]; Sodium amytal [5]

Cofactor(s)/prosthetic group(s)/activating agents
FAD (K_m: 0.009 mM in absence of phospholipids, 0.0002 mM in presence of phospholipids [1], K_m: 0.0004 mM [5], not replaceable by FMN [5]) [1, 5]; Phospholipids (activation [1], required by purified enzyme [5], K_m: 0.006 mM (phosphoethanolamine) [5]) [1, 5]; Quinone (required by purified enzyme, K_m: 0.02 mM (coenzyme Q9), 3.0 mM (vitamin K_3)) [5] Triton X-100 (activation) [1]

Metal compounds/salts

Turnover number (min^{-1})

Specific activity (U/mg)
92.8 [1]; 9.3 [5]; 0.108 [4]; More (assay method) [5]

K_m-value (mM)
0.45 (L-malate) [5]

pH-optimum

pH-range

Temperature optimum (°C)

Temperature range (°C)

---

3 ENZYME STRUCTURE

Molecular weight

Subunits

Glycoprotein/Lipoprotein
–

---

4 ISOLATION/PREPARATION

Source organism
E. coli (mutant strain lacking malate dehydrogenases EC 1.1.1.38 and EC 1.1.1.40) [1, 4]; Arthrobacter crystallopoietes [3]; Micrococcus lysodeikticus [2]; Pseudomonas ovalis (strain Chester, lacking malate dehydrogenase EC 1.1.1.37 and EC 1.1.1.40) [5]

Source tissue

Localisation in source
Cytoplasmic membrane (inside face) [1]; Cell-wall membrane [5]

Purification
Micrococcus lysodeikticus [2]; E. coli (partial) [1]; Pseudomonas ovalis [5]

Crystallization
–

Cloned
–

Renaturated
–

5 STABILITY

pH

Temperature (°C)
20 (24 h, 85% loss of activity) [5]

Oxidation

Organic solvent

General stability information
FAD and phospholipids together protect against proteolytic inactivation [1]

Storage
0°C, 24 h, 50% loss of activity [5]; –15°C, 24 h, 70% loss of activity [5]

6 CROSSREFERENCES TO STRUCTURE DATABANKS

PIR/MIPS code

Brookhaven code

7 LITERATURE REFERENCES

[1] Narindrasorasak, S., Goldie, A.H., Sanwal, B.D.: J. Biol. Chem.,254,1540–1545 (1979)
[2] Cohn, D.V.: J. Biol. Chem.,233,299–304 (1958)
[3] Meganathan, R., Ensign, J.C.: J. Gen. Microbiol.,94,90–96 (1976)
[4] Goldie, A.H., Narindrasorasak, S., Samwal, B.D.: Biochem. Biophys. Res. Commun.,83,421–426 (1978)
[5] Phizackerley, P.J.R.: Methods Enzymol.,13,135–140 (1969)

Enzyme Handbook © Springer-Verlag Berlin Heidelberg 1995
Duplication, reproduction and storage in data banks are only
allowed with the prior permission of the publishers

1 NOMENCLATURE

EC number
1.1.3.4

Systematic name
beta-D-Glucose:oxygen 1-oxidoreductase

Recommended name
Glucose oxidase

Synonymes
Oxidase, glucose
Corylophyline
Notatin
Penatin
Glucose aerodehydrogenase
Microcid
beta-D-Glucose oxidase
D-Glucose oxidase
D-Glucose-1-oxidase
beta-D-Glucose:quinone oxidoreductase
Glucose oxyhydrase
Deoxin-1
GOD [43]

CAS Reg. No.
9001-37-0

2 REACTION AND SPECIFICITY

Catalysed reaction
beta-D-Glucose + O_2 →
→ D-glucono-1,5-lactone + H_2O_2 (stereochemistry [23], mechanism [39, 40])

Reaction type
Redox reaction

Natural substrates

Substrate spectrum
1 beta-D-Glucose + O_2 (highly specific [22, 31]) [1–43]
2 beta-D-Glucose + benzoquinone [14]
3 alpha-D-Glucose + O_2 (very slow reaction) [22]
4 L-Sorbose + O_2 (5.8% of D-glucose reactivity) [1, 2]
5 D-Xylose + O_2 (4.8% of D-glucose reactivity) [1, 2]
6 D-Maltose + O_2 (4.5% of D-glucose reactivity) [1, 2]
7 2-Deoxy-D-glucose + O_2 [6, 22]

Product spectrum
1 D-Glucono-1,5-lactone + H_2O_2
2 D-Glucono-1,5-lactone + hydroquinone [14]
3 D-Glucono-1,5-lactone + H_2O_2
4 ?
5 ?
6 ?
7 2-Deoxy-D-glucono-1,5-lactone

Inhibitor(s)
Ag^+ [1, 2]; o-Phthalate [1, 2]; F^- (competitive, at low pH) [20]; Cl^- (competitive, at low pH) [20]; Br^- (competitive, at low pH) [20]; 8-Hydroxyquinoline [23]; Sodium nitrate [23]; Semicarbazide [23, 35]; Dimedon (partial) [23]; Phenylhydrazine (partial) [23, 31]; Hydrazine (partial) [23]; Hydroxylamine (partial) [23]; Sodium bisulfite (partial) [23]; $FeSO_4$ (inhibition of enzyme production) [31]; $HgCl_2$ [31]; $CuSO_4$ [31]; $NaHSO_3$ [31]; CN^- (honey enzyme only) [35]; Putrescine (i.e. 1,4-diaminobutane) [38]; Adenine nucleotides (inhibition of FAD-binding to apoprotein) [42]

Cofactor(s)/prosthetic group(s)/activating agents
FAD (2 mol per mol of enzyme [1], kinetic behavior [20], electron transfer between FAD centers and metal electrodes after chemical modification of enzyme [34], not detectable in honey enzyme [35]) [1, 9, 19, 20, 22, 34]; Flavin-hypoxanthine dinucleotide (FHD, can substitute FAD) [19]

Metal compounds/salts
Na^+ (required for maximal activity, honey enzyme) [35]

Turnover number (min^{-1})
20200 (glucose, native enzyme) [18]; 19400 (glucose, periodate-oxidized enzyme) [18]; More (pH-dependence) [41]

Specific activity (U/mg)
15.1 (oxidation of o-dianisidine at 37°C) [1, 2]; 172 (oxygen consumption/min/mg at 30°C); More (assay method [12, 13], continous flow determination [4]) [4, 8, 9, 12, 13, 22, 26, 35]

K_m-value (mM)
37–38 (D-glucose, native and deglycosylated enzyme [5]) [1, 2, 5]; 0.95
(O_2) [1, 2]; 11 (D-glucose, Penicillium amagasakiense, similar value [28])
[29]; 217. 4 (L-sorbose) [1, 2]; 105.2 (D-xylose) [1, 2]; 55.5 (D-maltose) [1,
2]; 30 (D-glucose, native enzyme) [33]; 35 (D-glucose, deglycosylated enzy-
me) [33]; 25.2 (D-glucose, soluble enzyme) [3]; 4.0–5.4 (D-glucose, immobilized enzyme) [5]; 33 (D-glucose, soluble enzyme) [10]; 44 (D-glucose, immobilized enzyme) [10]; 1.51–3.4 (D-glucose, depending on O_2-concentrati-
on, comparison of values with enzyme immobilized on various materials)
[16]; 26–28 (D-glucose, native and periodate-oxidized enzyme) [18];
0.18–0.2 (O_2, native and periodate-oxidized enzyme) [18]; More (values for
glyco-GOD and aglyco-GOD) [43]

pH-optimum
4.6–5.0 [1]; 5.0 [14, 26]; 5.0–6.0 (native and deglycosylated enzyme) [5];
5.2–7 (immobilized on silk fibroin) [3]; 5–6 [9]; 5.5–6 [33]; 5.5 (immobilized
on polyacrylamide [10], glyco-GOD [43]) [10, 16, 22, 43]; 5.6 [6, 23]; 5.6
(pH-dependence of individual reaction steps) [37]; 5.7 (soluble enzyme) [3];
5.8 [8]; 6.0 (aglyco-GOD) [43]; 6.1 [35]; 6.2 (immobilized on activated carbon) [16]; 5.9 [21]

pH-range
2.5–9 (immobilized on activated carbon) [16]; 3.4–7.5 (soluble enzyme)
[16]; 4.0–7.0 (more than 90% of activity maximum at pH 4.0 and 7.0) [33]

Temperature optimum (°C)
30 [8]; 40 [10, 21, 35]; 40–60 [33]; 45 [6]

Temperature range (°C)

3 ENZYME STRUCTURE

Molecular weight
175000–180000 (Phanerochaete chrysosporium, gel filtration [1, 2], Penicillium chrysogenum, gel filtration [9]) [1, 2, 9]
158000–160000 (Aspergillus niger, gel filtration [19], Penicillium amagasakiense, sedimemtation equilibrium centrifugation [26]) [19, 26]
150000–153000 (Aspergillus niger, sedimentation and diffusion data [18,
22], periodate-oxidized enzyme [18]) [18, 22]
130000–150000 (Penicillium amagasakiense, aglyco-GOD, gel filtration, native PAGE) [43]

3

Subunits
Tetramer (4 × 45000, Penicillium amagasakiense, 2 polypeptide chains linked by disulfide bond, sedimentation equilibrium centrifugation, treatment with guanidine-HCl and 2-mercaptoethanol) [26]
Dimer (2 × 72000–80000, identical, Phanerochaete chrysosporium, SDS-PAGE [1], Penicillium chrysogenum, SDS-PAGE [9], Aspergillus niger, SDS-PAGE [19], Penicillium vitale [27], 2 × 60000, Penicillium amagasakiense, aglyco-GOD, SDS-PAGE [43], 2 × 70000, Penicillium amagasakiense, glyco-GOD, SDS-PAGE [43]) [1, 9, 19, 27, 43]

Glycoprotein/Lipoprotein
Glycoprotein (N- and O-linked sugar chains [5], carbohydrate composition [18, 19], no significant differences in catalytic properties of glyco- and aglyco-GOD [43]) [5, 9, 18, 19, 22, 32, 43]; No glycoprotein [1, 2]

---

4 ISOLATION/PREPARATION

Source organism
Aspergillus niger (culture conditions [6]) [4–7, 10–13, 15, 16, 18–22, 24, 30, 32–34, 36, 38–42]; Phanerochaete chrysosporium [1, 2]; Bombyx mori [3]; Penicillium vitale [8, 27]; Penicillium chrysogenum [9, 28, 31]; Aspergillus sp. No.319 [17, 19]; Penicillium amagasakiense [20, 23, 25, 26, 32, 43]; Penicillium notatum [11, 31, 37]; Penicillium paxilli [14]; Penicillium sp. (8 species) [31]; Talaromyces stipitatus [31]; Penicillium purpurogenum [29, 31]; More (northeastern fall-flower honey) [35]

Source tissue
Mycelia [1, 2, 22]; Culture medium [26, 31]; Commercial preparations [12, 13, 15, 19, 26, 33, 37, 38]; Honey (northeastern fall-flower) [35]

Localisation in source
Periplasmic and peroxisome-like structures [1]; Extracellular (in all Penicillium sp.); Intracellular (in all Aspergillus sp.); Peroxysomes (Aspergillus) [30]

Purification
Phanerochaete chrysosporium [1, 2]; Aspergillus niger (partial [6], from commercial preparation [19, 33]) [6, 19, 22, 24, 33]; Penicillium chrysogenum [9, 28]; Penicillium amagasakiense (from commercial preparation [26], isolation of glyco-GOD and aglyco-GOD from tunicamycin containing growth medium [43]) [23, 25, 26, 43]; Penicillium purpurogenum (partial) [31]

Crystallization
[23, 25, 33]

Cloned
–

Renaturated

–

---

5 STABILITY

pH

1.9 (1 h, bound to Blue Dextran) [15]; 3.0–5.0 (deglycosylated enzyme, 4°C, 4 months, 65–75% activity retained) [33]; 3.5–7.5 [9]; 4.0 (rapid inactivation below) [6]; 5.0–7.0 [31]; 5–8 (native and deglycosylated form [5]) [3, 5]; 5.5 (50°C, most stable, immobilized enzyme) [10]; 6.0 (50°C, most stable, soluble enzyme) [10]; 6.0–8.0 [6]; 8 (unstable above) [23]; More (comparison of soluble and immobilized enzyme) [8]

Temperature (°C)

35 (stable up to) [43]; 40 (60 min [6], up to, soluble enzyme [10, 23]) [6, 10, 23]; 45 (bound to Blue Dextran, 3 h, 60% activity [15], 30 min, considerable inactivation [43]) [15, 43]; 50 (up to, immobilized enzyme [10], up to, native and deglycosylated enzyme [33]) [10, 33]; 55 (inactivation above) [5, 8]; 60 (complete inactivation) [35]; 65 (soluble enzyme: inactivation, mycelia-bound: 85% activity retained) [17]; 72.4 (denaturation point of native enzyme) [18]; 72.8 (denaturation point of periodate-oxidized enzyme) [19]; More (glucose: stabilization against heat inactivation [31], comparison of stability of enzyme from different sources [32]) [31, 32]

Oxidation

Organic solvent

General stability information

SDS, 5%, stable to [5]; Freezing/thawing, stable [5]; Urea: 7 M, 5 min, no inactivation [18]; Erythritol stabilizes [7]; Sorbitol stabilizes [7]; Xylitol stabilizes [7]; Polyethylene glycol stabilizes [7]; KCl stabilizes [7]; NaCl stabilizes [7]; D_2O stabilizes [7]; Comparison of stability immobilized on various materials [21]

Storage

Frozen, at least 1 week [1]; 4°C, pH 7, immobilized, 4 months [3]; 4–5°C, immobilized, 8 months, 90% activity retained [8]; 4°C, immobilized, 1 year, 9% loss of activity [10]; 4°C, immobilized, 3 months [17]; Lyophilized, over P_2O_5 [19]; 3°C, purified enzyme, several years [22]; –15°C crystalline suspension, 8 years [23]

6 CROSSREFERENCES TO STRUCTURE DATABANKS

PIR/MIPS code
PIR3:S14129 (Aspergillus niger); PIR2:A35459 (precursor Aspergillus niger);
PIR2:S05668 (precursor Aspergillus niger)
Brookhaven code
1GAL (Aspergillus niger)

7 LITERATURE REFERENCES

[1]   Kelley, R.L., Reddy, C.A.: Methods Enzymol.,161,307–316 (1988)
[2]   Kelley, R.L., Reddy, C.A.: J. Bacteriol.,166,269–274 (1986)
[3]   Demura, M., Asakura, T.: Biotechnol. Bioeng.,33,598–603 (1989)
[4]   Okuma, H., Sekimukai, S., Hoshi, M., Toyama, K., Watanabe, E.: Enzyme Microb.
      Technol.,11,824–829 (1989)
[5]   Takegawa, K., Fujiwara, K., Iwahara, S., Yamamoto, K., Tochikura, T.: Biochem. Cell
      Biol.,67,460–464 (1989)
[6]   Rogalski, J., Fiedurek, J., Szczordrak, J., Kapusta, K., Leonowicz, A.: Enzyme
      Microb. Technol.,10,508–511 (1988)
[7]   Ye, W.N., Combes, D., Monsan, P.: Enzyme Microb. Technol.,10,498–502 (1988)
[8]   Kozhukharova, A., Kirova, N., Popova, Y., Batsalova, K.: Biotechnol.
      Bioeng.,32,245–248 (1988)
[9]   Eriksson, K.-O., Kourteva, I., Yao, K., Liao, J.-L., Kilar, F., Hjerten, S.: J. Chromatogr.,
      397,239–249 (1987)
[10] Szajani, B., Molnar, A., Klamar, G., Kalman, M.: Appl. Biochem. Biotechnol.,14,
      37–47 (1987)
[11] Fiedurek, J., Rogalski, J., Ilczuk, Z., Leonowicz, A. : Enzyme Microb. Technol.,8,
      734–736 (1986)
[12] Woodward, J., Wagner, M., Lennon, K.W., Zanin, G., Scott, M.: Enzyme Microb.
      Technol.,7,449–453 (1985)
[13] Kunst, A., Draeger, B., Ziegenhorn, J. in "Methods Enzym. Anal.",3rd Ed
      (Bergmeyer, H.U., ed.) 6,178–185 (1984)
[14] Alberti, B.N., Klibanov, A.M.: Enzyme Microb. Technol.,4,47–49 (1982)
[15] Solomon, B., Lotan, N., Katchalski-Katzir, E.: J. Chromatogr.,215,121–129 (1981)
[16] Cho, Y.K., Bailey, J.E.: Biotechnol. Bioeng.,20,1651–1665 (1978)
[17] Karube, I., Hirano, K.-I., Suzuki, S.: Biotechnol. Bioeng.,19,1233–1238 (1977)
[18] Nakamura, S., Hayashi, S., Koga, K.: Biochim. Biophys. Acta,445,294–308 (1976)
[19] Tsuge, H., Natsuaki, O., Ohashhi, K.: J. Biochem.,78,835–843 (1975)
[20] Bright, H.J., Porter, D.J.T. in "The Enzymes",3rd Ed. (Boyer, P.D., ed.) 12,421–505
      (1975) (Review)
[21] Constantinides, A., Vieth, W.R., Fernandes, P.M.: Mol. Cell. Biochem.,1,127–133
      (1973)
[22] Pazur, J.H.: Methods Enzymol.,9,82–87 (1966)
[23] Bentley, R. in "The Enzymes",2nd Ed. (Boyer, P.D., Lardy, H., Myrbäck, K., eds.)
      7,567–586 (1963) (Review)
[24] Swoboda, B.E.P., Massey, V.: J. Biol. Chem.,240,2209–2215 (1965)

[25] Kusai, K., Sekuzu, I., Hagihara, B., Okumuki, K., Yamaguchi, S., Nakai, M.: Biochim. Biophys. Acta,40,555–557 (1960)
[26] Yoshimura, T., Isemura, T.: J. Biochem.,69,839–846 (1971)
[27] Abalikhina, T.A., Morozkin, A.D., Bogdanov, V.P., Kaverznera, E.D:: Biokhimiya,36, 191 ff. (1971)
[28] Chaga, S.: Nauchn. Tr. Nauchno-Issled. Inst. Konserv. Prom Plovdiv,10,173 (1973)
[29] Nakamatsu, T., Akamatsu, R., Miyajima, R., Shio, I.: Agric. Biol. Chem.,39,1803–1811 (1975)
[30] VanDijken, J.P., Veenhuis, M.: Eur. J. Appl. Microbiol. Biotechnol.,9,275–283 (1980)
[31] Nakamatsu, T., Akamatsu, T., Miyajima, R., Shio, I.: Agric. Biol. Chem.,39,1803–1811 (1975)
[32] Hayashi, S., Nakamura, S.: Biochim. Biophys. Acta,438,37–48 (1976)
[33] Kalisz, H.M., Hecht, H.-J., Schomburg, D., Schmid, R.D.: J. Mol. Biol.,213,207–209 (1990)
[34] Degani, Y., Heller, A.: J. Am. Chem. Soc.,110,2615–2620 (1988)
[35] Schepartz, A.I., Subers, M.H.: Biochim. Biophys. Acta,85,228–237 (1964)
[36] Sanner, C., Macheroux, P., Rüterjans, H., Müller, F., Bacher, A.: Eur. J. Biochem., 196,663–672 (1991)
[37] Bright, H.J., Appleby, M.: J. Biol. Chem.,244,3625–3634 (1969)
[38] Voet, J.G., Andersen, E.C.: Arch. Biochem. Biophys.,233,88–92 (1984)
[39] Chan, T.W., Bruice, T.C.: J. Am. Chem. Soc.,99,2387–2389 (1977)
[40] Duke, F.R., Weibel, M., Page, D.S., Bulgrin, V.G., Luthy, J.: J. Am. Chem. Soc.,91,3904–3909 (1969)
[41] Weibel, M.K., Bright, H.J.: J. Biol. Chem.,246,2734–2744 (1971)
[42] Swoboda, B.E.P.: Biochim. Biophys. Acta,175,380–387 (1969)
[43] Kim, J.M., Schmid, R.D.: FEMS Microbiol. Lett.,78,221–226 (1991)

1 NOMENCLATURE

EC number
1.1.3.5

Systematic name
D-Hexose:oxygen 1-oxidoreductase

Recommended name
Hexose oxidase

Synonymes

CAS Reg. No.
9028-75-5

2 REACTION AND SPECIFICITY

Catalysed reaction
beta-D-Glucose + O_2 →
→ D-glucono-1,5-lactone + H_2O_2

Reaction type
Redox reaction

Natural substrates
D-Glucose + O_2 [1–3]
D-Galactose + O_2 [1–3]

Substrate spectrum
1 D-Glucose + O_2 [1–4]
2 D-Galactose + O_2 [1–4]
3 Maltose + O_2 [1–4]
4 Cellobiose + O_2 [1–4]
5 Lactose + O_2 [1–4]
6 D-Glucose 6-phosphate + O_2 [1]
7 D-Mannose + O_2 [1, 3]
8 2-Deoxy-D-glucose + O_2 [1, 3]
9 2-Deoxy-D-galactose + O_2 [1]
10 D-Xylose + O_2 [3]
11 D-2-Glucosamine + O_2 [3]

Enzyme Handbook © Springer-Verlag Berlin Heidelberg 1995
Duplication, reproduction and storage in data banks are only
allowed with the prior permission of the publishers

Product spectrum
  1 D-Glucono-1,5-lactone + H_2O_2 [1–4]
  2 gamma-D-Galactonolactone + H_2O_2 [1–4]
  3 ?
  4 ?
  5 ?
  6 D-Glucono-1,5-lactone 6-phosphate + H_2O_2
  7 D-Mannono-1,5-lactone + H_2O_2
  8 2-Deoxy-D-glucono-1,5-lactone + H_2O_2
  9 2-Deoxy-D-galactono-1,5-lactone + H_2O_2
 10 ?
 11 D-Glucono-2-amine-1,5-lactone + H_2O_2

Inhibitor(s)
  Diethyldithiocarbamate [1, 2]; Cyanide [1, 2]; Hydroxylamine [1, 2]; Azide
  [1, 2]; Acetate [1, 2, 4]; Pyruvate [1, 2, 4]; Cu^{2+} [3]; Hg^{2+} [3]; Ag^{2+} [3]; Ba^{2+}
  [3]; Glucoronic acids [3]; Galacturonic acids [3]; Propionate [4]; Benzoate
  [4]

Cofactor(s)/prosthetic group(s)/activating agents

Metal compounds/salts
  Cu^{2+} (12 mol per mol enzyme) [1, 2]

Turnover number (min⁻¹)

Specific activity (U/mg)
  4.095 [1, 2]; 0.3 [4]

K_m-value (mM)
  2.5–4 (glucose) [1, 2, 4]; 5–8 (galactose) [1, 2, 4]

pH-optimum
  6.3 [1, 2]; 5.0 [4]

pH-range

Temperature optimum (°C)
  25 [1, 2]

Temperature range (°C)

3 ENZYME STRUCTURE

Molecular weight
  130000 (Chondrus crispus, gel filtration) [1, 2]

Subunits

Glycoprotein/Lipoprotein
Glycoprotein (70% carbohydrates, mainly galactose and xylose) [1, 2]

4 ISOLATION/PREPARATION

Source organism
Chondrus crispus (Irish moss, red alga) [1, 2]; Iridophycus flaccidum (red alga) [1–4]; Euthora cristata (red alga) [2]; Citrus sinensis (orange) [3]; Citrus macrophylla [3]; Citrus junios [3]; Citrus vulgaris [3]; Citrus macroptera [3]; Citrus aurantifolia [3]; Citrus aurantium [3]; Citrus limetta [3]

Source tissue
Fruits [3]

Localisation in source

Purification
Chondrus crispus [1, 2]; Iridophycus flaccidum (partially) [4]

Crystallization
–

Cloned
–

Renaturated
–

5 STABILITY

pH
7.5 (slow denaturation above) [4]

Temperature (°C)
50–60 (not stable above) [2]

Oxidation

Organic solvent

General stability information

Storage
–10°C, 7 months [4]

6 CROSSREFERENCES TO STRUCTURE DATABANKS

PIR/MIPS code

Brookhaven code

7 LITERATURE REFERENCES

[1] Ikawa, M.: Methods Enzymol.,89,145–149 (1982)
[2] Sullivan jr., J.D., Ikawa, M.: Biochim. Biophys. Acta,309,11–22 (1973)
[3] Bean, R.C., Porter, G.G., Steinberg, B.M.: J. Biol. Chem.,236,1235–1240 (1961)
[4] Bean, R.C., Hassid, W.Z.: J. Biol. Chem.,218,425–436 (1956)

1 NOMENCLATURE

EC number
1.1.3.6

Systematic name
Cholesterol:oxygen oxidoreductase

Recommended name
Cholesterol oxidase

Synonymes
Oxidase, cholesterol
Cholesterol-O_2 oxidoreductase [10]
3beta-Hydroxy steroid oxidoreductase [5, 22]
3beta-Hydroxysteroid:oxygen oxidoreductase [5]

CAS Reg. No.
9028-76-6

2 REACTION AND SPECIFICITY

Catalysed reaction
Cholesterol + O_2 →
→ cholest-4-en-3-one + H_2O_2 (mechanism [24])

Reaction type
Redox reaction

Natural substrates
Cholesterol + O_2 (first step in cholesterol degradation) [1, 24]

Substrate spectrum
1 Cholesterol + O_2 (r [14]) [1–24]
2 beta-Sitosterol + O_2 [5, 12, 13, 24]
3 Fucosterol + O_2 [13]
4 Dehydroepiandrosterone + O_2 (r [14]) [5, 7, 12, 14, 16]
5 Pregnenolone + O_2 (r [14]) [5, 7, 12, 14, 16]
6 Stigmasterol + O_2 [12, 17, 18, 24]
7 Lanosterol + O_2 (slow rate [16], not [17]) [16]
8 Ergosterol + O_2 (low activity [17], not [16]) [17, 18, 24]
9 Stigmastanol + O_2 [18]
10 Dihydrocholesterol + O_2 [18]
11 5alpha-Cholestan-3beta-ol + O_2 [18]
12 Desmosterol + O_2 [24]

13 Campesterol + O_2 [24]
14 7E,22-Ergostadien-3beta-ol + O_2 [24]
15 26-Hydroxycholesterol + O_2 [10]
16 25-Hydroxy-27-norcholesterol + O_2 [10]
17 5alpha-Cholestan-3beta-ol + O_2 [10, 24]
18 5alpha-Cholest-8(14)-en-3beta-ol + O_2 [10]
19 17beta-Hydroxymethyl-5-androsten-3beta-ol + O_2 [10]
20 5-Pregnen-3beta,20beta-diol + O_2 [10]
21 5-Cholen-3beta,24-diol + O_2 [10]
22 5,7-Cholestadien-3beta-ol + O_2 [10]
23 4,4-Dimethylcholesterol + O_2 [10]
24 23,24-Dinor-5-cholen-3beta,22-diol + O_2 [10]
25 4alpha-Methylcholesterol + O_2 [10]
26 5-Androstene-3beta,17beta-diol + O_2 [10]
27 20,25-Diazacholesterol + O_2 [10]
28 5-Cholene-3beta,24-diol + O_2 [10]
29 More (1,5-cholesten-3-one is an intermediate [14], enzyme specifically
     oxidizes 3beta-hydroxy groups in steroids [18], specificity in oxidation of
     3-hydroxy steroids [10], overview: substrate specificity [23, 24], size and
     shape of steroid 17beta-side-chain and oxygenation of nucleus [23], re-
     latively unreactive 17keto-DELTA[5]–3beta-hydroxysteroids are converted
     to satisfactory substrates by formation of isopentyloximes or benzyloxi-
     mes [23]) [18, 23, 24]

Product spectrum
1 Cholest-4-en-3-one + H_2O_2
2 Stigmast-5-en-3-one + H_2O_2
3 Stigmasta-5,24(28)-dien-3-one + H_2O_2
4 Androst-5-en-3,17-dione + H_2O_2
5 5-Pregnen-3,20-dione + H_2O_2
6 Stigmasta-5,22-dien-3-one + H_2O_2
7 Lanosta-8,24-dien-3-one + H_2O_2
8 Ergost-8(14)-en-3-one
9 ?
10 ?
11 5alpha-Cholestan-3-one + H_2O_2
12 Cholesta-5,24-dien-3-one + H_2O_2
13 (24R)-Ergost-5-en-3-one + H_2O_2
14 7E,22-Ergostadien-3-one + H_2O_2
15 26-Hydroxycholest-5-en-3-one + H_2O_2
16 ?
17 5alpha-Cholestan-3-one + H_2O_2
18 5alpha-Cholest-8(14)-en-3-one + H_2O_2
19 17beta-Hydroxymethyl-5-androsten-3-one + H_2O_2
20 5-Pregnen-20beta-ol-3-one + H_2O_2

21 5-Cholan-24-ol-3-one + H_2O_2
22 5,7-Cholestadien-3-one + H_2O_2
23 4,4-Dimethylcholest-4-en-3-one + H_2O_2
24 23,24-Dinor-5-cholen-22-ol-3-one + H_2O_2
25 4alpha-Methylcholest-4-en-3-one + H_2O_2
26 5-Androstan-17beta-ol-3-one + H_2O_2
27 20,25-Diazacholest-4-en-3-one + H_2O_2
28 5-Cholen-24-ol-3-one + H_2O_2
29 ?

Inhibitor(s)

p-Chloromercuribenzoate (not [7]) [5]; $AgNO_3$ [5, 7]; Cholest-4-en-3-one [1];
Trinitrobenzene sulfonate [7]; Sulfhydryl reagents [17]; $HgCl_2$ [1, 5, 7, 17];
Hg^{2+} [18]; Iodine [7]; Fe^{2+} [18]; $FeCl_3$ [7]; Zn^{2+} [18]; $CuSO_4$ (slight) [7]; KCN
[7]; NaN_3 [7]; N-Bromosuccinimide [7]; 5alpha-Cholestan-3beta-ol [7];
5alpha-Cholestan-3-one [7]; 5beta-Cholestan-3beta-ol [7];
5alpha-Cholestan-3beta,5alpha-diol [7]; 5alpha-Lanosta-8,24-dien-3beta-ol
[7]; 1-Fluoro-2,4-dinitrobenzene [7]; 5,10-Seco-19-nor-5-cholestyn-3,10-dione
[14]; Sodium deoxycholate [16]; Triton X-100 (stimulates at concentration up
to 0.2% [17], inactivates at a concentration of 0.02% [16]) [16]; More (overview: inhibition by steroids) [7]

Cofactor(s)/prosthetic group(s)/activating agents

Triton X-100 (stimulates at concentration up to 0.2% [17], inactivates at a
concentration of 0.02% [16]) [17]; Bile salts (activate) [19]; FAD (flavoprotein [5, 15, 16, 22], firmly bound [15, 16], 1 mol of FAD per mol of protein
[5], exogenous addition increases activity [18]) [5, 15, 16, 18, 22];
Long-chain alcohols (activate, maximum activation with decanol and undecanol, shorter and longer alcohols less effective) [12]; Cholic acid (stimulates at low concentrations of Triton X-100 without a significant change in K_m,
at high concentrations of Triton X-100 it changes the sigmoidal shape into
the normal Michaelis-Menten relationship with reduced K_m) [15]

Metal compounds/salts

Mn^{2+} (activates) [18]; No metal content [5]

Turnover number (min^{-1})

Specific activity (U/mg)

2.42 (inducible enzyme) [17]; 3.65 (constitutive enzyme) [17]; 0.6 (polyethylene glycol derivative-modified enzyme) [13]; 27.9 [18]; 17.0 [22]; 19.7 [7];
0.045–0.069 (cholesterol) [1]

K$_m$-value (mM)
0.0002 (cholesterol) [12, 18]; 0.00033 (cholesterol) [16]; 0.00045 (stigmasterol) [12]; 0.0005 (pregnenolone, beta-sitosterol) [12]; 0.001 (dehydroepiandrosterone) [12]; 0.0012 (26-hydroxycholesterol, 5-androstene-3beta,17beta-diol) [10]; 0.0014 (25-hydroxy-27-norcholesterol) [10]; 0.0017 (3beta-hydroxy-5-pregnen-20-one) [14]; 0.0023 (5alpha-cholestan-3beta-ol) [10, 24]; 0.0028 (5alpha-cholest-8(14)-en-3beta-ol) [10]; 0.0029 (cholesterol) [24]; 0.0031 (17beta-hydroxymethyl-5-androsten-3beta-ol) [10]; 0.00344 (cholesterol) [10]; 0.0036 (3beta-hydroxy-5-androsten-17-one) [14]; 0.0037 (5-pregnene-3beta,20beta-diol) [10]; 0.0048 (5-cholene-3beta,24-diol) [10]; 0.0051 (cholesterol) [14]; 0.0055 (5,7-cholestadien-3beta-ol) [10]; 0.0070 (5-cholesten-3-one) [14]; 0.0075 (4,4-dimethylcholesterol) [10]; 0.0093 (23,24-dinor-5-cholene-3beta,22-diol) [10]; 0.0107 (5alpha-cholest-8(14)en-3beta-ol) [10]; 0.0125 (5-pregnene-3,20-dione) [14]; 0.0126 (4alpha-methylcholesterol) [10]; 0.0284 (4beta-methylcholesterol) [10]; 0.124 (20,25-diazacholesterol) [10]; 0.548 (5-androstene-3,17-dione) [14]; More (K$_m$-values in presence and absence of undecanol [12], effect of bile salts on K$_m$ [19], increased K$_m$ of enzyme in AOT-isooctane reverse micelles [20], influence of cholic acid [15]) [7, 12, 15, 17, 19, 20, 23, 24]

pH-optimum
4–5 [16]; 6.0–8.0 (extracellular enzyme) [21]; 7.0 [1, 18]; 7.5 (enzyme after micellization [20]) [5, 7, 15, 17, 20]; 8.0 (membrane-bound enzyme) [21]

pH-range
4.5–8 (4.5: about 50% of activity maximum, 8: about 90% of activity maximum) [1]; 5.0–9.0 (33% of activity maximum at pH 5.0 and 9.0) [18]; 5–9.5 (about 50% of activity maximum at pH 5 and 9.5) [5]; 6.0–9.5 (6.0: 80% of activity maximum with inducible enzyme, 70% of activity maximum with constitutive enzyme, 9.5: about 50% of activity maximum) [17]

Temperature optimum (°C)
30 [1]; 40 (enzyme after micellization [20]) [20, 21]; 50 [7, 17]

Temperature range (°C)
10–80 (about 25% of activity maximum at 10°C and 80°C) [7]; 13–44 (about 50% of activity maximum at 13°C and 44°C) [1]; 20–70 (about 50% of activity maximum at 20°C and 70°C) [7]; 33–60 (33°C: about 50% of activity maximum, 60°C: about 50% of activity maximum of inducible enzyme, about 70% of activity maximum of constitutive enzyme) [17]

3 ENZYME STRUCTURE

Molecular weight
30000 (Streptomyces violascens, gel filtration) [7]
32500 (Brevibacterium sterolicum, gel filtration, sedimentation equilibrium method) [5, 15]
53000 (Schizophyllum commune, SDS-PAGE, sedimentation equilibrium, amino acid analysis) [16]
56000 (Pseudomonas sp., SDS-PAGE, gel filtration) [18]
65100 (Pseudomonas sp., analytical ultracentrifugation) [19]

Subunits
Monomer (1 × 53000, Schizophyllum commune, SDS-PAGE [16], 1 × 56000, Pseudomonas sp., SDS-PAGE [18]) [16, 18]

Glycoprotein/Lipoprotein
More (3 enzyme forms differ chiefly in the possession or absence of hydro-phobic anchor region connected by a trypsin-sensitive region, no phospholi-pids are extracted with the enzyme) [1]

---

4 ISOLATION/PREPARATION

Source organism
Rhodococcus equi [21]; Streptomyces violascens [7, 11, 24]; Schizophyl-lum commune [12, 16]; Nocardia rhodochrous (3 forms [1]) [1, 4]; Arthrob-acter simplex [2, 17]; Nocardia erythropolis [3, 10, 14, 20, 23, 24]; Brevibac-terium sterolicum [5, 15, 22, 24]; Streptomyces griseocarneus [6]; Coryne-bacterium sterolicium [9]; Pseudomonas sp. [18, 19]; Actinomyces lavendu-lae [8]; Nocardia sp. [13]

Source tissue
Fermentation broth [5, 7, 17]; Commercial product [20]

Localisation in source
Membrane (2 forms: membrane-bound and secreted [21], intrinsic membra-ne protein [1]) [1, 21]; Extracellular (2 forms: membrane-bound and secre-ted [21]) [11, 17, 21, 22]

Purification
Brevibacterium sterolicum [5, 22]; Schizophyllum commune [16]; Arthrobac-ter simplex (inducible and constitutive enzyme) [17]; More (overview: purifi-cation procedures) [24]

Crystallization
[22]

Cloned
–

Renaturated
–

---

5 STABILITY

pH
6–10 (30°C, 2 h, stable) [17]; 4.0–10.0 (37°C, 30 min, stable) [5]; 5.0–10.0
(37°C, 30 min, stable) [5]; 5.0–8.0 (60°C, 1 min, stable) [18]; 10 (60°C, 1 min,
25% loss of activity) [18]

Temperature (°C)
50 (60 h, about 25% loss of activity [1], pH 7.0, 30 min, stable [5]) [1, 5]; 60
(pH 7.0, 30 min, 19% loss of activity) [5]; 70 (5 min, no loss of activity [18],
30 min, 15% loss of activity [18], pH 7, 30 min, about 94% loss of activity
[5]) [5, 18]; 80 (5 min, about 70% loss of activity) [1]

Oxidation

Organic solvent

General stability information
Long-chain primary alcohols stabilize [12]; SDS, 0.2 mM, stabilizes [12]; Bile
salts stabilize [19]

Storage
0–5°C, crystallized enzyme in lyophilized state, in the dark, stable for at
least 10 months [5]

---

6 CROSSREFERENCES TO STRUCTURE DATABANKS

PIR/MIPS code
PIR3:PC2002 (Streptomyces sp. (fragment)); PIR2:JQ1193 (precursor Brevibacterium sterolicum (ATCC 21387)); PIR2:A32260 (precursor Streptomyces
sp.); PIR2:S15810 (precursor Streptomyces sp. (fragment))

Brookhaven code
3COX (Brevibacterium sterolicum)

7 LITERATURE REFERENCES

[1] Cheetham, P.S.J., Dunnill, P., Lilly, M.D.: Biochem. J.,201,515–521 (1982)
[2] Arima, K., Nagasawa, M., Bae, M., Tamura, G.: Agric. Biol. Chem.,33,1636–1643 (1969)
[3] Smith, A.G., Brooks, C.J.W.: J. Chromatogr.,101,373–378 (1974)
[4] Buckland, B.C., Richmond, W., Dunnill, P., Lilly, M.D. in "Industrial Aspects of Biochemistry" (Spencer, B., ed.) 65–79, FEBS, Amsterdam (1974)
[5] Uwajima, T., Yagi, H., Terado, O.: Agric. Biol. Chem.,38,1149–1156 (1974)
[6] Kerenyi, G., Szeentirmai, A., Natonek, M.: Acta Microbiol. Acad. Sci. Hung.,22,487–496 (1975)
[7] Tomioka, H., Kagawa, M., Nakamura, S.: J. Biochem.,79,903–915 (1976)
[8] Petrova, L.Y., Podsukhina, G.M., Dikun, T.A., Seleznava, A.A.: Appl. Biochem. Microbiol.,15,125–144 (1979)
[9] Shirokane, J., Nakamura, K., Mizusawa, K.: J. Ferment. Technol.,55,337–346 (1977)
[10] Smith, A.G., Brooks, C.J.W.: Biochem. Soc. Trans.,3,675–677 (1975)
[11] Fukuda, H., Kawakami, Y., Nakamura, S.: Chem. Pharm. Bull.,21,2057–2060 (1973)
[12] Abe, T., Nihira, T., Tanaka, A., Fukui, S.: Biochim. Biophys. Acta,749,69–76 (1983)
[13] Yoshimoto, T., Ritani, A., Ohwada, K., Takahashi, K., Kodera, Y., Matsushima, A., Saito, Y., Inada, Y.: Biochem. Biophys. Res. Commun.,148,876–882 (1987)
[14] Smith, A.G., Brooks, C.J.W.: Biochem. J.,167,121–129 (1977)
[15] Uwajima, T., Terada, O.: Agric. Biol. Chem.,42,1453–1454 (1978)
[16] Fukuyama, M., Miyake, Y.: J. Biochem.,85,1183–1193 (1979)
[17] Liu, W.-H., Meng, M.-H., Chen, K.-S.: Agric. Biol. Chem.,52,413–418 (1988)
[18] Lee, S., Rhee, H., Tae, W., Shin, J., Park, B.: Appl. Microbiol., Biotechnol.,31, 542–546 (1989)
[19] Cheillan, F., Lafont, H., Termine, E., Fernandez, F., Sauve, P., Lesgards, G.: Biochim. Biophys. Acta,999,233–238 (1989)
[20] Bru, R., Sanchez-Ferrer, A., Garcia-Carmona, F.: Biotechnol. Lett.,11,237–242 (1989)
[21] Johnson, T.L., Somkuti, G.A.: Biotechnol. Appl. Biochem.,13,196–204 (1991)
[22] Uwajima, T., Yagi, H., Nakamura, S., Terada, O.: Agric. Biol. Chem.,37,2345–2350 (1973)
[23] Brooks, C.J.W., Smith, A.G.: J. Chromatogr.,112,499–511 (1975)
[24] Smith, A.G., Brooks, C.J.W.: J. Steroid Biochem.,7,705–713 (1976) (Review)

1 NOMENCLATURE

EC number
1.1.3.7

Systematic name
Aryl-alcohol:oxygen oxidoreductase

Recommended name
Aryl-alcohol oxidase

Synonymes
Oxidase, aryl alcohol
Aryl alcohol oxidase
Veratryl alcohol oxidase
Arom. alcohol oxidase

CAS Reg. No.
9028-77-7

2 REACTION AND SPECIFICITY

Catalysed reaction
An aromatic primary alcohol + $O_2 \rightarrow$
$\rightarrow$ an aromatic aldehyde + H_2O_2

Reaction type
Redox reaction

Natural substrates
Aromatic primary alcohol + O_2 (lignin degradation) [2, 3, 5]

Substrate spectrum
1 Aromatic primary alcohol + O_2 [1–7]
2 Veratryl alcohol + O_2 (i.e. dimethoxybenzyl alcohol) [1–7]
3 Benzyl alcohol + O_2 [2, 5, 7]
4 Anisyl alcohol + O_2 (i.e. 4-methoxybenzyl alcohol) [2, 3, 5–7]
5 Cinnamyl alcohol + O_2 (i.e 3-phenyl-2-propen-1-ol) [5, 7]

6 3-Phenoxybenzyl alcohol + O_2 [5]
7 2,4-Dimethoxybenzyl alcohol + O_2 [5]
8 3-Hydroxy-4-methoxybenzyl alcohol + O_2 [5]
9 More (activity at least ten times greater with aromatic alcohols than with ethanol, propanol or butanol, no activity with aliphatic and secondary aromatic alcohols [2], not all aromatic alcohols are oxidized [5], not: methanol [2, 3, 7], ethanol [2, 3], butanol [2, 6], 1-(3,4-dimethoxyphenyl) ethanol [2, 6], glucose [3, 6], secondary phenylethanols [6], L-amino acids [6], cannot use artificial electron acceptor systems or pyridine nucleotides [7]) [2, 3, 5–7]

Product spectrum
1 Aromatic aldehyde + H_2O_2 [1, 2]
2 3,4-Dimethoxybenzaldehyde + H_2O_2
3 Benzaldehyde + H_2O_2
4 4-Methoxybenzaldehyde + H_2O_2
5 Cinnamaldehyde + H_2O_2
6 3-Phenoxybenzaldehyde + H_2O_2
7 2,4-Dimethoxybenzaldehyde + H_2O_2
8 3-Hydroxy-4-methoxybenzaldehyde + H_2O_2
9 ?

Inhibitor(s)
Ag^+ [2]; Pb^{2+} [2]; NaN_3 [2]; More (little affected by p-chloromercuribenzoate, cyanide or azide ions) [6]

Cofactor(s)/prosthetic group(s)/activating agents
FAD (enzyme contains FAD as prosthetic group) [3]; Flavin (contains a flavin prosthetic group) [5]

Metal compounds/salts

Turnover number (min^{-1})
4930 (veratryl alcohol, isozyme VAO I) [5]; 3860 (veratryl alcohol, isozyme VAO II) [5]

Specific activity (U/mg)
210 [3]; 30.7 [5]

K_m-value (mM)
0.41 (veratryl alcohol, isozyme VAO I) [5]; 1.2 (veratryl alcohol) [2]; 0.24 (anisyl alcohol) [2]; 0.46 (veratryl alcohol, isozyme VAO II) [5]

pH-optimum
4–7 (more than 90% of maximum activity between pH 4 and 7) [5]; 5.7 [3]; 6.0–6.5 [2, 6]

pH-range
2–10 (2: about 30% of activity maximum, 10: 18% of activity maximum) [2];
3–8 (3: about 30% of activity maximum, 8: about 60% of activity maximum)
[5]; 5.3–7.3 (5.3: 90% of activity maximum, 7.3: about 35% of activity maximum) [6]

Temperature optimum (°C)
45–50 [2]

Temperature range (°C)
20–65 (20°C: about 60% of activity maximum, 65°C: about 15% of activity
maximum) [2]

---

3 ENZYME STRUCTURE

Molecular weight

Subunits
? (x × 78000, Bjerkandera adusta, SDS-PAGE [3], x × 71000, Pleurotus
sajor-caju, SDS-PAGE [5]) [3, 5]

Glycoprotein/Lipoprotein
Glycoprotein [5]

---

4 ISOLATION/PREPARATION

Source organism
Phanerochaete chrysosporium (Sporotrichum pulverulentum, white-rot fungus) [4]; Chrysosporium pruinosum (white-rot fungus) [4]; Coriolus versicolor (Trametes versicolor, white-rot fungus) [4]; Pleurotus ostreatus (white-rot
fungus) [4]; Fomes lignosus (Rigidiporus microporus, white-rot fungus) [4];
Pleurotus sajor-caju (basidiomycete, 2 veratryl alcohol oxidases: VAO I, VAO
II) [5]; Bjerkandera adusta (white-rot fungus [1, 3, 4], 2 forms [3], i.e. Polyporus adusta [4]) [1, 3, 4]; Pleurotus eryngii (ligninolytic fungus) [2]; Poly-
stictus versicolor [6]; Arion ater (terrestrial slug) [7]

Source tissue
Mycelium [2]; Culture medium [2, 5]

Localisation in source
Intracellular [2]; Extracellular [2, 3]; Particulate (but not localized in mitochondria or peroxisomes) [7]

Purification
Bjerkandera adusta (white-rot fungus, 2 forms) [3]

Enzyme Handbook © Springer-Verlag Berlin Heidelberg 1995
Duplication, reproduction and storage in data banks are only
allowed with the prior permission of the publishers   3

Crystallization

–

Cloned

–

Renaturated

–

---

5 STABILITY

pH
    4.0–9.0 (room temperature, 24 h) [2]

Temperature (°C)
    35 (30 min, no inactivation) [3]; 45 (5 min, stable) [6]; 50 (30 min, stable [2],
    30 min, complete inactivation [3]) [2, 3]; 55 (5 min, complete inactivation)
    [6]; 60 (half-life: 30 min) [2]; 70 (10 min, total inactivation) [2]

Oxidation

Organic solvent

General stability information

Storage
    4°C, pH 5.7, weeks [3]

---

6 CROSSREFERENCES TO STRUCTURE DATABANKS

PIR/MIPS code

Brookhaven code

---

7 LITERATURE REFERENCES

[1] Muheim, A., Leisola, M.S.A., Schoemaker, H.E.: J. Biotechnol.,13,159–167 (1990)
[2] Guillen, F., Martinez, A.T., Martinez, M.J.: Appl. Microbiol. Biotechnol.,32,465–469
    (1990)
[3] Muheim, A., Waldner, R., Leisola, M.S.A., Fiechter, A. : Enzyme Microb. Technol.,
    12,204–209 (1990)
[4] Waldner, R., Leisola, M.S.A., Fiechter, A.: Appl. Microbiol. Biotechnol.,29,400–407
    (1988)
[5] Bourbonnais, R., Paice, M.G.: Biochem. J.,255,445–450 (1988)
[6] Farmer, V.C., Henderson, M.E.K., Russell, J.D.: Biochem. J.,74,257–262 (1960)
[7] Mann, V., Large, A., Khan, S., Malik, Z., Connock, M.J.: J. Exp. Zool.,251,265–274
    (1989)

1 NOMENCLATURE

EC number
1.1.3.8

Systematic name
L-Gulono-1,4-lactone:oxygen 2-oxidoreductase

Recommended name
L-Gulonolactone oxidase

Synonymes
Oxidase, L-gulonolactone
L-Gulono-gamma-lactone:O_2 oxidoreductase
L-Gulono-gamma-lactone oxidase [2]
L-Gulono-gamma-lactone:oxidoreductase [7]

CAS Reg. No.
9028-78-8

2 REACTION AND SPECIFICITY

Catalysed reaction
L-Gulono-1,4-lactone + O_2 →
→ L-xylo-hexulonolactone + H_2O (the product spontaneously isomerizes to
L-ascorbate [4])

Reaction type
Redox reaction

Natural substrates
L-Gulono-1,4-lactone + O_2 (last step of L-ascorbic acid biosynthesis in ani-
mals [4, 8], the product L-xylo-hexulonolactone isomerizes spontanouesly to
L-ascorbate [4]) [4, 8]

Substrate spectrum
1 L-Gulono-1,4-lactone + O_2 (similar enzyme: L-galactonolactone oxidase
  [5], not [11]) [1–10, 12–16]
2 L-Galactono-gamma-lactone + O_2 (similar enzyme: L-galactonolactone
  oxidase [5, 11]) [4, 5, 11, 16]
3 D-Mannono-gamma-lactone + O_2 [4]
4 D-Altrono-gamma-lactone + O_2 (similar enzyme: L-galactonolactone oxidase [5, 11]) [4, 5, 11]
5 L-Fucono-gamma-lactone + O_2 (similar enzyme: L-galactonolactone oxidase) [11]

6 L-Arabino-gamma-lactone + O_2 (similar enzyme: L-galactonolactone oxidase) [11]

7 D-Threo-1,4-lactone + O_2 (similar enzyme: L-galactonolactone oxidase) [11]

8 More (attacks semicarbazone, oxime and cyanohydrine of D-gulonono-gamma-lactone [4], phenazine methosulfate acts as electron acceptor in place of molecular oxygen [4]) [4, 11]

Product spectrum

1 L-Xylo-hexulonolactone + H_2O (the product spontaneously isomerizes to L-ascorbate [4]) [1–10, 12–16]

2 ?

3 ?

4 ?

5 ?

6 ?

7 ?

8 ?

Inhibitor(s)

Hg^{2+} (similar enzyme: L-galactonolactone oxidase [11]) [4, 11, 16]; Cd^{2+} (similar enzyme: L-galactonolactone oxidase [11], not [16]) [11]; p-Chloromercuribenzoate [4, 5]; p-Chloromercuriphenylsulfonate (similar enzyme: L-galactonolactone oxidase [11]) [4, 11]; N-Ethylmaleimide (similar enzyme: L-galactonolactone oxidase) [11]; Na_2S [16]; Na_2SO_3 [16]; p-Nitrothiophenol [16]; More (NaN_3 and KCN scarcely inhibit) [5]

Cofactor(s)/prosthetic group(s)/activating agents

FAD (flavoprotein [2–5, 8, 10], 8alpha-[N(1)-histidyl]riboflavin is the structure of the covalently bound flavin [2, 4, 10], flavin covalently bound to the enzyme [4, 5, 8, 15], flavin content in a purified preparation: 0.22 nmol per mg of protein [16]) [2–5, 8, 10, 15, 16]; More (specific functional pigment associated with the enzyme activity) [16]

Metal compounds/salts

No metal requirement [4, 5]

Turnover number (min⁻¹)

39 (L-gulono-1,4-lactone) [10]

Specific activity (U/mg)

0.540 (rat) [3, 4]; 1.1 [8]; 3.43 [10]; 2.050 (goat) [3, 4]

K_m-value (mM)
0.66 (L-gulono-gamma-lactone, rat) [3, 4]; 0.15 (L-gulono-gamma-lactone, goat) [3, 4]; 0.007 (L-gulono-gamma-lactone) [10]; 0.30 (L-galactonolactone, similar enzyme: L-galactonolactone oxidase) [11]; 0.36 (L-galactonolactone, similar enzyme: L-galactonolactone oxidase) [11]; 2.0 (D-altronolactone, similar enzyme: L-galactonolactone oxidase) [11]; 0.16 (D-arabinolactone, similar enzyme: L-galactonolactone oxidase) [11]; 15 (D-threonolactone, similar enzyme: L-galactonolactone oxidase) [11]

pH-optimum
7.0 (assay at) [10]; 7.4 (assay at, activity increases from pH 6.0–8.0 with no clear-cut optimum, slight decrease beyond pH 8.0) [7]; 8 (Tris-citrate buffer) [10]; 7.5 (phosphate-citrate buffer, higher activity than in Tris-citrate buffer) [10]

pH-range
6.0–8.0 (activity increases from pH 6.0–8.0 with no clear-cut optimum, slight decrease beyond pH 8.0) [7]

Temperature optimum (°C)
37 (assay at) [4, 5]

Temperature range (°C)

---

3 ENZYME STRUCTURE

Molecular weight
70000 (Saccharomyces cerevisiae, similar enzyme: L-galactonolactone oxidase, gel filtration in presence of deoxycholate) [11]
100000 (rat, gel filtration) [16]
290000 (Saccharomyces cerevisiae, gel filtration, similar enzyme: L-galactonolactone oxidase) [5]
400000 (chicken, gel filtration) [10]
500000 (rat, goat, gel filtration) [3, 4]
More (450000, rat, gel filtration, MW of the lipid-containing membrane subunit the enzyme is associated with) [1]

Subunits
Tetramer (4 × 18000, Saccharomyces cerevisiae, SDS-PAGE, similar enzyme: L-galactonolactone oxidase) [11]
Oligomer (x × 51000, rat, goat, SDS-PAGE [3, 4], x × 50000, chicken, SDS-PAGE [10], x × 56000, Saccharomyces cerevisiae, SDS-PAGE, similar enzyme: L-galactonolactone oxidase [5]) [3–5, 10]

Glycoprotein/Lipoprotein
–

Enzyme Handbook © Springer-Verlag Berlin Heidelberg 1995
Duplication, reproduction and storage in data banks are only
allowed with the prior permission of the publishers

4 ISOLATION/PREPARATION

Source organism
Tachyglossus aculeatus (prototheria, echidna) [9]; Ornithorhynchus anatinus (prototheria, platypus) [9]; Perameles nasuta (long-nosed bandicoot) [9]; Isoodon macrourus (brindled bandicoot) [9]; Macropus rufogriseus (red-necked wallaby) [9]; Thylogale thetis (red-necked pademelon) [9]; Limulus polyphemus [12]; Bullfrog [14]; Rat [1–4, 6, 8, 14–16]; Goat [3, 4]; Saccharomyces cerevisiae (similar enzyme: L-galactonolactone oxidase) [5, 11]; Chicken [10, 13, 14]; Mammals [7]; More (human, primates and guinea pig lack this enzyme) [4]

Source tissue
Muscle [12]; Gut [12]; Central nervous system [12]; Hepatopancreas [12]; Heart [12]; Yolk sac membrane [13]; Liver (not: prototheria (egg-laying mammals) [9]) [1, 3, 4, 6–9, 14–16]; Kidney [9, 10, 14]

Localisation in source
Microsomes [1, 3, 4, 8, 10, 15, 16]; Mitochondria (similar enzyme: L-galactonolactone oxidase) [5, 11]; Endoplasmic reticulum (enzyme is associated with a lipid-containing membrane subunit of the endoplasmic reticulum) [1]

Purification
Rat (partial [1]) [1, 3, 4, 8, 10, 15, 16]; Goat [3, 4]; Saccharomyces cerevisiae (similar enzyme: L-galactonolactone oxidase) [5, 11]; Chicken [10]

Crystallization
–

Cloned
(rat liver enzyme) [6]

Renaturated
–

5 STABILITY

pH

Temperature (°C)
24 (80 min, little loss of activity, solubilized preparation) [1]; 49 (10 min, 90% loss of activity) [1]; 70 (1 min, complete loss of activity) [1]

Oxidation

Organic solvent

General stability information
EDTA stabilizes [1]; Insulin stabilizes [1]; Albumin stabilizes [1]; Sodium deoxycholate diminishes stability at 37°C [12]

Storage
4°C, 1 mM EDTA, 1 mM mercaptoethanol, 2 weeks, 50% loss of activity [1];
-20°C, stable in frozen liver, rapid loss of activity in frozen homogenate [7]

6 CROSSREFERENCES TO STRUCTURE DATABANKS

PIR/MIPS code
PIR1:OXRTGU (rat); PIR2:A61199 (rat); PIR2:S06473 (rat (fragment));
PIR2:B61199 (rat (fragments))

Brookhaven code

7 LITERATURE REFERENCES

[1] Eliceiri, G.L., Lai, E.K., McCay, P.B.: J. Biol. Chem.,244,2641–2645 (1969)
[2] Kenney, W.C., Edmondson, D.E., Singer, T.P.: Biochem. Biophys. Res. Commun.,
     71,1194–1200 (1976)
[3] Nishikimi, M., Tolbert, B.M., Udenfriend, S.: Arch. Biochem. Biophys.,175,427–435
     (1976)
[4] Nishikimi, M.: Methods Enzymol.,62,24–30 (1979) (Review)
[5] Nishikimi, M., Noguchi, E., Yagi, K.: Arch. Biochem. Biophys.,191,479–486 (1978)
[6] Koshizaka, T., Nishikimi, M., Ozawa, T., Yagi, K.: J. Biol. Chem.,263,1619–1621
     (1988)
[7] Ayaz, K.M., Jenness, R., Birney, E.C.: Anal. Biochem.,72,161–171 (1976)
[8] Nakagawa, H., Asano, A., Sato, R.: J. Biochem.,77,221–232 (1975)
[9] Birney, E.C., Jenness, R., Hume, I.D.: Experientia,35,1425–1426 (1979)
[10] Kiuchi, K., Nishikimi, M., Yagi, K.: Biochemistry,21,5076–5082 (1982)
[11] Bleeg, H.S., Christensen, F.: Eur. J. Biochem.,127,391–396 (1982)
[12] Wallace, C.A., Jenness, R., Mullin, R.J., Herman, W. S.: Experientia,41,485–486
     (1985)
[13] Yew, M.S.: Experientia,41,943–944 (1985)
[14] Nishikimi, M., Yamauchi, N., Kiuchi, K., Yagi, K.: Experientia,37,479–480 (1981)
[15] Nishikimi, M., Kiuchi, K., Yagi, K.: FEBS Lett.,81,323–325 (1977)
[16] Nakagawa, H., Assano, A.: J. Biochem.,68,737–746 (1970)

1 NOMENCLATURE

EC number
1.1.3.9

Systematic name
D-Galactose:oxygen 6-oxidoreductase

Recommended name
Galactose oxidase

Synonymes
Oxidase, galactose
D-Galactose oxidase
beta-Galactose oxidase

CAS Reg. No.
9028-79-9

---

2 REACTION AND SPECIFICITY

Catalysed reaction
D-Galactose + O_2 →
→ D-galacto-hexodialdose + H_2O_2 (mechanism [7, 19, 21])

Reaction type
Redox reaction

Natural substrates

Substrate spectrum
1 D-Galactose + O_2 (alcohol oxidation step is not reversible [19]) [4, 5, 8–10, 12–14, 19, 20, 23, 26, 27, 29, 30]
2 Dihydroxyacetone + O_2 [4, 8, 10, 12]
3 Glycerol + O_2 [15]
4 Raffinose + O_2 [4, 8–10, 12, 26, 27]
5 Methyl-alpha-D-galactopyranose + O_2 [4, 8, 12, 26, 27]
6 Methyl-beta-D-galactopyranose + O_2 [12, 26–28]

Enzyme Handbook © Springer-Verlag Berlin Heidelberg 1995
Duplication, reproduction and storage in data banks are only
allowed with the prior permission of the publishers

7 More (other substrates: major glycolipid of human red cells [3], D-talose [12], 3-halo-1,2-propane-diols (R-isomer is a better substrate than S-isomer) [15], GM1 ganglioside [16], D-galactosamine [8, 26], lactose (low activity) [8, 26], melibiose [8, 10, 12, 26, 27], stachyose [8, 10, 12, 26, 27], guaran [8, 26, 27], desialyated glycoproteins (fetuin, mucin) [11], N-acetyl-D-galactosamine [12, 26], overview: snail galactans, galactosides and D-galactose-composed oligosaccharides [6], isopropyl-beta-D-thiogalactosylpyranoside [27], beta-thiodigalactoside [27], melibiitol [27], melibionic acid [27], 1,5-anhydrogalactitol [26], planteose [26, 27], 2-glycerol-alpha-D-galactopyranoside [27], galactobiose [27], beta-D-galactopyranosyl(1→ 6)beta-D-galactopyranosyl(1→ 4)-D-glucose [27], methyl-beta-D-thiogalactosylpyranoside [27], not: glycerol [4], beta-hydroxypyruvate [4], D-mannose [26], D-fructose [26], D-glucose [26], enzyme also functions as a superoxide dismutase [23], enzyme also catalyzes further oxidation to carboxyl group [9], acts on a specific subterminal D-galactosyl residue (2→ 1) D-Gal as well as upon terminal non-reducing galactosyl residue in Helix pomatia galactogen [13], stereospecificity [15], steric factors involved in enzyme action (overview) [17]) [3, 4, 6, 8–13, 15–17, 23, 26, 27]

Product spectrum
1 D-Galacto-hexodialdose + H_2O_2
2 3-Hydroxy-2-oxo-propionaldehyde + H_2O_2
3 S(-)-Glyceraldehyde + H_2O_2 [15]
4 6''-Carboxyraffinose + H_2O_2 [9]
5 Methyl-alpha-D-galacto-hexadialdose + H_2O_2
6 Methyl-beta-D-galacto-hexadialdose + H_2O_2
7 ?

Inhibitor(s)
CN^- [8, 26]; NaN_3 [8, 20]; Hydroxylamine [8]; Diethyldithiocarbamate [8, 20, 25, 26]; D-Galacturonic acid [12]; D-Galactono-gamma lactone [12]; H_2O_2 [9, 26]; $CuSO_4$ [20]; $FeSO_4$ (slight) [20]; $AgNO_3$ [20]; Iodoacetamide [20, 28]; N-Bromosuccinimide [20, 24]; Sodium dithionite [20]; Ascorbic acid [20]; Glutathione [20]; Dithiothreitol [20]; 2-Mercaptoethanol [20]; D-Galactosamine [25]; 2-Deoxy-D-galactosamine [25]; N-Acetyl-D-galactosamine [25]; TAY-SACHS ganglioside [25]; Bovine brain ganglioside [25]; Superoxide dismutase [22]

Cofactor(s)/prosthetic group(s)/activating agents
Pyrroloquinoline quinone (one molecule covalently bound to the enzyme, mechanistic role) [2]

Metal compounds/salts
Cu (0.95 gatom of Cu per 68000 g of protein [28], 1 gatom of Cu per 70000 g of protein [30], a unique mononuclear Cu site [1], Cu(II)-containing enzyme [14], contains about 1 atom each of Cu and Fe per molecule [20], the Cu(III)enzyme is the oxidant which converts the primary alcohol function of galactose to an aldehyde [21], enzyme contains a single nonblue Cu(II) atom [24], 1 equivalent of Cu per 40000–48000 g enzyme [26]) [1, 14, 20, 21, 24, 26, 28, 30]; Fe (contains about 1 atom each of Cu and Fe per molecule) [20]

Turnover number (min[-1])
19500 (O-methyl-beta-galactopyranoside) [7]; 390000 (galactose) [14]; 86100 (beta-methyl-D-galactopyranoside) [28]

Specific activity (U/mg)
116 [8]; 166 [18]; 3.32 [20]; More [5, 14, 19, 26, 28, 30]

K$_m$-value (mM)
35 (D-galactose, intracellular enzyme) [4]; 30 (D-galactose, extracellular enzyme) [4]; 20 (dihydroxyacetone) [8]; 3 (O$_2$) [14]; 175 (galactose) [14]; 0.24 (D-galactose) [27]; 0.45 (D-galactosamine) [27]; 0.025 (raffinose) [27]; 0.045 (melibiose) [27]; 0.013 (stachyose) [27]; 0.0003 (guaran) [27]; More (overview: K$_m$ of different free or surface-linked substrates (GM1 in micellar or vesicular dispersion) [16], K$_m$ of desialylated glycoproteins [11], nonlinear Lineweaver-Burk plot [12]) [11, 12, 16]

pH-optimum
6–7 [19]; 6.7 [7, 28]

pH-range

Temperature optimum (°C)
25 (assay at) [14]

Temperature range (°C)

---

3 ENZYME STRUCTURE

Molecular weight
42100 (Polyporus circinatus, determination with high speed osmometer) [25]
42400 (Polyporus circinatus, equilibrium centrifugation) [26]
68000 (Dactylium dendroides, extracellular enzyme, SDS-PAGE [5], Polyporus circinatus, sedimentation equilibrium, SDS-PAGE, gel filtration, osmometry [30]) [5, 30]
72000 (Dactylium dendroides, intracellular enzyme, gel filtration) [8]
90000 (Gibberella fujikuroi, gel filtration) [20]

Enzyme Handbook © Springer-Verlag Berlin Heidelberg 1995
Duplication, reproduction and storage in data banks are only
allowed with the prior permission of the publishers   3

Subunits
Monomer (1 × 68000, Polyporus circinatus, SDS-PAGE [30], 1 × 72000, Dactylium dendroides, intracellular enzyme, SDS-PAGE [8]) [8, 30]

Glycoprotein/Lipoprotein
Glycoprotein (carbohydrate content: intracellular enzyme 7.7%, extracellular enzyme 1.7% [4], intracellular enzyme: 7.7% neutral sugars, 1.7% aminosugars [8], 1% neutral carbohydrate [30]) [4, 8, 30]

---

4 ISOLATION/PREPARATION

Source organism
Dactylium dendroides [1–9, 11, 13, 14, 15, 17, 18, 28, 29]; Gibberella fujikuroi [10, 12, 20]; Polyporus circinatus [16, 19, 22–27, 30]

Source tissue
Culture medium [5, 14, 20, 30]; Cell [8]; Commercial preparation [13, 17, 29]

Localisation in source
Extracellular [1, 4, 5, 13, 14, 30]; Intracellular [4, 8]

Purification
Dactylium dendroides (extracellular enzyme [5, 14], intracellular enzyme [8]) [5, 8, 14, 18]; Polyporus circinatus [26, 27, 30]; Gibberella fujikuroi [20]

Crystallization
[1]

Cloned
–

Renaturated
[4]

---

5 STABILITY

pH
5.5 (1 h, intracellular enzyme: 30% loss of activity, extracellular enzyme: 50% loss of activity) [4]; 5.5–8 (slow inactivation below pH 5.5 and above 8) [19]

Temperature (°C)
23 (room temperature, native enzyme and apoenzyme are stable) [26]; 60 (transition temperature of intracellular enzyme) [4]

Oxidation
Exposure to galactose in air should be avoided [26]

Organic solvent

General stability information

Glycosylation protects against endogenous inactivation [4]; Lyophilization inactivates [14]; Repeated freezing and thawing inactivates [14]; Inactivated in dilute solutions, by shaking or if pigments have not been removed completely in the last purification step [26]; Autoinactivation of immobilized enzyme can be prevented by $K_3Fe(CN)_6$ [29]

Storage

−17°C, stable for at least 6 months, intracellular enzyme [8]; −70°C, 1 year [14]

6 CROSSREFERENCES TO STRUCTURE DATABANKS

PIR/MIPS Code

PIR2:A38084 (precursor fungus (Cladobotryum dendroides))

Brookhaven code

1GOF (Dactylium dendroides); 1GOG (Dactylium dendroides); 1GOH (Dactylium dendroides)

7 LITERATURE REFERENCES

[1] Ito, N., Phillips, S.E.V., Stevens, C., Ogel, Z.B., McPherson, M.J., Keen, J.N., Yadav, K.D.S., Knowles, P.F.: Nature,350,87–90 (1991)
[2] Van der Meer, R.A., Jongejan, J.A., Duine, J.A.: J. Biol. Chem.,264,7792–7794 (1989)
[3] Lampio, A., Siissalo, I., Gahmberg, C.G.: Eur. J. Biochem.,178,87–91 (1988)
[4] Mendonca, M.H., Zancan, G.T.: Arch. Biochem. Biophys.,266,427–434 (1988)
[5] Kelleher, F.M., Dubbs, S.B., Bhavanandan, V.P.: Arch. Biochem. Biophys.,263, 349–354 (1988)
[6] Bretting, H., Jacobs, G.: Biochim. Biophys. Acta,913,342–348 (1987)
[7] Driscoll, J.J., Kosman, D.J.: Biochemistry,26,3429–3436 (1987)
[8] Mendonca, M.H., Zancan, G.T.: Arch. Biochem. Biophys.,252,507–514 (1987)
[9] Kelleher, F.M., Bhavanandan, V.P.: J. Biol. Chem.,261,11045–11048 (1986)
[10] Aisaka, K., Uwajima, T., Terada, O.: Agric. Biol. Chem.,49,1201–1202 (1985)
[11] Avigad, G.: Arch. Biochem. Biophys.,239,531–537 (1985)
[12] Aisaka, K., Uwajima, T., Terada, O.: Agric. Biol. Chem.,48,1425–1431 (1984)
[13] Goudsmit, E.M., Matsuura, F., Blake, D.A.: J. Biol. Chem.,259,2875–2878 (1984)
[14] Tressel, P.S., Kosman, D.J.: Methods Enzymol.,89,163–171 (1982) (Review)
[15] Klibanov, A.M., Alberti, B.N., Marletta, M.A.: Biochem. Biophys. Res. Commun., 108,804–808 (1982)
[16] Masserini, M., Sonnino, S., Ghidoni, R., Chigorno, V. : Biochim. Biophys. Acta,688,333–340 (1982)
[17] Cathmann, W.D., Aminoff, D.: Biochem. Biophys. Res. Commun.,103,68–76 (1981)
[18] Tressel, P., Kosman, D.J.: Anal. Biochem.,105,150–153 (1980)

[19] Hamilton, G.A., Adolf, P.K., De Jersey, J., DuBois, G.C., Dyrkacz, G.R., Libby, R.D.: J. Am. Chem. Soc.,100,1899–1912 (1978)
[20] Aisaka, K., Terada, O.: Agric. Biol. Chem.,46,1191–1197 (1982)
[21] Hamilton, G.A., Libby, R.D.: Biochem. Biophys. Res. Commun.,55,333–340 (1973)
[22] Kwiatowski, L.D., Kosman, D.J.: Biochem. Biophys. Res. Commun.,53,715–721 (1973)
[23] Cleveland, L., Davis, L.: Biochim. Biophys. Acta,341,517–523 (1974)
[24] Weiner, R.E., Ettinger, M.J., Kosman, D.J.: Biochemistry,16,1602–1606 (1977)
[25] Yip, M.C.M., Dain, J.A.: Enzymologia,35,368–376 (1968)
[26] Amaral, D., Kelly-Falcoz, F., Horecker, B.L.: Methods Enzymol.,9,87–92 (1966)
[27] Avigad, G., Amaral, D., Asensio, C., Horecker, B.L.: J. Biol. Chem.,237,2736–2743 (1962)
[28] Kwiatowski, L.D., Siconolfi, L., Weiner, R.E., Giordano, R.S., Bereman, R.D., Ettinger, M.J., Kosman, D. J.: Arch. Biochem. Biophys.,182,712–722 (1977)
[29] Dahodwala, S.K., Weibel, M.K., Humphrey, A.E.: Biotechnol. Bioeng.,18,1679–1694 (1976)
[30] Kosman, D.J., Ettinger, M.J., Weiner, R.E., Massaro, E.J.: Arch. Biochem. Biophys., 165,456–467 (1974)

1 NOMENCLATURE

EC number
1.1.3.10

Systematic name
Pyranose:oxygen 2-oxidoreductase

Recommended name
Pyranose oxidase

Synonymes
Glucose 2-oxidase
Pyranose-2-oxidase

CAS Reg. No.
37250-80-9

2 REACTION AND SPECIFICITY

Catalysed reaction
D-Glucose + $O_2 \rightarrow$
$\rightarrow$ 2-dehydro-D-glucose + H_2O_2

Reaction type
Redox reaction

Natural substrates
D-Glucose + O_2 [4]

Substrate spectrum
1 D-Glucopyranose + O_2 (other electron acceptors: 2,6-dichlorophenolindo-phenol, cytochrome c [7]) [1–9]
2 L-Sorbose + O_2 [1, 3, 6–9]
3 D-Xylose + O_2 [1, 3, 4, 6–9]
4 D-Mannose + O_2 [6]
5 D-Galactose + O_2 [3, 6]
6 delta-D-Gluconolactone + O_2 [1, 3, 4, 7–9]
7 6-Deoxy-D-glucose + O_2 [1, 3]
8 2-Deoxy-D-glucose + O_2 [3]

Product spectrum
1 D-Glucosone + H_2O_2 (D-glucosone is identical with D-arabino-2-hexosulo-se) [2–4, 7, 9]
2 5-Keto-D-fructose + H_2O_2 [7, 9]
3 D-Xylosone + H_2O_2 [7, 9]
4 ? + H_2O_2 [6]
5 ? + H_2O_2 [3, 6]
6 2-Keto-D-gluconate + D-araboascorbate + H_2O_2 [7, 9]
7 ? + H_2O_2 [3]
8 ? + H_2O_2 [3]

Inhibitor(s)
Ag^+ [1, 3, 6]; Hg^{2+} [1, 3]; Cu^{2+} [1, 6]; Ni^{2+} [6]; Co^{2+} [6]; p-Chloromercuriben-zoate [3, 6]; o-Phenanthroline [1, 6]; 8-Hydroxyquinoline [1, 6]; H_2O_2 [6]

Cofactor(s)/prosthetic group(s)/activating agents
FAD (covalently bound) [1, 6]

Metal compounds/salts
Fe^{2+} (stimulation) [1]

Turnover number (min^{-1})

Specific activity (U/mg)
30.9–33.3 [1, 6]; 12.8–15.5 [3, 7, 8]

K_m-value (mM)
1.0 (alpha-D-glucose) [5]; 0.57 (beta-D-glucose) [5]; 0.83–3.1 (D-glucose) [1, 3, 4, 6]; 20 (D-xylose) [3, 4]

pH-optimum
5.5–8.0 [1]; 7.5 [3, 4]; 6.2 [6]; 6.0–8.0 [7, 8]

pH-range

Temperature optimum (°C)
60 [1]; 50 [6]

Temperature range (°C)

3 ENZYME STRUCTURE

Molecular weight
300000 (unidentified basidiomycete, gel filtration) [1]
220000 (Coriolus versicolor, gel filtration) [6]

Subunits
Tetramer (4 × 69000, unidentified basidiomycete, SDS-PAGE [1], 4 × 68000, Coriolus versicolor, SDS-PAGE [6]) [1, 6]

Glycoprotein/Lipoprotein
Glycoprotein (0.7% carbohydrate) [1]

4 ISOLATION/PREPARATION

Source organism
Coriolus versicolor [2, 3, 5, 6]; Trametes cinnabarinus [2]; Auricularia polytrica [2]; Irpex lacteus [2]; Polyporus optusus [2, 3, 5–9]; Phanerochaete chrysosporium [3, 4]; Coriolus hirsutus [6]; Oudemansiella mucida [5]; Daedaleopsis styracina [6]; Gloeophyllum separium [2, 6]; Unidentified basidiomycete [1]; More (genera of basidiomycetes) [6]

Source tissue

Localisation in source

Purification
Unidentified basidiomycete [1]; Phanerochaete chrysosporium [3, 4]; Coriolus versicolor [6]; Polyporus optusus [7, 8]

Crystallization
–

Cloned
–

Renaturated
–

5 STABILITY

pH
5.5–9.0 [1]; 5.0–7.4 [6]; 4.0–9.0 [7, 8]

Temperature (°C)
50 (not stable above) [1]; 60 (not stable above) [4]; 70 (not stable above) [6]

Oxidation

Organic solvent

General stability information

Storage
4°C, 30 mM sodium phosphate buffer, pH 6.5, 3 weeks [3]; 4°C, over anhydrous silica gel, 1 year [7]

6 CROSSREFERENCES TO STRUCTURE DATABANKS

PIR/MIPS code

Brookhaven code

7 LITERATURE REFERENCES

[1] Izumi, Y., Furuya, Y., Yamada, H.: Agric. Biol. Chem.,54,1393–1399 (1990)
[2] Izumi, Y., Furuya, Y., Yamada, H.: Agric. Biol. Chem.,54,799–801 (1990)
[3] Volc, J., Eriksson, K.E.: Methods Enzymol.,161,316–322 (1988)
[4] Eriksson, K.E., Pettersson, B., Volc, J., Musilik, V.: Appl. Microbiol. Biotechnol.,23,
    257–262 (1986)
[5] Taguchi, T., Ohwaki, K., Okuda, J.: J. Appl. Biochem.,7,289–295 (1985)
[6] Machida, Y., Nakanishi, T.: Agric. Biol. Chem.,48,2463–2470 (1984)
[7] Janssen, F.W., Ruelius, H.W.: Methods Enzymol.,41 B,170–173 (1975)
[8] Ruelius, H.W., Kerwin, R.M., Janssen, F.W.: Biochim. Biophys. Acta,167,493–500
    (1968)
[9] Janssen, F.W., Ruelius, H.W.: Biochim. Biophys. Acta,167,501–510 (1968)

1 NOMENCLATURE

EC number
1.1.3.11

Systematic name
L-Sorbose:oxygen 5-oxidoreductase

Recommended name
L-Sorbose oxidase

Synonymes
Oxidase, sorbose

CAS Reg. No.
37250-81-0

2 REACTION AND SPECIFICITY

Catalysed reaction
L-Sorbose + O_2 →
→ 5-dehydro-D-fructose + H_2O_2

Reaction type
Redox reaction

Natural substrates
L-Sorbose + O_2 [1]

Substrate spectrum
1 L-Sorbose + O_2 (O_2 can be substituted by 2,6-dichloroindophenol) [1]
2 D-Glucose + O_2 [1]
3 D-Galactose + O_2 [1]
4 D-Xylose + O_2 [1]
5 More (no reaction with D-fructose) [1]

Product spectrum
1 5-Dehydro-D-fructose + H_2O_2 [1]
2 2-Dehydro-D-glucose + H_2O_2
3 2-Dehydro-D-galactose + H_2O_2
4 ?
5 ?

Inhibitor(s)
F⁻ (slight) [1]; Mn^{2+} (slight) [1]; Ag^+ [1]; Hg^{2+} [1]; Pb^{2+} (slight) [1]; Cu^{2+} (slight) [1]; Ni^{2+} (slight) [1]; Cyanide (10 mM, slight) [1]; More (not: p-chloromercuribenzoate, phenylmercuric nitrate) [1]

Cofactor(s)/prosthetic group(s)/activating agents

Metal compounds/salts

Turnover number (min^{-1})

Specific activity (U/mg)
More [1]

K_m-value (mM)
22 (L-sorbose (+ O_2)) [1]; 100 (L-sorbose (+ 2,6-dichlorophenolindophenol)) [1]

pH-optimum
5.7–6.0 (potassium phosphate buffer) [1]

pH-range
5.1–7.0 (5.1: about 85% of activity maximum, 7.0: about 40% of activity maximum) [1]

Temperature optimum (°C)
30 (assay at) [1]

Temperature range (°C)

---

3 ENZYME STRUCTURE

Molecular weight

Subunits

Glycoprotein/Lipoprotein
–

---

4 ISOLATION/PREPARATION

Source organism
Trametes sanguinea (basidiomycete) [1]

Source tissue
Cell [1]

Localisation in source

Purification
  Trametes sanguinea (basidiomycete) [1]

Crystallization
  –

Cloned
  –

Renaturated
  –

---

5 STABILITY

pH

Temperature (°C)
  45–60 (5 min, stable) [1]; 75 (5 min, complete loss of activity) [1]

Oxidation

Organic solvent

General stability information

Storage
  0°C, 1 week stable [1]

---

6 CROSSREFERENCES TO STRUCTURE DATABANKS

PIR/MIPS code

Brookhaven code

---

7 LITERATURE REFERENCES

[1] Yamada, Y., Iizuka, K., Aida, K., Uemura, T.: J. Biochem.,62,223–229 (1967)

1 NOMENCLATURE

EC number
1.1.3.12

Systematic name
Pyridoxine:oxygen 4-oxidoreductase

Recommended name
Pyridoxine 4-oxidase

Synonymes
Oxidase, pyridoxol 4-
Pyridoxin 4-oxidase
Pyridoxol 4-oxidase

CAS Reg. No.
37250-82-1

2 REACTION AND SPECIFICITY

Catalysed reaction
Pyridoxine + O_2 →
→ pyridoxal + H_2O_2

Reaction type
Redox reaction

Natural substrates

Substrate spectrum
1 Pyridoxine + O_2 (other acceptor: 2,6-dichlorophenolindophenol, high degree of specificity for pyridoxine) [1]
2 5-Deoxypyridoxine + O_2 [1]
3 2-Dimethyl-2-ethylpyridoxine + O_2 [1]

Product spectrum
1 Pyridoxal + H_2O_2 [1]
2 5-Deoxypyridoxal + H_2O_2 [1]
3 2-Dimethyl-2-ethylpyridoxal + H_2O_2

Inhibitor(s)
1,10-Phenanthroline [1]; Chloromercuribenzoate (o-, m-, and p-substituted, inhibits in absence but not in presence of cyanide) [1]; alpha,alpha'-Dipyridyl [1]

Enzyme Handbook © Springer-Verlag Berlin Heidelberg 1995
Duplication, reproduction and storage in data banks are only
allowed with the prior permission of the publishers

Cofactor(s)/prosthetic group(s)/activating agents
FAD (flavoprotein, FAD-dependent) [1]

Metal compounds/salts

Turnover number (min⁻¹)

Specific activity (U/mg)
More [1]

K_m-value (mM)
0.43 (pyridoxine) [1]; 0.0083 (2,6-dichloroindophenol) [1]

pH-optimum
7.5–8.0 [1]

pH-range
6–9 [1]

Temperature optimum (°C)
26 (assay at) [1]

Temperature range (°C)

---

3 ENZYME STRUCTURE

Molecular weight

Subunits

Glycoprotein/Lipoprotein
–

---

4 ISOLATION/PREPARATION

Source organism
Pseudomonas sp. MA-1 [1]

Source tissue
Cell [1]

Localisation in source

Purification
Pseudomonas sp. MA-1 [1]

Crystallization
–

Cloned
–

Renaturated

–

---

5 STABILITY

pH

Temperature (°C)

Oxidation

Organic solvent

General stability information

Storage

---

6 CROSSREFERENCES TO STRUCTURE DATABANKS

PIR/MIPS code

Brookhaven code

---

7 LITERATURE REFERENCES

[1] Sundaram, T.K., Snell, E.E.: J. Biol. Chem.,244,2577–2584 (1969)

1 NOMENCLATURE

EC number
1.1.3.13

Systematic name
Alcohol:oxygen oxidoreductase

Recommended name
Alcohol oxidase

Synonymes
Oxidase, alcohol
Ethanol oxidase
More (the enzyme is probably identical with methanol oxidase EC 1.1.3.31)

CAS Reg. No.
9073-63-6

2 REACTION AND SPECIFICITY

Catalysed reaction
A primary alcohol + $O_2 \rightarrow$
$\rightarrow$ an aldehyde + H_2O_2 (mechanism [9, 12, 16], ping-pong mechanism [10])

Reaction type
Redox reaction

Natural substrates
Methanol + O_2 (key enzyme of methanol metabolism) [1–8, 14, 15]

Substrate spectrum
1 Methanol + O_2 (best substrate) [1–17]
2 Ethanol + O_2 (oxidation at 28% [1], 82% [3], 92% [7, 14, 15] the rate of methanol oxidation) [1–4, 6–8, 14, 15]
3 n-Propanol + O_2 (oxidation at 74% [7], 53% [1], 38% [3], 34.4% [14, 15] the rate of methanol oxidation) [1, 3–8, 14, 15]
4 n-Butanol + O_2 (oxidation at 2.1% [1], 10.6% [14, 15], 27% [3], 52% [7] the rate of methanol oxidation) [1, 3, 4, 7, 8, 14, 15]
5 1-Pentanol + O_2 (i.e. n-amyl alcohol, oxidation at 1% [1], 21% [3], 30% [7] the rate of methanol oxidation) [1, 3, 7]
6 1-Hexanol + O_2 (oxidation at 4% the rate of methanol oxidation) [7]
7 2-Propen-1-ol + O_2 (i.e. allyl alcohol, oxidation at 17% [1], 82% [3] the rate of methanol oxidation) [1, 3, 8, 10]

8 Isopropanol + O_2 (oxidation at 21% [3], 1.7% [15] the rate of methanol oxidation, oxidation only at high concentrations [4]) [3, 4, 15]

9 2-Buten-1-ol + O_2 [8]

10 Chloroethanol + O_2 (oxidation at 6% [1], 70% [7] the rate of methanol oxidation) [1, 7, 8]

11 Bromoethanol + O_2 (oxidation at 4% the rate of methanol oxidation) [1]

12 3-Chloro-1-propanol + O_2 (oxidation at 22% the rate of methanol oxidation) [7]

13 4-Chloro-1-butanol + O_2 (oxidation at 11% the rate of methanol oxidation) [7]

14 Isobutanol + O_2 (poor substrate) [7]

15 Formaldehyde + O_2 (oxidation at 15% [7], 23% [3] the rate of methanol oxidation, no substrate for Phanerochaete chrysosporium [15]) [3, 7, 8, 11]

16 2-Propyn-1-ol + O_2 (i.e. propargyl alcohol, oxidized at 45% the rate of methanol oxidation [1], oxidized to mechanism-based irreversible inactivator: propynal) [1, 4, 5]

17 1,4-Butynediol + O_2 (oxidized to mechanism-based irreversible inactivator: 4-hydroxy-2-butynal) [5]

18 Ethylene glycol + O_2 (oxidation at 1.7% the rate of methanol oxidation) [15]

19 Ethylene glycol mono-methylether + O_2 (oxidation at 10% the rate of methanol oxidation) [15]

20 More (lower primary alcohols are oxidized to the corresponding aldehydes [1–10, 14, 15], unsaturated [1, 2] and halogenated alcohols [1, 2, 10] are also good substrates, isopropanol, benzyl alcohol and 2-mercaptoethanol are oxidized only at high concentrations [4], 2-propin-1-ol and methylene cyclopropyl alcohol are not suicide substrates [9], no substrates: longer-chain [8], branched-chain [1, 2], secondary [1, 2, 7, 8], tertiary [7, 8], cyclic [7], aromatic [7, 8], C-2-substituted alcohols [1], diols [7], aldehydes [7], DL-serine, propane diols [1], glycerol, vanillyl alcohol, formaldehyde [15]) [1–10, 14, 15]

Product spectrum

1 Formaldehyde + H_2O_2 [1–11, 14, 15]

2 Acetaldehyde + H_2O_2 [1, 7]

3 n-Propanal + H_2O_2 [1, 7]

4 n-Butanal + H_2O_2 [1, 7]

5 n-Pentanal + H_2O_2

6 n-Hexanal + H_2O_2

7 2-Propen-1-al + H_2O_2 (leads in vivo to cell intoxination) [10]

8 Isopropanal + H_2O_2

9 2-Buten-1-al + H_2O_2

10 ?
11 ?
12 3-Chloro-1-propanal + H_2O_2
13 4-Chloro-1-butanal + H_2O_2
14 Isobutanal + H_2O_2
15 Formate + H_2O_2 [3, 7, 8, 11]
16 Propyn-1-al + H_2O_2 (mechanism-based inhibitor) [5]
17 4-Hydroxy-2-butyn-1-al + H_2O_2 (mechanism-based inhibitor) [5]
18 ?
19 ?
20 ?

Inhibitor(s)

PCMB (complete inhibition [3, 4, 7, 8, 10], $t_{1/2}$: 11 min, beta-mercaptoacetic acid restores, substrates or products slow the inactivation rate, not tert-butanol [12]) [3, 4, 7, 8, 10, 12]; Mercuric acetate (strong) [3]; Mercuric chloride [7]; 5,5-Dithiobis-2-nitrobenzoate [7]; Phenylhydrazine (strong) [3]; o-Phenanthroline [3]; KCN [3, 10]; Iodoacetate (time-dependent inactivation, reversible on removal of excess inhibitor [12]) [7, 12]; Hg^{2+} (strong [3], reversible by mercaptoethanol [12]) [3, 12]; Mo^{6+} (strong) [3]; Cu^{2+} (reversible by EDTA [12]) [3, 7, 8, 12]; Sodium acetate (competitive) [12]; Cd^{2+} [3]; Ag^{2+} [12]; NaN_3 (strong, competitive to methanol [4]) [4, 10]; H_2O_2 (reversible [6], irreversible, pseudo-first order saturation kinetic, methanol or catalase partially protects [9]) [6, 7, 9, 10]; Diethyldicarbonate [12]; Urea (6 M at 30°C for 1–2 h: partial inactivation, urea + KBr: complete inactivation) [10]; KBr (3.5 M at 30°C for 1–2 h: partial inactivation, KBr + urea: complete inactivation) [10]; Formaldehyde (Hansenula polymorpha [11]) [10, 11]; 2-Aminoethanol [7]; Propynal (GSH or anaerobiosis prevents [5]) [5, 10]; 4-Hydroxy-2-butynal (GSH or anaerobiosis prevents [5]) [5, 10]; Hydroxylamine (slight [7], Hansenula polymorpha [11]) [7, 11]; Thiosemicarbazide (slight) [7]; Imidazol (slight) [7]; Metal chelators (slight) [7]; Cyclopropanol (inactivation, suicide substrate) [10, 17]; Cyclopropanone (suicide substrate [10], mechanism [13]) [10, 13]; Anaerobiosis (Hansenula polymorpha) [11]; More (no inhibition by NaCN [4], NEM, N-butylmaleimide, acrylonitrile, DTNB, methylmethanesulfonate [12], 2-propin-1-ol and methylenecyclopropyl alcohol are not suicide substrates [9]) [4, 9, 12]

Cofactor(s)/prosthetic group(s)/activating agents

FAD (requirement, flavoprotein, 1 mol FAD per mol subunit [3, 4, 8], 5.62 mol per mol enzyme [14, 15], 6 FAD per octamer, mechanism [16], 5 oxygen-stable red flavin semiquinones (ESR-spectroscopy) plus 2 oxidized flavins per octamer, the reconstituted enzyme which lacks the flavin semiquinones is still half as active as the native enzyme [17], noncovalently bound [10]) [1–11, 14–17]

Metal compounds/salts
More (no iron cofactor [4], no metal requirement [8]) [4, 8]

Turnover number (min^{-1})
3100 (1H-methanol) [9]; 6000 (2H-methanol) [9]

Specific activity (U/mg)
More [1, 9]; 2.8 [3]; 4.5 [9]; 6.6 [7]; 9.5 [2]; 10.94 [8]; 11.9 (Pichia pastoris) [6]; 16.5 [15]; 17.3 [14]; 20.7 [4]; 23.2 [10]

K_m-value (mM)
0.019 (methanol) [3]; 0.13 (ethanol) [3]; 0.2 (methanol) [4]; 0.4–0.44 (O_2 [6, 10], 4-hydroxy-2-butynal [5]) [5, 6, 10]; 0.5 (methanol) [7]; 0.7 (O_2 (+ 0.01 M methanol)) [6]; 0.785 (methanol) [15]; 1.0 (ethanol [4], O_2 (+ 100 mM methanol [6]) [4, 6]; 1.4 (methanol (+ 0.19 mM O_2)) [6]; 1.97 (ethanol) [15]; 2.6 (formaldehyde) [10]; 3.0 (methanol) [5]; 3.1 (methanol (+ 0.93 mM O_2)) [6]; 3.2 (2-propin-1-ol) [4]; 3.5 (formaldehyde) [7]; 6.1 (formaldehyde) [4]; 8.3 (n-propanol) [4]; 10.0 (2-propin-1-ol) [5]; 13.8 (n-propanol) [15]; 21.3 (n-butanol) [4]; 25.0 (isopropanol) [4]; 27.2 (2-mercaptoethanol) [4]; 35.9 (n-butanol) [15]; 36.0 (1,4-butynediol) [5]

pH-optimum
More (pI: 4.4–4.5 [3], pI: 5.4 [14, 15], pI: 6.2 [10], pI: 6.3 [6]) [3, 6, 10, 14, 15]; 5.5–8.5 (purified) [10]; 6.0 (in vivo, owing to acidic nature of peroxisomal matrix) [10]; 6.0–10.5 (plateau) [14, 15]; 6.3–9.0 (plateau) [1]; 7.5 (Pichia pastoris [6]) [3, 6]; 8.5 (Hansenula polymorpha) [6]; 7.5–9.0 [8]; 9.0 [7]

pH-range
5.5–11.0 (6% of maximal activity at pH 5.5 and 12% at pH 11.0) [15]; 6.0 (activity drops rapidly below) [14]; 6.0–10.0 (about 65% of maximal activity at pH 6.0 and 85% at pH 10.0) [7]; 6.5–8.3 (about half-maximal activity at pH 6.5 and 8.3, Pichia pastoris) [6]; 6.7–9.8 (about half-maximal activity at pH 6.7 and 9.8, Hansenula polymorpha) [6]; 7.0 (below, activity drops rapidly) [8]

Temperature optimum (°C)
25 (assay at) [1, 4]; 30 [8]; 37.0 (Pichia pastoris) [6]; 37.5 [3]

Temperature range (°C)
18–45 (half-maximal activity at 18°C and 45°C, Pichia pastoris) [6]

3 ENZYME STRUCTURE

Molecular weight
300000 (Pichia sp., gel filtration) [7]
310000 (Phanerochaete chrysosporium, gel filtration) [14, 15]
520000 (Candida sp., sedimentation equilibrium centrifugation) [3]
600000 (Candida boidinii, sedimentation equilibrium [8], Hansenula polymorpha, gel filtration [10]) [8, 10]
610000 (Poria contigua, sedimentation equilibrium ultracentrifugation) [4]
675000 (Pichia pastoris, meniscus depletion method) [6]

Subunits
Tetramer (4 × 75000, Phanerochaete chrysosporium, SDS-PAGE) [14, 15]
Octamer (8 × 65000, Candida sp., SDS-PAGE [3], 8 × 74000, Candida boidinii, SDS-PAGE [8], Hansenula polymorpha, calculated from amino acid sequence derived from nucleotide sequence of a structural gene [10],
8 × 79000, Poria contigua, SDS-PAGE [4], 8 × 80000, Pichia pastoris,
SDS-PAGE [6]) [3, 4, 6, 8, 10]
More (electron microscopic analysis: the complete structure is an octad aggregate composed of 2 tetragons face-to-face [4], the subunits are noncovalently associated [8], minimal molecular weight calculated from amino
acid analysis: 81000 [6]) [4, 6, 8]

Glycoprotein/Lipoprotein
No carbohydrates [1]

---

4 ISOLATION/PREPARATION

Source organism
Basidiomycete (strain B191039 [2]) [1, 2]; Polyporus obtusus [2]; Radulum
casearium [2]; Candida sp. (strain 25-A) [3]; Poria contigua (brown rot fungus) [4]; Candida boidinii [5, 8, 9, 11–13, 16]; Pichia pastoris (strain IFP 206
[6]) [6, 11]; Pichia sp. [7]; Hansenula polymorpha (strain DL-1 [17]) [6, 10,
11, 17]; Phanerochaete chrysosporium (syn. Sporotrichum pulverulentum,
strain K-3) [14, 15]

Source tissue
Mycelium [1, 2, 4]; Cell [3, 5–13]

Localisation in source
Cytoplasm [4, 6]; Microbodies [4, 8]; Peroxisomes (synthesis of enzyme monomers in cytoplasm) [10, 11]

Purification
Basidiomycete [1, 2]; Candida sp. [3]; Poria contigua [4]; Pichia pastoris
[6]; Pichia sp. [7]; Candida boidinii [8]; Phanerochaete chrysosporium [14,
15]

Crystallization
(Basidiomycete [1], Candida sp. [3], Pichia sp. [7]) [1, 3, 7]

Cloned
–

Renaturated
–

---

5 STABILITY

pH
More (unstable at acidic pH-values) [6]; 5.0 (complete inactivation after 3 h at 25°C) [1]; 6.0–8.0 (45 min stable at 30°C) [3]; 6.5 (below: rapid inactivation) [1, 2]; 6.5–8.5 (stable, below: rapid loss of activity, even at 0°C) [12]; 7.0–8.0 (stable, Pichia pastoris) [6]; 7.0–9.0 (stable) [1]; 7.0–10.0 (stable) [8]; 8.0–10.0 (stable, Hansenula polymorpha) [6]; 9.0 (and above: inactivation, Pichia pastoris) [6]

Temperature (°C)
20 (24 h stable) [10]; 25 (complete inactivation at pH 5.0 after 3 h) [1]; 30 (retains full activity for 10 min [3], activation occurs at pH 7.5–8.5 [10], about 30% loss of activity after 15 min [15]) [3, 10, 15]; 38 (50% loss of activity) [6]; 40 (about 85% loss of activity after 15 min) [15]; 45 (1 h stable) [10]; 50 (complete inactivation after 15 min) [15]; 60 (inactivation after 10 min) [3]

Oxidation

Organic solvent
Acetonitril, 2% v/v, up to 2 h stable at 0°C [12]

General stability information
Polyethylene glycol stabilizes [2]; 2-Mercaptoethanol does not stabilize, it is oxidized as substrate analogue [10]; Sucrose stabilizes during freezing and thawing [10]; Freezing leads to loss of activity [2]; Freezing at –20°C and thawing leads to 25% loss of activity [4]; Lyophilization leads to buffer-insoluble, powdered protein [6]; 0°C, stable in 2% v/v acetonitril up to 2 h [12]

Storage
–20°C, stable [8]; –20°C, 25% loss of activity after thawing [4]; 0°C, stable at pH 6.5–8.5 [12]; 4°C, purified, stable for 2 weeks [2, 4]; 4°C, lyophilized, 20% loss of activity within a year [2]; Crystalline stable for at least 5 months in 0.05 M phosphate buffer, pH 7.7 [1]; Crystalline or precipitate, in phosphate buffer, pH 7.5 with 60–80% ammonium sulfate, loss of 20% activity per month [10]

6 CROSSREFERENCES TO STRUCTURE DATABANKS

PIR/MIPS code
PIR1:OXHQAP (yeast (Hansenula polymorpha)); PIR2:A23483 (yeast (Pichia pastoris) (fragment)); PIR2:JC1117 (precursor yeast (Candida boidinii))

Brookhaven code

7 LITERATURE REFERENCES

[1] Janssen, F.W., Ruelius, H.W.: Biochim. Biophys. Acta,151,330–342 (1968)
[2] Janssen, F.W., Kerwin, R.M., Ruelius, H.W.: Methods Enzymol.,41B,364–369 (1975) (Review)
[3] Yamada, H., Shin, K.-C., Kato, N., Shimizu, S., Tani, Y.: Agric. Biol. Chem.,43,877–878 (1979)
[4] Bringer, S., Sprey, B., Sahm, H.: Eur. J. Biochem.,101,563–570 (1979)
[5] Nichols, C.S., Cromartie, T.H.: Biochem. Biophys. Res. Commun.,97,216–221 (1980)
[6] Couderc, R., Baratti, J.: Agric. Biol. Chem.,44,2279–2289 (1980)
[7] Patel, R.N., Hou, C.-T., Laskin, A.I., Derelanko, P.: Arch. Biochem. Biophys.,210, 481–488 (1981)
[8] Sahm, H., Schütte, H., Kula, M.-R.: Methods Enzymol.,89,424–428 (1982) (Review)
[9] Geissler, J., Ghisla, S., Kroneck, P.M.H.: Eur. J. Biochem.,160,93–100 (1986)
[10] Van der Klei, I.J., Bystryck, L.V., Harder, W.: Methods Enzymol.,188,420–427 (1990)
[11] Bystryck, L.V., Dijkhuizen, L., Harder, W.: J. Gen. Microbiol.,137,2381–2386 (1991)
[12] Cromartie, T.H.: Biochemistry,20,5416–5423 (1981)
[13] Cromartie, T.H.: Biochem. Biophys. Res. Commun.,105,785–790 (1982)
[14] Eriksson, K.-E., Nishida, A.: Methods Enzymol.,161,322–326 (1988)
[15] Nishida, A., Eriksson, K.-E. : Biotechnol. Appl. Biochem.,9,325–338 (1987)
[16] Geissler, J., Hemmerich, P.: FEBS Lett.,126,152–156 (1981)
[17] Mincey, T., Tayrtien, G., Mildvan, A.S., Abeles, R.H.: Proc. Natl. Acad. Sci. USA,77,7099–7101 (1980)

Enzyme Handbook © Springer-Verlag Berlin Heidelberg 1995
Duplication, reproduction and storage in data banks are only
allowed with the prior permission of the publishers

1 NOMENCLATURE

EC number
1.1.3.14

Systematic name
Catechol:oxygen oxidoreductase (dimerizing)

Recommended name
Catechol oxidase (dimerizing)

Synonymes

CAS Reg. No.
37250-83-2

2 REACTION AND SPECIFICITY

Catalysed reaction
4 Catechol + 3 O_2 →
→ 2 dibenzo[1,4]dioxin-2,3-dione + 6 H_2O

Reaction type
Redox reaction

Natural substrates
Catechol + O_2 [1, 2]

Substrate spectrum
1 Catechol + O_2 [1, 2]

Product spectrum
1 Dibenzo[1,4]dioxin-2,3-dione + H_2O [1, 2]

Inhibitor(s)
Cyanide [2]; Hg^{2+} [2]; Co^{2+} [2]; $FeSO_4$ [2] Atebrin [2]; Ascorbate [2]

Cofactor(s)/prosthetic group(s)/activating agents
FAD (increases activity) [2]

Metal compounds/salts

Turnover number (min^{-1})

Specific activity (U/mg)
36.36 [2]

K_m-value (mM)
  0.5 (catechol) [2]

pH-optimum
  7.4 [2]

pH-range

Temperature optimum (°C)
  30 [2]

Temperature range (°C)

---

3 ENZYME STRUCTURE

Molecular weight

Subunits

Glycoprotein/Lipoprotein
  –

---

4 ISOLATION/PREPARATION

Source organism
  Spinacea oleracea (spinach) [1, 2]

Source tissue
  Leaf [1, 2]

Localisation in source

Purification
  Spinacea oleracea [2]

Crystallization
  –

Cloned
  –

Renaturated
  –

5 STABILITY

pH

Temperature (°C)

Oxidation

Organic solvent

General stability information

Storage
  -20°C, 1 month [2]

6 CROSSREFERENCES TO STRUCTURE DATABANKS

PIR/MIPS code

Brookhaven code

7 LITERATURE REFERENCES

[1] Uchiyama, S., Tamata, M., Tofuku, Y., Suzuki, S.: Anal. Chim. Acta,208,287–290 (1988)
[2] Nair, P.M., Vining, L.C.: Arch. Biochem. Biophys.,106,422–427 (1964)

1 NOMENCLATURE

EC number
1.1.3.15

Systematic name
(S)-2-Hydroxy-acid:oxygen 2-oxidoreductase

Recommended name
(S)-2-Hydroxy-acid oxidase

Synonymes
Glycolate oxidase
Hydroxy-acid oxidase A
Hydroxy-acid oxidase B
Oxidase, L-2-hydroxy acid
Hydroxyacid oxidase A
L-alpha-Hydroxy acid oxidase
L-2-Hydroxy acid oxidase
EC 1.1.3.1 (formerly)

CAS Reg. No.
9037-63-2

2 REACTION AND SPECIFICITY

Catalysed reaction
(S)-2-Hydroxy acid + O_2 →
→ 2-oxo acid + H_2O_2 (zero-order kinetics [3], carbanion formation is the first step in the enzyme catalyzed dehydrogenation reactions of alpha-hydroxy acids and alpha-amino acids [11], mechanism: possibly ternary complex involved [16])

Reaction type
Redox reaction
Oxidative deamination (rat kidney enzyme, hydroxyacid oxidase B) [1]

Natural substrates
L-alpha-Hydroxy acids + O_2 [5]
Thiol glyoxylate adducts + O_2 (glyoxylate thiohemiacetals [2]) [2, 12]
More (true physiological substrate remains to be found) [11]

Substrate spectrum
1 alpha-Hydroxy-n-butyrate analogues + O_2 [4]
2 alpha-Hydroxy-propionate analogues + O_2 (aliphatic and aromatic) [4]

Enzyme Handbook © Springer-Verlag Berlin Heidelberg 1995
Duplication, reproduction and storage in data banks are only
allowed with the prior permission of the publishers

3 Glycolate + O_2 (highly specific for glycolate [3], preferentially oxidized by isozyme A [1], best substrate for short-chain oxidase [20], best substrate for rat liver and leaf peroxisomal enzyme [19], oxidation at 1% the rate of L-alpha-hydroxyisocaproate by long-chain oxidase [20]) [1, 3, 6, 9, 14, 17, 19, 20]

4 L-alpha-Hydroxyisocaproate + O_2 (racemate [17], best substrate for long-chain oxidase, oxidation at 32% the rate of glycolate by short-chain oxidase [20]) [6, 11, 14, 17, 19, 20]

5 L-alpha-Hydroxy-n-caproate + O_2 (racemate [17], isozyme B [1], oxidation of the racemate at 5% the oxidation rate of glycolate by short-chain oxidase [20], oxidation at 25% the oxidation rate of L-alpha-hydroxyisocaproate by long-chain oxidase [20]) [1, 6, 11, 13, 17, 20]

6 L-alpha-Hydroxy-n-butyrate + O_2 (oxidation at 8% the oxidation rate of L-alpha-hydroxyisocaproate, long-chain oxidase [20]) [2, 6, 11, 13, 15, 17, 20]

7 L-alpha-Hydroxyvalerate + O_2 (oxidation of the racemate at 30% the rate of L-alpha-hydroxyisocaproate by long-chain oxidase and 10% the rate of glycolate by short-chain oxidase [20], racemate [17]) [6, 11, 13, 15, 17, 20]

8 L-alpha-Hydroxyisovalerate + O_2 (oxidation at 23% the oxidation rate of L-alpha-hydroxyisocaproate by long-chain oxidase [20]) [6, 13, 20]

9 L-alpha-Aminomonocarboxylic acids + H_2O + O_2 (oxidative deamination, the reaction proceeds more slowly than the oxidation of alpha-hydroxy acids [11]) [4, 11, 19]

10 L-Leucine + H_2O + O_2 [4, 11, 13]

11 L-Tryptophan + H_2O + O_2 (rat kidney enzyme [19]) [11, 13, 19]

12 L-Methionine + H_2O + O_2 [11, 13]

13 L-Lysine + H_2O + O_2 [13]

14 Phenyllactate + O_2 [11, 13]

15 beta-Phenyllactate + O_2 (oxidation at 94% the rate of L-alpha-hydroxyisocaproate oxidation by long-chain oxidase and 32% the rate of glycolate oxidation by short-chain oxidase) [20]

16 L-alpha-Hydroxyphenyllactate + O_2 [13]

17 p-Hydroxy-beta-phenyllactate + O_2 (oxidation at 90% the rate of L-alpha-hydroxyisocaproate oxidation by long-chain oxidase) [20]

18 L-alpha-Hydoxy-beta-methylvalerate + O_2 (oxidation at 81% the rate of L-alpha-hydroxyisocaproate oxidation by long-chain oxidase, poor substrate for short-chain oxidase) [20]

19 alpha-Hydroxy-gamma-methylmercaptobutyrate + O_2 (oxidation of the racemate at 42% the rate of L-alpha-hydroxyisocaproate oxidation by long-chain oxidase, poor substrate for short-chain oxidase) [20]

20 L-Lactate + O_2 (greater affinity than for L-alpha-hydroxybutyrate [17], poor substrate for both oxidases from pig kidney [20]) [4, 6, 11, 13, 15, 17, 19, 20]

21 beta-Chlorolactate + O_2 [10, 11, 13]
22 3-Indolelactate + O_2 (oxidation of the racemate at 38% the rate of
    L-alpha-hydroxyisocaproate oxidation by long-chain oxidase) [20]
23 Thiol-glyoxylate adducts + O_2 (i.e. glyoxylate thiohemiacetals, possible
    physiological substrates for the enzyme [2]) [2, 12]
24 2-Hydroxy-3-butenoate + O_2 (i.e. vinylglycolate) [10]
25 2-Hydroxy-3-butynoate + O_2 [10]
26 2-Hydroxy-3-pentynoate + O_2 [10]
27 2-Hydroxy-3-hexynoate + O_2 [10]
28 2-Hydroxy-3-heptynoate + O_2 [10]
29 2-Hydroxy-3-octynoate + O_2 [10]
30 L-Mandelate + O_2 (oxidation at 6% the rate of L-alpha-hydroxyisocaproa-
    te by long-chain oxidase, poor substrate for short-chain oxidase [20])
    [11, 20]
31 alpha-Hydroxyoctanoate + O_2 (oxidation at 72% the rate of L-alpha-hy-
    droxyisocaproate by long-chain oxidase, poor substrate for short-chain
    oxidase) [20]
32 alpha-Hydroxydecanoate + O_2 (oxidation at 62% the rate of L-alpha-hy-
    droxyisocaproate by long-chain oxidase, poor substrate for short-chain
    oxidase) [20]
33 Glyoxylate + O_2 [19]
34 More (hydroxy acid oxidase exists as two isozymes: hydroxy acid oxida-
    se A oxidizes preferentially short-chain aliphatic hydroxy acids, hydroxy
    acid oxidase B prefers long-chain aliphatic and aromatic hydroxy acids
    [6], the rat hydroxy acid oxidase B also acts as L-amino acid oxidase,
    no substrate of rat liver enzyme, hydroxy acid oxidase A: D-lactate,
    phenyllactate, indol-beta-lactate, mandelate, hydroxyphenyllactate,
    beta-hydroxybutyrate, alpha-hydroxyisobutyrate, L-leucine, glycine [6],
    L-lactate, L-phenylglycolate [3], equally active with C_3- and C_6-hydroxy
    acids, much less with C_4- and C_5-hydroxy acids [19]. No substrate of rat
    kidney enzyme, hydroxy acid oxidase B: DL-aminobutyrate, nitroethane,
    glycolate [13], L-serine, L-threonine, L-aspartate, L-glutamate, L-lysine,
    D-leucine, D-lactate, glycine and alpha-hydroxyisobutyrate [4]. No sub-
    strates of spinach enzyme: tartronic acid, malate, (+)-tatrate, (-)-tatrate
    and meso-tartrate [8]. No substrates for pig kidney oxidases:
    D-alpha-hydroxyisocaproate, D-alpha-hydroxy-beta-methylvalerate,
    D-alpha-hydroxyisovalerate, D-lactate, glucuronate, glucuronolactone,
    3-phospho-D-glycerate, DL-allo-isocitrate, L-malate, D-malate, (+)-, (-)
    and meso-tartrate, D-alanine, L-leucine, DL-serine, L-phenylalanine, gly-
    cerate, DL-alpha-hydroxyaspartate, DL-alpha-hydroxyisobutyrate,
    L-alpha-hydroxyglutarate [20], bromopyruvate is metabolized in vitro to
    pyruvate and bromolactate in a ratio of 10:1 [11]. No substrate for proto-
    zoon enzyme: phenyllactate [17]. The reversibility of the oxidation reac-
    tion has not been demonstrated [20]) [3, 4, 6, 8, 11, 13, 17, 19, 20]

Enzyme Handbook © Springer-Verlag Berlin Heidelberg 1995
Duplication, reproduction and storage in data banks are only
allowed with the prior permission of the publishers   3

Product spectrum

1 alpha-Keto-n-butanoate analogues + H_2O_2 [4]
2 alpha-Keto-propanoate analogues + H_2O_2 (aliphatic and aromatic) [4]
3 Glyoxylate + H_2O_2 [1, 3, 6, 9, 14, 17, 19, 20]
4 L-alpha-Ketoisocaproate + H_2O_2 [6, 11, 14, 17, 20]
5 L-alpha-Ketocaproate + H_2O_2 [1, 6, 11, 13, 17, 20]
6 L-alpha-Ketobutyrate + H_2O_2 [2, 6, 11, 13, 15, 17, 20]
7 L-alpha-Ketovalerate + H_2O_2 [6, 11, 13, 15, 17, 20]
8 L-alpha-Ketoisovalerate + H_2O_2 [6, 13, 20]
9 alpha Oxo acids + NH_3 + H_2O_2 [4, 11, 19]
10 L-alpha-Hydroxyisocaproate + NH_3 + H_2O_2 [4, 11, 13, 20]
11 alpha-Keto-1H-imidazole-3-propanoate + H_2O_2 + NH_3
12 2-Oxo-4-(methylthio)butanoate + H_2O_2 + NH_3
13 ?
14 ?
15 2-Oxo-3-phenylpropanoate + H_2O_2
16 ?
17 ?
18 2-Oxo-3-methylvalerate + H_2O_2
19 2-Oxo-4-methylmercaptobutyrate + H_2O_2
20 Pyruvate + H_2O_2 [4, 6, 13, 15, 20]
21 3-Chloropyruvate + H_2O_2 [10]
22 ?
23 Oxalyl thioester + H_2O_2 [2, 12]
24 2-Oxo-3-butenoate + H_2O_2 [10]
25 2-Oxo-3-butynoate + H_2O_2 [10]
26 2-Oxo-3-pentynoate + H_2O_2 [10]
27 2-Oxo-3-hexynoate + H_2O_2 [10]
28 2-Oxo-3-heptynoate + H_2O_2 [10]
29 2-Oxo-3-octynoate + H_2O_2 [10]
30 2-Oxobenzeneacetate + H_2O_2
31 2-Oxooctanoate + H_2O_2
32 2-Oxodecanoate + H_2O_2
33 Oxalate + H_2O_2 (enzyme of rat liver and leaves) [19]
34 ?

Inhibitor(s)

Iodoacetate (inhibits the recombination of apoenzyme with flavin nucleotide, no inhibition of native enzyme [3], inhibits short-chain oxidase at higher concentrations [20]) [3, 8, 20]; Iodoacetamide [8]; alpha-Oxophenazine [3]; Cu^{2+} (long-chain oxidase [20]) [3, 6, 14, 20]; Heavy metal ions [20]; Aliphatic aldehydes [3]; Aromatic aldehydes [3]; Diethyldithiocarbamate (competitive [14]) [6, 14]; Hydroxylamine (competitive to glycolate [3]) [3, 14]; Atebrin (flavin enzyme inhibitor, inhibits long-chain oxidase) [20]; Diphenylglycolate (competitive to glycolate [3]) [3, 14]; KCN [6, 14]; PCMB (non-competitive [20]) [6, 8, 14, 20]; o-Iodosobenzoate (inhibits short-chain oxidase at higher concentrations) [6, 14, 20]; DL-Lipoate (non-competitive) [20]; L-beta-Phenyllactate (competitive to glycolate) [20]; Glycolate (at concentrations above 1.7 mM) [16]; alpha-Hydroxybutyrate [20]; L-alpha-Hydroxyisocaproate (high concentrations) [20]; Monocarboxylic acids (e.g. acetate, propionate, butyrate, valerate, hexanoate, heptanoate, non-competitive to glycolate or glyoxylate/dichlorophenolindophenol as electron acceptor) [16]; Dicarboxylic acids (e.g. oxalate, succinate, malonate, competitive to glycolate/dichlorophenolindophenol as electron acceptor and with respect to glyoxylate/oxygen as electron acceptor, non-competitive inhibitors with respect to glycolate/oxygen and to glyoxylate/dichlorophenolindophenol as electron acceptors) [16]; Pyruvate (competitive to lactate) [13]; Oxalate [16]; 2-Hydroxy-3-butynoate [10, 13]; 2-Keto-3-butenoate [10]; Phosphate (with ferricyanide as electron acceptor, phosphate concentrations above 100 mM inhibit [16]) [11, 16]; Arsenate (with ferricyanide as electron acceptor) [16]; alpha-Hydroxycaproate [15]; alpha-Hydroxyvalerate [15]; alpha-Ketobutyrate (competitive) [15]; alpha-Ketovalerate (competitive) [15]; alpha-Ketoisovalerate (competitive) [15]; alpha-Ketoisocaproate (competitive) [15]; Dihydrolipoic acid (competitive to alpha-hydroxybutyrate) [12]; L-Leucine (competitive to alpha-hydroxybutyrate [13]) [13, 20]; L-Phenylalanine [20]; Cysteine [14]; More (short-chain oxidase is not inhibited by KCN, NaN$_3$, 2,2'-bipyridyl, CuSO$_4$ [20], dicarboxylic acids: 1 mol inhibitor per active site, inhibition studies [16]) [16, 20]

Cofactor(s)/prosthetic group(s)/activating agents

FMN (0.5 FMN-residues per subunit of isozyme B [4, 13], 1.0 FMN-residue per subunit [1, 11], 2 mol FMN per enzyme molecular weight of 100000 [7], one active site per flavin, i.e. one per dimer [10], inactivation without FMN [8, 20], K_m: 0.00113 mM [14], 0.02 mM [3]) [1, 3–5, 7–15, 19, 20]; FAD (can replace FMN [3], K_m: 0.05 mM [3], no FAD detectable [20]) [3]; Pyocyanine (activation, oxygen or ferricyanide as terminal electron acceptor) [3]; 2-Amino-3-hydroxyphenazine (activation in the presence of flavin nucleotides, cannot replace FMN or FAD) [3]; Isatin-6-carboxylate (activation in the presence of flavin nucleotides, cannot replace FMN or FAD) [3]; Ascorbate (slight activation) [6]; Green chromophore (FMN-free apoprotein is of green colour) [7]; Thiols (such as ethanethiol, 1-propanethiol, 2-mercaptoethanol, N-acetylcysteamine, propane-1,3-dithiol, dihydrolipoic acid, Coenzyme A, D-phosphopantetheine, pantetheine, activation, increase of rate of glyoxylate oxidation, not very reactive are GSH, L-cysteine and cysteamine [12]) [2, 12]; 2,6-Dichlorophenolindophenol (can act as electron acceptor in vitro) [16]; EDTA (activation) [20]; Ferricyanide (can act as electron acceptor) [7]; More (riboflavin cannot replace FMN [3], modified apoenzyme is not reactivated by FMN and/or FAD, EDTA, Mo, Cd, Fe, Cu, cysteine or lipoic acid [20], no NAD$^+$ or NADP$^+$ requirement [20], apoenzyme of short-chain oxidase exhibits atypical flavin spectrum [20]) [3, 20]

Metal compounds/salts
Phosphate (5–100 mM phosphate activate glycolate oxidase with $K_3Fe(CN)_6$ or 2,6-dichlorophenolindophenol as electron acceptor and glycolate or glyoxylate as substrate, formation of spectrophotometrically detectable enzyme/phosphate complex) [16]; Arsenate (5–100 mM activate glycolate oxidase with $K_3Fe(CN)_6$ or 2,6-dichlorophenolindophenol as electron acceptor and glycolate or glyoxylate as substrate) [16]; More (no metal cofactor requirement [6, 7, 20], Cl^- or HEPES does not activate [16]) [6, 7, 16, 20]

Turnover number (min^{-1})
6.3 (L-leucine, pH 8.85) [4]; 26 (L-lactate) [4]; 51.6 (L-mandelate) [11]

Specific activity (U/mg)
0.416 (soluble enzyme) [13]; 0.45 (mitochondrial enzyme) [13]; 0.63 [14]; 0.9 (isozyme B) [1]; 1.05 (leucine oxidation) [4]; 1.2–1.5 [7, 11]; 1.625 (L-alpha-hydroxyisocaproate oxidation) [6]; 2.08 [9]; 3.66 (isozyme A) [1]

K_m-value (mM)
0.007 (glyoxylate thiohemiacetal of DL-dihydrolipoate) [2]; 0.021 (glyoxylate thiohemiacetal of propane-1,3-dithiol) [2]; 0.03 (propane-1,3-dithiol/glyoxylate adduct) [12]; 0.1 (glycolate) [9, 19]; 0.14 (2-hydroxy-3-octynoate) [10]; 0.22 (glycolate) [14]; 0.28 (2,6-dichlorophenolindophenol) [16]; 0.3 (O_2, (+ alpha–hydroxybutyrate)) [12]; 0.31 (glycolate, short-chain oxidase) [20]; 0.38 (2-hydroxy-3-heptynoate) [10]; 0.40 (N-acetylcysteamine/glyoxylate adduct) [12]; 0.42 (glyoxylate thiohemiacetal of N-acetylcysteamine [2], glycolate [16]) [2, 16]; 0.46 (O_2, (+ propane-1,3-dithiol/glyoxylate adduct)) [12]; 0.5 (glycolate) [6]; 0.59 (glyoxylate thiohemiacetal of 2-mercaptoethanol) [2]; 0.6 (alpha-hydroxybutyrate) [2]; 0.67 (glyoxylate thiohemiacetal of D-pantetheine) [2]; 0.68 (L-alpha-hydroxyisocaproate) [20]; 0.70 (D-pantetheine/glyoxylate adduct) [12]; 0.75 (2-mercaptoethanol/glyoxylate adduct) [12]; 1.2 (alpha-hydroxybutyrate) [12]; 1.24 (alpha-hydroxyisocaproate) [13, 14]; 1.34 (alpha-hydroxycaproate) [6]; 1.65 (L-alpha-hydroxyisocaproate) [6]; 1.78 (glyoxylate) [14]; 1.9 (L-alpha-phenyllactate) [13]; 2.0 (glyoxylate thiohemiacetal of D-phosphopantetheine) [2]; 2.1 (glycolate, under 100% O_2-atmosphere) [3]; 2.2 (coenzyme A/glyoxylate adduct [12], L-beta-phenyllactate, long-chain hydroxylase [20]) [12, 20]; 2.4 (L-alpha-hydroxy-beta–methylvalerate) [20]; 2.5 (L-alpha-hydroxyisocaproate, long-chain oxidase) [20]; 3.0 (glyoxylate thiohemiacetal of coenzyme A) [2]; 3.2 (alpha-hydroxycaproate) [13]; 4.0 (2-hydroxy-3-butynoate) [10]; 4.68 (L-lactate) [4]; 7.0 (2-hydroxy-3-hexynoate) [10]; 8.0 (alpha-hydroxyisovalerate) [13]; 8.5 (L-lactate) [6]; 9.0 (2-hydroxy-3-pentynoate) [10]; 10.0 (2-hydroxy-3-butenoate, i.e. vinylglycolate) [10]; 12.7 (alpha-hydroxybutyrate) [6, 8]; 13.0 (alpha-hydroxyvalerate) [13]; 13.1 (L-leucine) [4]; 14.0 (alpha-hydroxybutyrate) [13];

15.0 (L-leucine) [13]; 27.0 (lactate) [13]; 28.0 (beta-chlorolactate) [10, 13];
40.0 (L-tryptophane) [13]; 53.0 (L-methionine) [13]; 71.0 (phenyllactate)
[13]; 90.0 (L-lysine) [13]; More (different K_m-values at suboptimal pH-values
in phosphate or imidazole buffer) [11]

pH-optimum
7.0–9.0 (alpha-hydroxybutyrate) [17]; 7.7 (short-chain oxidase) [20]; 7.9
(long-chain oxidase) [20]; 8.0 (L-lactate [4], L-lactate in 0.033 M phosphate
buffer [11]) [4, 11]; 8.2 (glycerate) [17]; 8.4 [11, 13]; 8.5 (approximately, with
three different substrates) [6]; 8.6–8.9 (0.033 M Tris buffer [11], L-alpha-hydroxybutyrate [15]) [11, 15]; 8.7 (L-leucine) [4]; 8.8 [3]; 8.8–9.0 (L-lactate or
glycolate) [17]; 8.9–9.4 (0.033 M Tris buffer, L-alpha-hydroxyisocaproate)
[15]

pH-range
6.0–9.0 (about 70% of maximal activity at pH 6.0 and 100% at pH 9.0,
alpha-hydroxybutyrate) [17]; 6.0–9.4 (about 50% of maximal activity at pH
6.0 and 9.4, L-lactate) [6]; 7.0–9.0 (about 60% of maximal activity at pH 7.0
and 100% at pH 9.0, L-lactate or glycolate) [17]; 7.0–10.0 (about 50% of
maximal activity at pH 7.0 and 10.0, L-alpha-hydroxyisocaproate) [6];
7.3–8.6 (about 60% of maximal activity at pH 7.3 and 8.6, L-lactate) [4];
7.5–9.0 (about 50% of maximal activity at pH 7.5 and 9.0) [13]; 8.0–9.4
(about 66% of maximal activity at pH 8.0 and 9.4, L-leucine) [4]

Temperature optimum (°C)
25 (assay at) [2, 7, 12, 13]; 30 (assay at) [1, 8, 9, 11]; 37 (assay at) [4, 6, 18,
20]

Temperature range (°C)

---

3 ENZYME STRUCTURE

Molecular weight
140000 (spinach, ultracentrifugation) [8]
150000 (rat, meniscus depletion method) [5]
150000–180000 (rat, gel exclusion chromatography) [5]
169000 (chicken, gel electrophoresis) [9]
190000 (rat, gel filtration) [13]
210000 (rat, gel filtration, SDS-gel electrophoresis of enzyme cross-linked
with dimethylsuberimidate) [13]
250000 (rat, sedimentation equilibrium) [13]
270000 (spinach, velocity sedimentation) [8]
300000 (rat, gel filtration) [6]

Enzyme Handbook © Springer-Verlag Berlin Heidelberg 1995
Duplication, reproduction and storage in data banks are only
allowed with the prior permission of the publishers

Subunits
Tetramer (4 × 37000, mouse, SDS-PAGE [1], 4 × 37760, chicken, amino acid analysis, 4 × 39000, chicken, SDS-PAGE [9], 4 × 39000 rat, SDS-PAGE [11], 4 × 40000, rat, SDS-PAGE [1], 4 × 43000, rat, SDS-PAGE [5], 4 × 47500, rat, SDS-PAGE [13]) [1, 5, 9, 11, 13]
? (x × 51500, pig, amino acid analysis [7], x × 66800–71900, spinach, minimal molecular weight calculation from spectrophotometric and fluorometric data [8]) [7, 8]
More (both isozymes tend to self-associate at protein concentrations above 0.5 mg/ml, following a monomer-dimer-tetramer model [5], N-terminus of polypeptide chain of isozyme A blocked [9]) [5, 9]

Glycoprotein/Lipoprotein
–

---

4 ISOLATION/PREPARATION

Source organism
Chicken (white leghorn pullet [18]) [9, 18]; Mouse [1]; Pig [7, 16, 20]; Rat (Wistar strain [1, 6], Sprague-Dawley [4]) [1–6, 10–15, 19]; Spinach [8]; Rana pipiens (frog) [18]; Plants (green) [19]; Yeast [19]; Mammals [19]; Algae (multicellular) [19]; Tetrahymena pyriformis (ciliated protozoon) [17]

Source tissue
Kidney (cortex [20]) [1, 2, 4, 5, 10–13, 15, 18–20]; Leaf (C$_3$-plants) [8, 19]; Liver [1, 3, 5–7, 9, 14, 16, 18–20]; Seeds (germinating) [19]; Bundle-sheath cells (C$_4$-plants) [19]

Localisation in source
Mitochondria (long- and short-chain oxidase [20]) [4, 6, 13, 15, 20]; Peroxisomes (mainly [18]) [2, 6, 9, 11, 12, 18, 19]; Cytoplasm [3, 4, 6–8, 10, 13, 15, 18]; Microsomes (long- and short-chain oxidase) [20]; Lysosomes [18]

Purification
Chicken [9]; Mouse [1]; Pig [7, 20]; Rat (hydroxyapatite chromatography [11]) [1, 3–6, 11, 13, 15]; Spinach (preparative electrophoresis) [8]

Crystallization
(pig [7], rat [4], spinach [8]) [4, 7, 8]

Cloned
–

Renatured
–

5 STABILITY

pH
6.0 (and below, denaturation under acidic conditions) [13]; 7.0 (stable in
1.0 M potassium phosphate buffer) [20]; 7.0–8.8 (stable, outside this range
increasing inactivation rate) [8]; 8.4 (stable, when frozen in 0.1 M Tris/HCl
buffer) [11]

Temperature (°C)
25 (stable for 24 h) [13]; 60 (5 min: mitochondrial enzyme retains 63% activity, soluble enzyme retains 97% activity) [15]; 65 (5 min: mitochondrial enzyme retains 16% activity, soluble enzyme 45% activity) [15]

Oxidation

Organic solvent

General stability information
Dilution inactivates, addition of crystalline human albumin protects [6]; Free-
ze-thawing inactivates dilute enzyme solutions [6]; Shaking inactivates dilute
solutions [6]; Ammonium sulfate stabilizes during storage [7]; Light leads to
photodecomposition [7]; Increase of ionic strength inactivates crude and
partially purified preparations, FMN retards rate of this inactivation [8]; High
salt concentration stabilizes crystalline enzyme [8]; FMN stabilizes [8]; Dialysis, against potassium bromide for two days, stable [13]; Lyophilization,
stable to [13]; Redissolution, stable to [13]; Freezing in 0.1 M Tris/HCl buffer,
pH 8.4, stable [11]; EDTA stabilizes and doubles activity of homogenate during purification [20]; Glycerol stabilizes during storage [7]

Storage
–80°C, unstable [9]; –20°C, unstable [9]; –20°C, stable as ammonium sulfate suspension in 0.1 M sodium phosphate without deterioration [7]; –20°C,
crystalline enzyme stable for several months without loss of activity in the
presence of glycerol or ammonium sulfate [7]; –20°C, long-chain oxidase,
hydroxyapatite purified preparation stable in 1.0 M potassium phosphate
buffer, pH 7 [20]; –20°C, purified short-chain oxidase stable for several
months in 1.0 M potassium phosphate buffer, pH 7.0 [20]; Frozen, with or
without EDTA stable for at least 2 months without loss of activity [6]; Frozen,
unstable [9]; –5 – 2°C, purified preparation, stable for many months in the
dark [8]; 0°C, stable for at least a week [20]; 0 – 3°C, dialyzed enzyme loses
10% of its initial activity in 2 weeks [4]; 1°C, in the dark, in the presence of
FMN: 97% of initial activity after 2 days, 89% after 5 days, in the absence of
FMN: 31% of initial activity after 2 days and 18% after 5 days [8]; 4°C, in the
dark stable for several months [13]; 4°C, quite stable for several weeks in
buffer [9]; 4°C, stable for months as ammonium sulfate precipitate [11]; 4°C,
in ammonium sulfate, FMN-loss more pronounced for isozyme A, 70% loss
of activity, than for isozyme B, 30% loss of activity [1]

6 CROSSREFERENCES TO STRUCTURE DATABANKS

PIR/MIPS code
PIR3:S29407 (Arabidopsis thaliana (fragment)); PIR2:A40838 (rat);
PIR3:S33322 (rat); PIR2:A31138 (rat (fragment)); PIR2:S00621 (spinach);
PIR1:OXSPH (peroxisomal spinach)

Brookhaven code

---

7 LITERATURE REFERENCES

[1] Duley, J.A., Holmes, R.S.: Eur. J. Biochem.,63,163–173 (1976)
[2] Gunshore, S., Brush, E.J., Hamilton, G.A.: Bioorg. Chem.,13,1–13 (1985)
[3] Kun, E., Dechary, J.M., Pitot, H.C.: J. Biol. Chem.,210,269–280 (1954)
[4] Nakano, M., Danowski, T.S.: J. Biol. Chem.,241,2075–2083 (1966)
[5] Phillips, D.R., Duley, J.A., Fennell, D.J., Holmes, R.S.: Biochim. Biophys.
     Acta,427,679–687 (1976)
[6] Nakano, M., Ushijima, Y., Saga, M., Tsutsumi, Y., Asami, H.: Biochim. Biophys.
     Acta,167,9–22 (1968)
[7] Schuman, M., Massey, V.: Biochim. Biophys. Acta,227,500–520 (1971)
[8] Frigerio, N.A., Harbury, H.A.: J. Biol. Chem.,231,135–157 (1958)
[9] Dupuis, L., DeCaro, J., Brachet, P., Puigserver, A.: FEBS Lett.,266,183–186 (1990)
[10] Cromartie, T.H., Walsh, C.: Biochemistry,14,3482–3490 (1975)
[11] Urban, P., Chirat, I., Lederer, F.: Biochemistry,27,7365–7371 (1988)
[12] Brush, E.J., Hamilton, G.A.: Biochem. Biophys. Res. Commun.,103,1194–1200
     (1981)
[13] Cromartie, T.H., Walsh, C.: Biochemistry,14,2588–2596 (1975)
[14] Ushijima, Y.: Arch. Biochem. Biophys.,155,361–367 (1973)
[15] Domenech, C.E., Machado, E.E., Blanco, A.: Biochim. Biophys. Acta,321,54–63
     (1973)
[16] Schuman, M., Massey, V.: Biochim. Biophys. Acta,227,521–537 (1971)
[17] Eichel, H.J.: Biochim. Biophys. Acta,128,183–186 (1966)
[18] Scott, P.J., Visentin, L.P., Allen, J.M: Ann. N. Y. Acad. Sci.,168,244–264 (1969)
[19] Tolbert, N.E.: Annu. Rev. Biochem.,50,133–157 (1981) (Review)
[20] Robinson, J.C., Keay, L., Molinari, R., Sizer, I.W.: J. Biol. Chem.,237,2001–2010
     (1962)

1 NOMENCLATURE

EC number
1.1.3.16

Systematic name
Ecdysone:oxygen 3-oxidoreductase

Recommended name
Ecdysone oxidase

Synonymes
Oxidase, ecdysone
beta-Ecdysone oxidase

CAS Reg. No.
56803-12-4

2 REACTION AND SPECIFICITY

Catalysed reaction
Ecdysone + O_2 →
→ 3-dehydroecdysone + H_2O_2

Reaction type
Redox reaction

Natural substrates
Ecdysone + O_2 (first step in reaction sequence for the conversion of ecdysone and 20-hydroxyecdysone to their 3(alpha)-epimers [1], probably part of the catabolic system that inactivates the molting hormones [3]) [1, 3]
20-Hydroxyecdysone + O_2 (first step in reaction sequence for the conversion of ecdysone and 20-hydroxyecdysone to their 3(alpha)-epimers [1], probably part of the catabolic system that inactivates the molting hormones [3]) [1, 3]

Substrate spectrum
1 Ecdysone + O_2 (ir [3], i.e.
2beta,3beta,14alpha,22,25-pentahydroxy-5beta-cholest-7-en-6-one [3], 2,6-dichloroindophenol can act as acceptor [3, 6], artificial electron acceptors in vitro: ferricyanide < methylene blue < 2,6-dichloroindophenol < 5-methazine methyl sulfate [3, 7]) [1, 3, 6, 7]

Enzyme Handbook © Springer-Verlag Berlin Heidelberg 1995
Duplication, reproduction and storage in data banks are only
allowed with the prior permission of the publishers

2 20-Hydroxyecdysone + O_2 [1]
3 Ecdysteroids + O_2 (e.g. inokosterone [3, 6], makisterone [3, 6], 2-deoxy-20-hydroxyecdysone [3, 6], cyasterone [3, 6], 2-deoxyecdysone [3, 6], structural requirements: alpha,beta-unsaturated keto function at C-6 and several hydroxyl functions including the hydroxyls at C-3 and at C-22 [3]) [3, 6]

Product spectrum
1 3-Dehydroecdysone + H_2O_2 (ir [3], i.e. 2beta,14alpha,22,25-tetrahydroxy-5beta-cholest-7-en-3,6-dione [3]) [1, 3]
2 3-Dehydro-20-hydroxyecdysone + H_2O_2 [1]
3 3-Dehydroecdysteroids + H_2O_2 [3, 6]

Inhibitor(s)
Methanol [1, 6]; $HgCl_2$ [6]

Cofactor(s)/prosthetic group(s)/activating agents

Metal compounds/salts

Turnover number (min^{-1})

Specific activity (U/mg)
More [3, 7]

K_m-value (mM)
0.013 (ecdysone) [1]; 0.098 (ecdysone, radio assay) [3, 6]; 0.041 (ecdysone, optical assay) [3]; 0.031 (20-hydroxyecdysone) [3, 6]; 0.007 (2,6-dichloroindophenol) [3]

pH-optimum
6.5 [3, 7]

pH-range
5.3–8.6 (5.3: about 85% of activity maximum, 8.6: about 15% of activity maximum) [7]

Temperature optimum (°C)
45 [7]

Temperature range (°C)

3 ENZYME STRUCTURE

Molecular weight
240000 (Calliphora vicina, gel filtration [3], Calliphora erythrocephala, gel filtration [7]) [3, 7]

Subunits

Glycoprotein/Lipoprotein

–

---

4 ISOLATION/PREPARATION

Source organism
Manduca sexta (tobacco hornworm) [1]; Spodoptera littoralis (sixth instar) [2]; Calliphora vicina (blue blowfly) [3, 5, 6]; Pieris brassicae (lepidoptera) [4]; Calliphora erythrocephala [7]; Periplaneta americana (dictyoptera) [5]; Daphnis nerei (lepidoptera) [5]; Drosophila melanogaster (diptera) [5]; More (reaction product of ecdyson oxidase detected in: Locusta migratoria, Aeshna cyanea, Choristoneura fumiferana, Drosophila hydei, Tenebrio molitor, Gryllus bimaculatus) [3]

Source tissue
Larvae [1, 2]; Prepupae [7]; Pupae [3, 5]; Midgut [1, 2]; Wings [4]; Tegument [4]; Eggs [5]; Gut [4, 5]; Fat body [4, 5]

Localisation in source
Cytoplasm [1, 3, 5]

Purification
Calliphora vicina (blue blowfly) [3]; Calliphora erythrocephala [7]

Crystallization

–

Cloned

–

Renaturated

–

---

5 STABILITY

pH

Temperature (°C)

Oxidation

Organic solvent

General stability information
Stable to repeated freezing and thawing [3]; Partial loss of activity by lyophilization [6]; Stable to precipitation with ammonium sulfate and redissolvation or dialysis [6]

Enzyme Handbook © Springer-Verlag Berlin Heidelberg 1995
Duplication, reproduction and storage in data banks are only
allowed with the prior permission of the publishers   3

Storage
−25°C, more than 3 months [3, 6]; 0°C, several days [3]

6 CROSSREFERENCES TO STRUCTURE DATABANKS

PIR/MIPS code

Brookhaven code

7 LITERATURE REFERENCES

[1] Weinrich, G.F., Thompson, M.J., Svoboda, J.A.: Arch. Insect Biochem. Physiol.,12, 201–218 (1989)
[2] Milner, N.P., Rees, H.H.: Biochem. J.,231,369–374 (1985)
[3] Koolman, J.: Methods Enzymol.,111,419–429 (1985) (Review)
[4] Blais, C., Lafont, R.: Hoppe-Seyler's Z. Physiol. Chem.,365,809–817 (1984)
[5] Koolman, J.: Hoppe-Seyler's Z. Physiol. Chem.,359,1315–1321 (1978)
[6] Koolman, J., Karlson, P.: Eur. J. Biochem.,89,453–460 (1978)
[7] Koolman, J., Karlson, P.: Hoppe-Seyler's Z. Physiol. Chem.,356,1131–1138 (1975)

1 NOMENCLATURE

EC number
1.1.3.17

Systematic name
Choline:oxygen 1-oxidoreductase

Recommended name
Choline oxidase

Synonymes
Oxidase, choline
Choline dehydrogenase
Choline-cytochrome c reductase
Dehydrogenase, choline

CAS Reg. No.
9028-67-5 (identical with CAS Reg. No. of EC 1.1.99.1)

---

2 REACTION AND SPECIFICITY

Catalysed reaction
Choline + O_2 →
→ betaine aldehyde + H_2O_2

Reaction type
Redox reaction

Natural substrates
Choline + O_2 [1]

Substrate spectrum
1 Choline + O_2 [1–7]
2 Betaine aldehyde + O_2 + H_2O (46% of the activity with choline [1]) [1, 4]
3 N,N-Dimethylaminoethanol + O_2 (5.2% of the activity with choline [1]) [1, 4]
4 Triethanolamine + O_2 (2.6% of the activity with choline) [1]
5 Diethanolamine + O_2 (0.8% of the activity with choline) [1]
6 More (not: monoethanolamine, N-methylaminoethanol, methanol, ethanol, propanol, formaldehyde, acetaldehyde, propionaldehyde [1], photoreaction with covalently bound flavin [6]) [1, 6]

Product spectrum
1 Betaine aldehyde + H_2O_2 [1–7]
2 Betaine + H_2O_2 [1]
3 ?
4 ?
5 ?
6 ?

Inhibitor(s)
Ag^+ [4]; Hg^{2+} [4]; Cu^{2+} [4]; Zn^{2+} [4]; p-Chloromercuribenzoate [4]

Cofactor(s)/prosthetic group(s)/activating agents
FAD (flavoprotein [1, 3, 5], covalently bound flavin is
8alpha-[N(3)-histidyl]FAD [3, 5, 7], 2 mol FAD per mol of enzyme [4], 1 mol
of FAD per mol of enzyme [5], photoreaction with covalently bound flavin
[6]) [1, 3–7]

Metal compounds/salts

Turnover number (min^{-1})

Specific activity (U/mg)
More [1, 2]

K_m-value (mM)
0.87 (choline) [5]; 1.2 (choline) [1]; 1.3 (choline) [4]; 5.8 (betaine aldehyde)
[4]; 6.2 (betaine aldehyde) [5]; 8.7 (betaine aldehyde) [1]; 14 (N,N-dimethyl-
aminoethanol) [4]

pH-optimum
7.5 [1]; 9.0 [4]

pH-range
6–10 (about 60% of activity maximum at pH 6 and 10) [1]

Temperature optimum (°C)
40 [4]

Temperature range (°C)

3 ENZYME STRUCTURE

Molecular weight
72000 (Alcaligenes sp., HPLC gel filtration) [5]
83000 (Arthrobacter globiformis, gel filtration) [1]
120000 (Cylindrocarpon didymum M-1, gel filtration) [4]
145000 (Cylindrocapron didymum M-1, sedimentation velocity method) [4]

Subunits
Monomer (1 × 71000, Arthrobacter globiformis, SDS-PAGE [1], 1 × 66000, Alcaligenes sp., SDS-PAGE [5]) [1, 5]
Dimer (2 × 64000, Cylindrocarpon didymum M-1, SDS-PAGE in presence of 2-mercaptoethanol) [4]

Glycoprotein/Lipoprotein
–

4 ISOLATION/PREPARATION

Source organism
Arthrobacter globiformis (B-0577, NRRLB-11097 [3]) [1, 3]; Achromobacter cholinophagum [2]; Cylindrocarpon didymum M-1 [4, 7]; Alcaligenes sp. [5, 6]

Source tissue
Cell [1, 4, 7]

Localisation in source

Purification
Arthrobacter globiformis [1]; Cylindrocarpon didymum M-1 [4, 7]; Achromobacter cholinophagum [2]

Crystallization
–

Cloned
–

Renaturated
–

5 STABILITY

pH
6.0 (10 min, 40°C, 50% loss of activity) [4]; 7.0–9.0 (10 min, 40°C, stable) [4]; 9.5 (10 min, 40°C, 50% loss of activity) [4]

Temperature (°C)
40 (10 min, 20% loss of activity) [4]; 45 (10 min, 96% loss of activity) [4]

Oxidation

Organic solvent

General stability information
Glycerol, 10%, stabilizes against heat inactivation [4]

Storage

6 CROSSREFERENCES TO STRUCTURE DATABANKS

PIR/MIPS code
PIR2:A15398 (Alcaligenes sp. (fragment))

Brookhaven code

7 LITERATURE REFERENCES

[1] Ikuta, S., Imamura, S., Misaki, H., Horiuti, Y.: J. Biochem.,82,1741–1749 (1977)
[2] Tani, Y., Mori, N., Ogata, K.: Agric. Biol. Chem.,41,1101–1102 (1977)
[3] Ohishi, N., Yagi, K.: Biochem. Biophys. Res. Commun.,86,1084–1088 (1979)
[4] Yamada, H., Mori, N., Tani, Y.: Agric. Biol. Chem.,43,2173–2177 (1979)
[5] Ohta-Fukuyama, M., Miyake, Y., Emi, S., Yamano, T.: J. Biochem.,88,197–203 (1980)
[6] Ohta, M., Miura, R., Yamano, T., Miyake, Y.: J. Biochem.,94,879–892 (1983)
[7] Mori, N., Tani, Y., Yamada, H., Hayashi, R.: Agric. Biol. Chem.,45,539–540 (1981)

1 NOMENCLATURE

EC number
1.1.3.18

Systematic name
Secondary-alcohol:oxygen oxidoreductase

Recommended name
Secondary-alcohol oxidase

Synonymes
Polyvinyl alcohol oxidase
Secondary alcohol oxidase

CAS Reg. No.
71245-08-4

---

2 REACTION AND SPECIFICITY

Catalysed reaction
Secondary alcohol + O_2 →
→ ketone + H_2O_2

Reaction type
Redox reaction

Natural substrates
Polyvinyl alcohol + O_2 [1–10]

Substrate spectrum
1 Vinyl alcohol oligomers + O_2 (molecular weight of vinyl alcohol oligomers: 220–1500 [1, 8], other electron acceptors: 2,6-dichlorophenolindo-
   phenol, nitro blue tetrazolium [2, 4, 7]) [1–10]
2 5-Hydroxy-3-heptanone + O_2 (0.36% of polyvinyl alcohol activity) [1]
3 4-Hydroxy-2-nonanone + O_2 (6.02% of polyvinyl alcohol activity) [1]
4 3-Hydroxy-5-nonanone + O_2 (5.0% of polyvinyl alcohol activity) [1]
5 6-Hydroxy-4-nonanone + O_2 (2.89% of polyvinyl alcohol activity) [1]
6 7-Hydroxy-5-dodecanone + O_2 (3.12% of polyvinyl alcohol activity) [1]
7 8-Hydroxy-6-tridecanone + O_2 (3.4% of polyvinyl alcohol activity) [1]
8 2-Pentanol + O_2 (18.2% of polyvinyl alcohol activity) [2]
9 3-Pentanol + O_2 (10.1% of polyvinyl alcohol activity) [2]
10 2-Hexanol + O_2 (16.8% of polyvinyl alcohol activity) [2]
11 3-Hexanol + O_2 (17.4% of polyvinyl alcohol activity) [2]
12 2-Heptanol + O_2 (14.4% of polyvinyl alcohol activity) [2]

13 3-Heptanol + O_2 (10.7% of polyvinyl alcohol activity) [2]
14 4-Heptanol + O_2 (27.6% of polyvinyl alcohol activity) [2]
15 4-Octanol + O_2 (11.8% of polyvinyl alcohol activity) [2]
16 4-Nonanol + O_2 (10.4% of polyvinyl alcohol activity) [2]
17 More (not : 4-hydroxy-2-pentanone [1], overview primary alcohols, secondary alcohols [2, 7]) [2, 7]

Product spectrum
1 Corresponding ketones + H_2O_2 [1–10]
2 Heptan-3,5-dione + H_2O_2
3 Nonan-2,4-dione + H_2O_2
4 Nonan-3,5-dione + H_2O_2
5 Nonan-4,6-dione + H_2O_2
6 Dodecan-5,7-dione + H_2O_2
7 Tridecan-6,8-dione + H_2O_2
8 2-Pentanone + H_2O_2
9 3-Pentanone + H_2O_2
10 2-Hexanone + H_2O_2
11 3-Hexanone + H_2O_2
12 2-Heptanone + H_2O_2
13 3-Heptanone + H_2O_2
14 4-Heptanone + H_2O_2
15 4-Octanone + H_2O_2
16 4-Nonanone + H_2O_2
17 ?

Inhibitor(s)
Hg^{2+} [2, 4, 7]; Ni^{2+} [4, 8, 10]; Mn^{2+} [4]; Ba^{2+} [4]; Pb^{2+} [7]; Zn^{2+} [7]; Co^{2+} [8, 10]; Hydroxylamine [8, 10]; EDTA [8, 10]; Salicylaldoxime [8, 10]

Cofactor(s)/prosthetic group(s)/activating agents

Metal compounds/salts
Non-heme iron (one atom per molecule) [2, 7, 8]

Turnover number (min^{-1})

Specific activity (U/mg)
3.80 [2]; 1.38 [4]; 3.0 [7]

K_m-value (mM)
0.4–1.0 (polyvinyl alcohol) [4, 8]; 11 (4-heptanol) [4]

pH-optimum
7.0 [2, 7]; 8.0–8.4 [4]; 6.5–7.0 [9] 9.0 [10]

pH-range

Temperature optimum (°C)
45 [2, 4, 9]; 50 [7]; 40 [10]

Temperature range (°C)

---

3 ENZYME STRUCTURE

Molecular weight
49500 (Pseudomonas sp., sedimentation equilibrium) [7]
38500 (unidentified bacteria, gel filtration, sedimentation equilibrium) [2]
26000 (Pseudomonas sp., gel filtration) [8, 10]

Subunits
Monomer (1 × 45000, unidentified bacteria, SDS-PAGE [2], 1 × 50500, Pseudomonas sp., SDS-PAGE [7]) [2, 7]

Glycoprotein/Lipoprotein
–

---

4 ISOLATION/PREPARATION

Source organism
Pseudomonas sp. [1–10]; Unidentified bacteria (mixed culture) [2, 5]

Source tissue

Localisation in source
Extracellular [3, 4, 6, 10]; Cytosol [3, 4, 6]; Membranes [4, 6]

Purification
Bacterial mixed culture [2, 5]; Pseudomonas sp. [4, 7, 8, 10]

Crystallization
–

Cloned
–

Renaturated
–

---

5 STABILITY

pH
5.0–9.0 [2]; 4.5–9.0 [7]; 5.0–11.0 [10]

Temperature (°C)

45 (not stable above) [2]; 50 (not stable above) [7, 10]

Oxidation

Organic solvent

General stability information

Storage

---

6 CROSSREFERENCES TO STRUCTURE DATABANKS

PIR/MIPS code

Brookhaven code

---

7 LITERATURE REFERENCES

[1] Sakai, K., Hamada, N., Watanabe, Y.: Agric. Biol. Chem.,50,989–996 (1986)
[2] Sakai, K., Hamada, N., Watanabe, Y.: Agric. Biol. Chem.,49,817–825 (1985)
[3] Shimao, M., Nishimura, Y., Kato, N., Sakazawa, C.: Appl. Environ. Microbiol.,49, 8–10 (1985)
[4] Shimao, M., Tsuda, T., Takahashi, M., Kato, N., Sakazawa, C.: FEMS Microbiol. Lett.,20,429–433 (1983)
[5] Sakai, K., Hamada, N., Watanabe, Y.: Agric. Biol. Chem.,47,153–155 (1983)
[6] Shimao, M., Taniguchi, Y., Shikata, S., Kato, N., Sakazawa, C.: Appl. Environ. Microbiol.,44,28–32 (1982)
[7] Morita, M., Hamada, N., Sakai, K., Watanabe, Y.: Agric. Biol. Chem.,43,1225–1235 (1979)
[8] Suzuki, T.: Agric. Biol. Chem.,42,1187–1194 (1978)
[9] Morita, M., Watanabe, Y.: Agric. Biol. Chem.,41,1535–1537 (1977)
[10] Suzuki, T.: Agric. Biol. Chem.,40,497–504 (1976)

1 NOMENCLATURE

EC number
1.1.3.19

Systematic name
(S)-2-Hydroxy-2-(4-hydroxyphenyl)acetate:oxygen 1-oxidoreductase

Recommended name
4-Hydroxymandelate oxidase

Synonymes
L-4-Hydroxymandelate oxidase (decarboxylating)

CAS Reg. No.

2 REACTION AND SPECIFICITY

Catalysed reaction
(S)-2-Hydroxy-2-(4-hydroxyphenyl)acetate + O_2 →
→ 4-hydroxybenzaldehyde + CO_2 + H_2O_2

Reaction type
Redox reaction

Natural substrates
(S)-2-Hydroxy-2-(4-hydroxyphenyl)acetate + O_2 (enzyme of degradation
pathway from mandelic acid to benzoic acid [1]) [1, 2]

Substrate spectrum
1 (S)-2-Hydroxy-2-(4-hydroxyphenyl)acetate + O_2 (absolute requirement for
  molecular oxygen, specific for L-isomer of 4-hydroxymandelic acid) [1]

Product spectrum
1 4-Hydroxybenzaldehyde + CO_2 + H_2O_2

Inhibitor(s)
Reduced glutathione [1]; 2-Mercaptoethanol [1]; Dithiothreitol [1]; Cysteine
[1]; SDS [1]; Guanidine hydrochloride [1]; DL-3,4-Dihydroxymandelic acid
[1]; Atebrin [1]; Ag^+ [1]; Hg^{2+} [1]; 8-Hydroxyquinoline [1]; Potassium cyanide
[1]; Diethyldithiocarbamate [1]; Fe^{2+} [1]; Fe^{3+} [1]; Co^{2+} [1]; Cu^{2+} [1]; Zn^{2+} [1];
Hydroxylamine [1]; 2,4-Dinitrophenylhydrazine [1]; EDTA [1]

Cofactor(s)/prosthetic group(s)/activating agents
FAD (flavoprotein, FAD required) [1, 2]; FMN (74% of the activity with FAD)
[1]

Metal compounds/salts
  Mn^{2+} (required) [1, 2]

Turnover number (min^{-1})

Specific activity (U/mg)
  More [1]

K_m-value (mM)
  0.44 (DL-4-hydroxymandelate) [1]; 0.038 (FAD) [1]

pH-optimum
  6.6 [1]

pH-range

Temperature optimum (°C)
  55 [1]

Temperature range (°C)

3 ENZYME STRUCTURE

Molecular weight
  155000 (Pseudomonas convexa, gel filtration) [1]

Subunits

Glycoprotein/Lipoprotein
  –

4 ISOLATION/PREPARATION

Source organism
  Pseudomonas convexa [1, 2]

Source tissue
  Cell [1]

Localisation in source
  Membrane (bound) [1]

Purification
  Pseudomonas convexa (partial) [1]

Crystallization
  –

Cloned
  –

Renaturated

–

---

5 STABILITY

pH

Temperature (°C)

Oxidation

Organic solvent

General stability information
After solubilization the enzyme gradually loses activity when kept at 5°C, full reactivation by freezing and thawing [1]

Storage
–20°C, particulate fraction suspended in 0.025 M sodium phosphate buffer, pH 7.0, more than 6 months [1]; –20°C, 2 months, solubilized enzyme [1]

---

6 CROSSREFERENCES TO STRUCTURE DATABANKS

PIR/MIPS code

Brookhaven code

---

7 LITERATURE REFERENCES

[1] Bhat, S.G., Vaidyanathan, C.S.: Eur. J. Biochem.,68,323–331 (1976)
[2] Bhat, S.G., Vaidyanathan, C.S.: J. Bacteriol.,127,1108–1118 (1976)

1 NOMENCLATURE

EC number
1.1.3.20

Systematic name
Long-chain-alcohol:oxygen oxidoreductase

Recommended name
Long-chain-alcohol oxidase

Synonymes
Oxidase, long-chain fatty alcohol
Fatty alcohol oxidase [3, 5, 6]
Fatty alcohol:oxygen oxidoreductase [3]
Long-chain fatty acid oxidase

CAS Reg. No.
129430-50-8

2 REACTION AND SPECIFICITY

Catalysed reaction
2 Long-chain alcohol + O_2 →
→ 2 long-chain aldehyde + 2 H_2O_2 (ping-pong mechanism [8]) [1, 2, 4–8];
Long-chain alcohol + O_2 →
→ long-chain aldehyde + H_2O [3]

Reaction type
Redox reaction

Natural substrates
Long-chain alcohols + O_2 (reaction in wax ester catabolism in the storage wax bodies of jojoba plant during germination [2, 3], reaction in n-alkane catabolism [4], substrates are primary alcohols with a chain-length of C_6-C_{10} [1], monounsaturated long-chain alcohols [2], n-alkan-inducible enzyme [4, 6], constitutive enzyme, perhaps two enzymes [5]) [1–8]

Substrate spectrum
1 n-Octanol + O_2 (oxidation at 34% the rate of dodecanol oxidation [8]) [1, 4, 5, 8]
2 1-Dodecanol + O_2 (i.e. lauryl alcohol, preferred substrate [2–4, 7, 8]) [2–5, 7, 8]
3 Hexanol + O_2 [4]

4 1-Decanol + O_2 (best substrate [5], oxidation at 51% [3], 70% [8] the rate of dodecanol oxidation) [2–5, 8]

5 1-Tetradecanol + O_2 (i.e. myristyl alcohol, best substrate [5], oxidation at 74% [3], 75% [8] the rate of dodecanol oxidation) [2–5, 8]

6 1-Hexadecanol + O_2 (i.e. palmityl or cetyl alcohol, oxidation at 26% [3], 68% [7], 19% [8] the rate of dodecanol oxidation) [2–5, 7, 8]

7 1-Octadecanol + O_2 (i.e. stearyl alcohol, oxidation at 16% [3], 12% [8] the rate of dodecanol oxidation [3]) [2–4, 8]

8 1-Eicosanol + O_2 (i.e. arachidyl alcohol, poor substrate) [2, 7]

9 1-Docosanol + O_2 (i.e. behenyl alcohol, poor substrate) [2]

10 Eicosenol + O_2 (higher oxidation rate than saturated analog [2], oxidation at 19% [3], 1.9% [7] the rate of dodecanol oxidation [3]) [2, 3, 7]

11 Docosenol + O_2 (higher oxidation rate than saturated analog [2], oxidation at 13% the rate of dodecanol oxidation [3]) [2, 3]

12 cis-11-Hexadecenol + O_2 [4]

13 12-Bromododecanol + O_2 (oxidation at the same rate as dodecanol [5], oxidation at 92% [7] the rate of dodecanol oxidation) [5, 7]

14 Hex-trans-2-ene-1-ol + O_2 (preferred substrate, the 2-ene-group enhances activity) [1]

15 Geraniol + O_2 (i.e. 3,7-dimethylocta-trans-2,6-dien-1-ol, better substrate than cis-isomer, the 2-ene-group enhances activity) [1]

16 alpha,omega-Alkanediols + O_2 (from C_{10} to C_{16}) [5, 8]

17 Diols + O_2 (e.g. hexadecan-1,16-diol (oxidation at 31% the rate of dodecanol oxidation [7]), poor substrates [1]) [1, 7]

18 Terpenol + O_2 (poor substrate) [1]

19 16-Hydroxyhexadecanoic acid + O_2 [4, 7]

20 Farnesol + O_2 [7]

21 Chrysanthemyl alcohol + O_2 [7]

22 12-Hydroxydodecanoic acid + O_2 (poor substrate) [8]

23 2-Dodecanol + O_2 [8]

24 More (substrates are primary alcohols with a chain-length of C_6-C_{10}, optimal chain-length: C_8-C_{10} [1], weak activity towards fatty aldehydes [8], no substrates for Tanacetum enzyme: secondary, tertiary alcohols and Tris [1], no substrates for Simmondsia: ethanol [2, 3, 8], palmitic acid [2, 3], no substrates for Candida: dodecanal, omega-hydroxydodecanoate [4], omega-hydroxy fatty acids [5], geraniol, hexadecan-2-ol, dodecan-5-ol, 2-adamantol, 5-phenylpentan-1-ol, 4-cyclohexylbutan-1-ol, 1-adamantane ethanol [7], dodecan-4-ol, dodecan-5-ol [8], perhaps two long-chain alcohol oxidases in Candida bombicola: the two best substrates are decanol and tetradecanol [5]) [1–5, 7, 8]

Product spectrum
1 Octanal + H_2O_2 [1, 4]
2 1-Dodecanal + H_2O_2 [2, 4]
3 Hexanal + H_2O_2 [4]
4 1-Decanal + H_2O_2 [2]
5 1-Tetradecanal + H_2O_2 [2, 4]
6 Hexadecanal + H_2O_2 [2, 4]
7 1-Octadecanal + H_2O_2 [2, 4]
8 ?
9 ?
10 ?
11 ?
12 ?
13 ?
14 Hex-trans-2-ene-1-al + H_2O_2 [1]
15 ?
16 ?
17 ?
18 ?
19 ?
20 ?
21 ?
22 ?
23 ?
24 ?

Inhibitor(s)
Anaerobiosis (reversible by introduction of oxygen) [4]; p-Chloromercuriben-zoate (completely reversible with dithiothreitol) [2]; O_2 (together with light of 405 nm photochemical inactivation) [6]; Light (photochemical inactivation at wavelength of 450 nm, i.e. blue-light region of the spectrum of visible light [6, 8], complete inactivation, $t_{1/2}$: 7 min [8], enzyme from Yarrowia more sensitive to light than Candida tropicalis, inactivation diminished under anaerobic conditions, light of wavelength longer than 480 nm and shorter than 430 nm does not inactivate [6]) [6, 8]; More (no inhibitors: iodoacetate, $MnSO_4$, EDTA, p-hydroxymercuribenzoate, 2,2'-dipyridyl, 1,10-phenanthroline, 8-hydroxyquinoline, fluoride, azide [1], cyanide [1–3], pyrazole, decanethiol [4]) [1–4]

Cofactor(s)/prosthetic group(s)/activating agents
Flavin (not FAD, FMN, riboflavin, different absorption and fluorescence emission spectra) [8]; Detergents (e.g. cholate, activation, purified preparations do not need detergents for activation) [8]; More (no cofactor: riboflavin, 2,6-dichlorophenolindophenol/phenazine methosulfate, potassium ferricyanide, cytochrome c, nitro blue tetrazolium, tetramethyl-p-phenylenediamine [8], FAD, FMN [1, 8], NAD(P)$^+$ [1, 3]) [1, 3, 8]

Metal compounds/salts

Turnover number (min^{-1})

Specific activity (U/mg)
0.042 (glucose grown, tetradecanol) [5]; 0.07 [6]; 0.14 (decanol, in octane/aqueous phase 99:1) [7]; 0.144 (n-hexane grown, tetradecanol) [5]; 0.218 (dodecanol) [4]; 0.22 (decanol, in aqueous phase) [7]; 0.28 (Candida tropicalis) [6]

K$_m$-value (mM)
0.0015 (tetradecanol) [8]; 0.005 (hexadecanol) [8]; 0.0061 (dodecanol) [8]; 0.008 (dodecanol, aqueous phase) [7]; 0.042 (decanol) [8]; 0.061 (O$_2$) [8]; 0.19 (hex-trans-2-ene-1-ol) [1]; 0.49 (above, n-octanol) [1]; 1.0 (octanol) [8]; 1.56 (geraniol) [1]; 40.0 (dodecanol, octane/aqueous phase 99:1) [7]

pH-optimum
5.5–9.0 (plateau) [1]; 7.6–8.8 (decanol) [5]; 8.1–8.3 (tetradecanol) [5]; 9.0 (1-dodecanol) [2, 3]

pH-range
4.8–10.0 (about half-maximal activity at pH 4.8 and 10.0) [1]; 6.6–9.3 (about half-maximal activity at pH 6.6 and 9.3) [2]

Temperature optimum (°C)
18–22 [1]; 20 (below) [7]; 30 [5]

Temperature range (°C)
0–45 (about 60% of maximal activity at 0°C and about 30% of maximal activity at 45°C) [1]; 18–35 (about 60% of maximal activity at 18°C and 35°C, decanol) [5]; 20–38 (about half-maximal activity at 20 and 38°C, tetradecanol) [5]

3 ENZYME STRUCTURE

Molecular weight
145000 (Candida tropicalis, gel filtration) [8]
180000 (Tanacetum vulgare, gel filtration) [1]

Subunits
Dimer (2 × 70000, Candida tropicalis, SDS-PAGE [8], 1 × 94000 + 1 × 75000, Tanacetum vulgare, SDS-PAGE [1]) [1, 8]

Glycoprotein/Lipoprotein
–

4 ISOLATION/PREPARATION

Source organism
Simmondsia chinensis (jojoba) [2, 3]; Candida tropicalis [4, 6–8]; Candida bombicola (syn. Torulopsis bombicola) [5]; Yarrowia lipolytica [6]; Tanacetum vulgare [1]

Source tissue
Wax bodies [2, 3]; Cell [4–7]; Leaf [1]

Localisation in source
Membrane [2–8]; Glyoxysomes [2]; Microsomes [6, 8]

Purification
Candida tropicalis (partial [6] [6, 8]; Simmondsia chinensis (partial) [3]; Tanacetum vulgare [1]; Yarrowia lipolytica (partial) [6]

Crystallization
–

Cloned
–

Renaturated
–

5 STABILITY

pH
6.0 (inactivation after 3 min) [8]; 7.0 (and below unstable) [8]; 9.0 (at least 15 min stable) [8]

Temperature (°C)
25 (at least 15 min stable [5], inactivation overnight [8]) [5, 8]; 30 (inactivation after 10 min) [5]; 37 (inactivation after 1 min) [5]

Oxidation
Oxygen together with light of 450 nm inactivates [6]

Organic solvent

Acetone, stable to [1]; Octane, enzyme stable in 99% octane/1% H_2O [7]

General stability information

Prolonged dialysis against EDTA, stable to [1]; Freezing leads to 50% loss of activity overnight [3]; Light of 450 nm inactivates, Yarrowia lipolytica enzyme is more sensitive than Candida tropicalis enzyme, inactivation diminished under anaerobic conditions [6]; Lyophilization, membrane preparation stable to [7]; Phenylmethylsulfonyl fluoride stabilizes during purification and storage [8]; CHAPS, 0.5%, stabilizes during solubilization [8]; Cholate, 1%, solubilizes and stabilizes [8]

Storage

0°C, in the dark stable [6]; 0°C, in the dark stable for at least 10 h [8]

6 CROSSREFERENCES TO STRUCTURE DATABANKS

PIR/MIPS code

Brookhaven code

7 LITERATURE REFERENCES

[1] Banthorpe, D.V., Cardemil, E., Del Carmen Contraras, M.: Phytochemistry,15, 391–394 (1976)
[2] Moreau, R.A., Huang, A.H.C.: Arch. Biochem. Biophys.,194,422–430 (1979)
[3] Moreau, R.A., Huang, A.H.C.: Methods Enzymol.,71,804–813 (1981)
[4] Kemp, G.D., Dickinson, F.M., Ratledge, C.: Appl. Microbiol. Biotechnol.,29,370–374 (1988)
[5] Hommel, R., Ratledge, C.: FEMS Microbiol. Lett.,70,183–186 (1990)
[6] Kemp, G.D., Dickinson, F.M., Ratledge, C.: Appl. Microbiol. Biotechnol.,32,461–464 (1990)
[7] Kemp, G.D., Dickinson, F.M., Ratledge, C.: Appl. Microbiol. Biotechnol.,34,441–445 (1991)
[8] Dickinson, F.M., Wadforth, C.: Biochem. J.,282,325–331 (1992)

1 NOMENCLATURE

EC number
1.1.3.21

Systematic name
sn-Glycerol-3-phosphate:oxygen 2-oxidoreductase

Recommended name
Glycerol-3-phosphate oxidase

Synonymes
Oxidase, glycerol phosphate
Glycerol-1-phosphate oxidase
Glycerol phosphate oxidase
L-alpha-Glycerophosphate oxidase
alpha-Glycerophosphate oxidase
L-alpha-Glycerol-3-phosphate oxidase

CAS Reg. No.
9046-28-0

2 REACTION AND SPECIFICITY

Catalysed reaction
sn-Glycerol 3-phosphate + O_2 →
→ glycerone phosphate + H_2O_2 (mechanism [3])

Reaction type
Redox reaction

Natural substrates

Substrate spectrum
1 sn-Glycerol 3-phosphate + O_2 (high specificity for L-alpha-glycerophosphate [2, 4], other electron acceptors: ferricyanide (in absence of O_2 [3])
[3, 4], 2,6-dichloroindophenol (at lower rate than O_2) [4], methylene blue
and cytochrome c (at very low rate) [4]) [1–4]

Product spectrum
1 Glycerone phosphate + H_2O_2 (i.e. dihydroxyacetone phosphate) [1, 3, 4] -

Inhibitor(s)
Fructose 6-phosphate [2]; Fructose 1-phosphate [2]; Iodoacetate (weak) [4];
Azide [4]; Acriflavine [4]; Atebrin [4]

Cofactor(s)/prosthetic group(s)/activating agents
FAD (prosthetic group [1–4], 2 mol bound per mol of enzyme [2, 3]) [1–4]

Metal compounds/salts
CaCl$_2$ (activation) [2]; MgCl$_2$ (activation) [2]; ZnCl$_2$ (activation) [2]; MnCl$_2$ (activation) [2]; CoCl$_2$ (activation) [2]; NaCl (activation) [2]; KCl (activation) [2]; More (at 10 mM DL-alpha-glycerophosphate activity is increased from 2.6- to 10-fold by increasing buffer concentration from 0.01 to 0.1 M, potassium phosphate and other anionic buffers) [2]

Turnover number (min^{-1})

Specific activity (U/mg)
More [1, 3, 4]

K$_m$-value (mM)
4 (L-alpha-glycerophosphate) [4]; 26 (sn-glycerol 3-phosphate) [2]; 6 (sn-glycerol 3-phosphate) [1]

pH-optimum
5.8 [4]; 7 [1, 2]

pH-range
5.0–6.7 (5.0: about 50% of activity maximum, 6.7: about 30% of activity maximum) [4]

Temperature optimum (°C)
37 (assay at) [1]

Temperature range (°C)

3 ENZYME STRUCTURE

Molecular weight
131000 (Streptococcus faecium, gel filtration, sucrose density centrifugation) [2]

Subunits
Dimer (2 × 72000, Streptococcus faecium, SDS-PAGE) [2]

Glycoprotein/Lipoprotein
–

4 ISOLATION/PREPARATION

Source organism
Streptococcus faecium (ATCC 12755 [2], F24 [1]) [1–3]; Streptococcus faecalis 10C1 [4]

Source tissue
  Cell [1–4]

Localisation in source

Purification
  Streptococcus faecalis [4]; Streptococcus faecium [1–3]

Crystallization
  –

Cloned
  –

Renaturated
  –

---

5 STABILITY

pH

Temperature (°C)

Oxidation

Organic solvent

General stability information

Storage
  2°C, several weeks, slow decline of activity [4]

---

6 CROSSREFERENCES TO STRUCTURE DATABANKS

PIR/MIPS code

Brookhaven code

---

7 LITERATURE REFERENCES

[1] Koditschek, L.K., Umbreit, W.W.: J. Bacteriol.,98,1063–1068 (1969)
[2] Esders, T.W., Michrina, C.A.: J. Biol. Chem.,254,2710–2715 (1979)
[3] Claiborne, A.: J. Biol. Chem.,261,14398–14407 (1986)
[4] Jacobs, N.J., VanDemark, P.J.: Arch. Biochem. Biophys.,88,250–255 (1960)

1 NOMENCLATURE

EC number
1.1.3.22

Systematic name
Xanthine:oxygen oxidoreductase

Recommended name
Xanthine oxidase

Synonymes
Hypoxanthine:oxygen oxidoreductase
Oxidase, xanthine
Hypoxanthine oxidase
Schardinger enzyme
Xanthine oxidoreductase
Hypoxanthine-xanthine oxidase
Xanthine:O_2 oxidoreductase [1]
Xanthine:xanthine oxidase
EC 1.2.3.2. (formerly)
More (see EC 1.1.1.204, the enzyme from animal tissues can be interconverted to EC 1.1.1.204, that from liver exists in vivo mainly as the dehydrogenase form, but can be converted into the oxidase form by storage at –20°C, by treatment with proteolytic enzymes or with organic solvents, or by thiol reagents such as Cu^{2+}, N-ethylmaleimide or 4-hydroxymercuribenzoate, the effect of the thiol reagents can be reversed by thiols such as 1,4-dithioerythritol, in other animal tissues the enzyme exists almost entirely as EC 1.1.3.22 but can be converted into the dehydrogenase form by 1,4-dithioerythritol) [1, 4, 28]

CAS Reg. No.
9002-17-9

2 REACTION AND SPECIFICITY

Catalysed reaction
Xanthine + H_2O + O_2 →
→ urate + H_2O_2 (mechanism [5, 6])

Reaction type
Redox reaction

Natural substrates

Aldehyde + H_2O + O_2 (enzyme is implicated in the control of various redox reactions in the cell, in milk: assures absorption of iron from the gut, coupling antibacterial effect via the lactoperoxidase system) [1]

Pteridines + H_2O + O_2 (enzyme is implicated in the control of various redox reactions in the cell, in milk: assures absorption of iron from the gut, coupling antibacterial effect via the lactoperoxidase system) [1]

Purines + H_2O + O_2 (enzyme is implicated in the control of various redox reactions in the cell, in milk: assures absorption of iron from the gut, coupling antibacterial effect via the lactoperoxidase system) [1]

Substrate spectrum

1 Xanthine + H_2O + O_2 [1–28]
2 Hypoxanthine + H_2O + O_2 [1–28]
3 Purine + H_2O + O_2 (and derivatives) [5, 25, 27]
4 6,8-Dihydroxypurine + H_2O + O_2 [5]
5 1-Methyl-2-hydroxypurine + H_2O + O_2 [5]
6 6-Cyanopurine + H_2O + O_2 [5]
7 1-Methylxanthine + H_2O + O_2 [5]
8 6-Thioxanthine + H_2O + O_2 [5]
9 3-Methylhypoxanthine + H_2O + O_2 [5]
10 Adenine + H_2O + O_2 (not [25]) [5, 27]
11 Pyrimidine derivatives + H_2O + O (e.g. 2-hydroxypyrimidine [5], 6-hydroxy-6-aminepyrimidine [5], 7-hydroxy-(1,2,5)-thiadiazolo(3,4-d)-pyrimidine [5]) [5]
12 N^1-Methylnicotinamide + H_2O + O_2 [5]
13 NADH + H_2O + O_2 [5]
14 Allopurinol + H_2O + O_2 [5]
15 Pteridine + H_2O + O_2 (and derivatives, e.g.: 4-amino-7-hydroxy pteridine, 4-hydroxy-7-azapteridine) [5]
16 Acetaldehyde + H_2O + O_2 [5, 17]
17 Salicylaldehyde + H_2O + O_2 [5]
18 Pyridine-2-aldehyde + H_2O + O_2 [17]
19 Pyrimidine-3-aldehyde + H_2O + O_2 [17]
20 Indole-3-aldehyde + H_2O + O_2 [17]
21 Guanine + H_2O + O_2 [25]
22 More (the enzyme from animal tissues can be interconverted to EC 1.1.1.204, that from liver exists in vivo mainly as the dehydrogenase form, but can be converted into the oxidase form by storage at –20°C, by treatment with proteolytic enzymes or with organic solvents, or by thiol reagents such as Cu^{2+}, N-ethylmaleimide or 4-hydroxymercuribenzoate, the effect of the thiol reagents can be reversed by thiols such as 1,4-dithioerythritol, in other animal tissues the enzyme exists almost entirely as EC 1.1.3.22 but can be converted into the dehydrogenase form by 1,4-dithioerythritol [1, 4, 28], bovine milk enzyme: conversion from

oxidase into dehydrogenase by treatment with dithioerythritol or dihydro-lipoic acid [8], oxidase is converted into an irreversible oxidase form by pretreatment with chymotrypsin, papain or subtilisin, but only partial with trypsin [8], rat liver enzyme: is unstable as dehydrogenase and is gra-dually converted to oxidase [21], low specificity to substrate [5, 6, 27], enzyme also oxidizes hypoxanthine, some other purines, pterines and aldehydes [5] (i.e. possesses the activity of EC 1.2.3.1), probably acts on the hydrated derivatives of these substrates, Arthrobacter S-2 enzy-me: relatively specific [23], specificity for electron acceptor is low [5, 6]: O_2 [1–28], methylene blue [5, 6], 2,6-dichlorophenolindophenol [5, 6, 23, 25], ferricyanide [5, 6, 23, 25], quinones [5], NAD^+ (not [23, 25]) [5, 6], triphenyltetrazolium chloride [6], phenazine methosulfate [6], nitrate [6], cytochrome c [6], ferritin [6], reversibility: enzyme couples through the iron-containing protein, ferredoxin, and through hydrogenase, directly to molecular hydrogen, so that uric acid can be reduced with uptake of H_2 [6], Micrococcus enzyme: can use ferredoxin as acceptor) [1–28]

Product spectrum

1   Uric acid + H_2O_2 (under some conditions the product is mainly superoxi-de rather than peroxide: $RH + H_2O + 2\,O_2 \rightarrow ROH + 2\,H^+ + 2\,O_2^-$) [1–28]
2   Xanthine + H_2O_2
3   ?
4   ?
5   ?
6   ?
7   1-Methylurate + H_2O_2
8   6-Thiourate + H_2O_2
9   3-Methylxanthine + H_2O_2
10 ?
11 ?
12 ?
13 ?
14 ?
15 ?
16 Acetic acid + H_2O_2
17 Salicylic acid + H_2O_2
18 ?
19 ?
20 ?
21 ?
22 More (reversibility: enzyme couples through the iron-containing protein, ferredoxin, and through hydrogenase, directly to molecular hydrogen, so that uric acid can be reduced with uptake of H_2) [6]

Enzyme Handbook © Springer-Verlag Berlin Heidelberg 1995
Duplication, reproduction and storage in data banks are only
allowed with the prior permission of the publishers

Inhibitor(s)

Arsenite [3, 5, 7]; Cyanide [3, 5, 7]; Methanol [3, 5]; 2,4-Dinitrofluorobenzene [5]; Formaldehyde [5]; Urea [5]; Guanine [5]; Salicylate [5, 6]; Thiocyanate [5]; 2,4-Diamino-6-hydroxy-s-triazine [5]; 2-Amino-4-hydroxypterine-6-aldehyde [5]; Purine-6-aldehyde [5]; Allopurinol [5]; Hydroxylamine [6, 7]; Semicarbazide [6]; Phenylhydrazine [6]; Myoglobin (inhibits reaction with cytochrome c as acceptor) [6]; Purines [6]; Pterines (or other heterocyclic compounds, which are either not oxidized or oxidized rather slowly) [6]; 2-Amino-4-hydroxy-6-formylpterine [6, 7]; 3,3',4,4'-Tetrahydroxychalcone [6]; Copper [6, 7]; p-Aminophenol quinimine [6, 7]; Dinitrophenol quinimine [6]; 1,2-Dihydroxybenzene-3,5-disulfonic acid (inhibits reaction with cytochrome c) [6]; Ascorbic acid [7, 25]; Tetraethylthiuram disulfide [7]; Borate [7]; Imidazotriazines [7]; Chalcones [7]; Isatin [7]; Ninhydrin [7]; Alloxan [7]; 8-Hydroxyquinoline-7-sulfonic acid [7]; p-Chloromercuribenzoate [7, 25]; Azaguanine [9]; Aldehydes (e.g. formaldehyde, 4-pyridinecarboxaldehyde, propionaldehyde, glycolaldehyde) [16]; Ag^{2+} [25]; Cu^{2+} [25]; Fe^{3+} [25]; Co^{2+} [25]; H_2O_2 [25]; Xanthine (substrate inhibition at high concentration [22], no substrate inhibition [23]) [22]

Cofactor(s)/prosthetic group(s)/activating agents

FAD (flavoprotein [1–28], contains FAD, molybdenum and iron in the ratio 1:4:4 [5, 6], bovine liver: 1 FAD per subunit (2 per molecule) [15]) [1–28]

Metal compounds/salts

Iron (iron-molybdenum protein [1–28], contains FAD, iron and molybdenum in the ratio 1:4:4 [5, 6], bovine liver: 2 Fe-S centres per subunit (4 per molecule) [15]) [1–28]; Molybdenum (an iron-molybdenum protein [1–28], contains FAD, molybdenum and iron in the ratio 1:4:4 [5, 6], purification of the molybdenum cofactor from milk xanthine oxidase [13], molybdenum cofactor is a complex between molybdenum and molybdopterin (a 6-alkylpterin with 4-carbon side chain which has an enedithiol at carbon 1' and 2', a hydroxyl at carbon 3', and a terminal phosphate group [14]) [18], bovine liver: 1 molybdenum atom per subunit (2 per molecule) [15]) [1–28]

Turnover number (min⁻¹)

970 (xanthine, pH 8.2, 23.5°C) [5]; 70 (aldehyde (+ indophenol)) [7]; 261 (hypoxanthine (+ indophenol)) [7]; 295 (xanthine (+ O_2)) [7]; 315 (xanthine (+ indophenol)) [7]; 154 (NADPH (+ indophenol)) [7]; 77 (NADPH (+ cytochrome c)) [7]

Specific activity (U/mg)

7.8 (bovine milk) [1]; 2.04 (human colostrum) [1]; 123 (goat milk) [1]; 2.4–5.1 [5]; 0.278 [12]; 230 [23]; 15.2 [25]; 1.8 [27]

K_m-value (mM)
More [1, 5, 6, 9, 10, 12, 17, 22, 23, 27]; 0.0279 (xanthine, free enzyme, human) [1]; 0.0264 (xanthine, membrane-bound enzyme, human) [1]; 0.0134 (hypoxanthine, free enzyme, human) [1]; 0.00924 (hypoxanthine, membrane-bound enzyme, human) [1]; 0.08 (O_2 (+ xanthine), pH 10.0) [5]; 0.05 (O_2 (+ xanthine), pH 8.5 [5], xanthine (+ O_2) [7]) [5, 7]; 0.088 (xanthine (+ indophenol)) [7]; 0.0475 (hypoxanthine (+ indophenol)) [7]; 20 (aldehyde (+ indophenol)) [7]; 0.0005 (NADPH (+ indophenol)) [7]; 0.00045 (NADPH (+ cytochrome c)) [7]; 0.00143 (2-amino-4-hydroxypteridine (+ methylene blue), locust) [9]; 0.0067 (2-amino-4-hydroxypteridine (+ methylene blue), Drosophila) [9]; 161.5 (formaldehyde) [17]; 130 (acetaldehyde) [17]; 430 (propionaldehyde) [17]; 142 (butyraldehyde) [17]; 0.36 (pyridine 2-aldehyde) [17]; 0.046 (pyridine 3-aldehyde) [17]; 1.7 (pyridine 4-aldehyde) [17]; 1.03 (o-hydroxybenzaldehyde) [17]; 0.068 (2,5-dihydroxybenzaldehyde) [17]; 0.085 (indole 3-aldehyde) [17]; 0.3 (indole 3-acetaldehyde) [17]; 0.56 (4-hydroxyphenylglycolaldehyde) [17]; 1 (succinate semialdehyde) [17]; 2 (glyceraldehyde 3-phosphate) [17]; 48 (N-methylnicotinamide) [27]; 0.004 (allopurinol) [27]; 0.55 (formycin B) [27]; 0.93 (benzaldehyde) [27]

pH-optimum
7.0 (locust) [9]; 7.1 (Enterobacter cloacae) [25]; 8.0 (Lens esculenta) [2]; 8.1 (rabbit liver) [12]; 8.2 (human) [1]; 8.3 [5]; 8.35 (goat) [1]

pH-range
More [1]; 5–9 (5: about 10% of activity maximum, 9: about 25% of activity maximum) [25]; 6–9 (6.0: about 25% of activity maximum, 9.0: about 20% of activity maximum) [22]; 7.7–9.1 (at pH 7.7 and 9.1: about 30% of activity maximum) [12]

Temperature optimum (°C)
35–45 (Enterobacter cloacae) [25]; 23–30 (assay at) [1]; 25 (assay at) [10]; 37 (assay at) [25]

Temperature range (°C)
5–60 (5°C: about 10% of activity maximum, 60°C: about 50% of activity maximum) [25]

---

3 ENZYME STRUCTURE

Molecular weight
100000 (around, bovine milk, ultrafiltration) [24]
128000 (Enterobacter cloacae, gel filtration) [25]
146000 (Arthrobacter sp. S-2, gel filtration) [23]
275000 (bovine, gel filtration) [15]
280000 (rabbit, gel filtration) [12]

300000 (human liver, FPLC size exclusion chromatography [27], mouse,
FPLC gel filtration [10]) [10, 27]
303000 (bovine, sedimentation equilibrium) [20]
310000 (human, gel filtration) [1]
More [5, 7]

Subunits

Dimer (2 × 69000, Enterobacter cloacae, SDS-PAGE [25], 2 × 80000, Arthrobacter sp. S-2, SDS-PAGE [23], 2 × 120000, chicken liver [5],
2 × 130000–140000, Drosophila melanogaster [5], 2 × 150000, bovine milk,
SDS-PAGE [5, 20], mouse, SDS-PAGE [10], human liver, SDS-PAGE [27]) [5,
10, 20, 23, 25, 27]
Oligomer (x × 52000 + x × 99000, rabbit, SDS-PAGE) [12]

Glycoprotein/Lipoprotein
–

---

4 ISOLATION/PREPARATION

Source organism

Bovine [1–4, 6, 8, 11, 13–20, 24, 26]; Goat [1]; Human [1, 6, 27]; Sheep [1];
Mouse (mouse embryo cell lines: 3T3, 3T6, B-3T3, 3T12 [29]) [1, 9, 10, 29];
Rat [1, 4, 7, 21, 28]; Dog [1]; Patas monkey [1]; Guinea pig [1]; Rabbit [1,
12]; Donkey [1]; Horse [1]; Cat [1]; Chicken [5, 7]; Drosophila melanogaster
[5, 9]; Locust [9]; Lens esculenta (lentil) [22]; Arthrobacter sp. S-2 [23]; Enterobacter cloacae (KY 3074) [25]; More (the enzyme from animal tissues
can be interconverted to EC 1.1.1.204, that from liver exists in vivo mainly as
the dehydrogenase form, but can be converted into the oxidase form by
storage at –20°C, by treatment with proteolytic enzymes or with organic solvents, or by thiol reagents such as Cu^{2+}, N-ethylmaleimide or 4-hydroxymer-
curibenzoate, the effect of the thiol reagents can be reversed by thiols such
as 1,4-dithioerythritol, in other animal tissues the enzyme exists almost entirely as EC 1.1.3.22 but can be converted into the dehydrogenase form by
1,4-dithioerythritol) [1, 4, 28]

Source tissue

Milk [1–3, 5–8, 11, 13–20, 24, 26]; Colostrum [1]; Liver [1, 4, 6, 9, 10, 12, 15,
21, 27, 28]; Duodenum [9]; Pancreas [9]; Kidney [9]; Lung [9]; Spleen [9];
Cell [23]; Seedlings [22]; Cell culture (mouse embryo cell lines: 3T3, 3T6,
B-3T3, 3T12) [29]

Localisation in source

Membrane (enzyme is associated with bovine milk-fat-globule membrane
[19, 24], membrane-bound [1]) [1, 19, 24]; Microsomes (enzyme from milk
and mammary gland) [6, 26]; Lipid globules [19, 24, 26]; More (liver enzyme: in supernatant fraction [6], distribution of enzyme in milk globules,
mammary gland and liver [26]) [6, 26]

Purification
Bovine (purification of the molybdenum cofactor of milk xanthine oxidase
[13]) [1, 13, 15, 20, 24, 26]; Human [1, 27]; Mouse [10]; Rabbit [12]; Rat
[21]; Arthrobacter sp. S-2 [23]; Enterobacter cloacae [25]

Crystallization
[2]

Cloned
–

Renaturated
–

---

5 STABILITY

pH
6.2–6.8 (60°C, 30 min) [25]

Temperature (°C)
60 (pH 6.2–6.8, 30 min stable) [25]; 70 (30 min, complete loss of activity)
[25]

Oxidation
Photooxidation, protection by competitive inhibitors [6]

Organic solvent

General stability information
Phosphate stabilizes [6]; Low operational stability by immobilization [11];
Stable for more than 1 year with alternate freezing and thawing [23]; More
(the enzyme from animal tissues can be interconverted to EC 1.1.1.204, that
from liver exists in vivo mainly as the dehydrogenase form, but can be converted into the oxidase form by storage at –20°C, by treatment with proteolytic enzymes or with organic solvents, or by thiol reagents such as Cu^{2+},
N-ethylmaleimide or 4-hydroxymercuribenzoate, the effect of the thiol
reagents can be reversed by thiols such as 1,4-dithioerythritol, in other animal tissues the enzyme exists almost entirely as EC 1.1.3.22 but can be
converted into the dehydrogenase form by 1,4-dithioerythritol [1, 4, 28], rat
liver enzyme is unstable as dehydrogenase and is gradually converted to
oxidase [21]) [1, 4, 21, 28]

Storage
4°C, 31% loss of activity after 6 days, 54% loss of activity after 12 days,
72% loss of activity after 16 days, goat enzyme [1]; –20°C, 27% loss of activity after 2 weeks, 51% loss of activity after 4 weeks, 89% loss of activity
after 12 weeks, goat enzyme [1]; –75°C, more than 1 year with alternate
freezing and thawing, stable [23]

6 CROSSREFERENCES TO STRUCTURE DATABANKS

PIR/MIPS code

Brookhaven code

7 LITERATURE REFERENCES

[1] Zikakis, J.P., Dressel, M.A., Silver, M.R. in " Instrum. Anal. Foods, Recent Prog. Proc. Symp. Int. Flavor Conf.",3rd Ed. (Charalambous, G., Inglett, G., Eds.) 2,243–303 (1983) (Review)

[2] Avis, P.G., Bergel, F., Bray, R.C.: J. Chem. Soc.,2 (Lond.) ,1100–1105 (1955)

[3] Coughlan, M.P., Rajagopalan, K.V., Handler, P.: J. Biol. Chem.,244,2658–2663 (1969)

[4] Stirpe, F., Della Corte, E.: Biochim. Biophys. Acta,212,195–197 (1970)

[5] Bray, R.C. in "The Enzymes",3rd Ed. (Boyer, P.D., Ed.) 12,299–412 (1975) (Review)

[6] Bray, R.C. in "The Enzymes",2nd Ed. (Boyer, P.D., Lardy, H., Myrback, K., Eds.) 7,533–556 (1963) (Review)

[7] De Renzo, E.C.: Adv. Enzymol. Relat. Subj. Biochem.,17,293–328 (1956) (Review)

[8] Batelli, M.G., Lorenzoni, E., Stirpe, F.: Biochem. J.,131,191–198 (1973)

[9] Hayden, T.J., Ryan, J.P., Duke, E.J.: Biochem. Soc. Trans.,1,247–250 (1973) (Review)

[10] Carpani, G., Racchi, M., Ghezzi, P., Terao, M., Garattini, E.: Arch. Biochem. Biophys.,279,237–241 (1990)

[11] Tramper, J.: Methods Enzymol.,136,254–262 (1987) (Review)

[12] Catigani, G.L., Chytil, F., Darby, W.J.: Biochim. Biophys. Acta,377,34–41 (1975)

[13] van Spanning, R.J.M., Wansell-Bettenhaussen, C.W., Oltmann, L.F., Stouthamer, A.H.: Eur. J. Biochem.,169,349–352 (1987)

[14] Kramer, S.P., Johnson, J.L., Ribeiro, A., Millington, D.S., Rajagopalan, K.V.: J. Biol. Chem.,262,16357–16363 (1987)

[15] Cabre, F., Canela, E.I.: Biochem. Soc. Trans.,15,511–512 (1987)

[16] Morpeth, F.F., Bray, R.C.: Biochemistry,23,1322–1338 (1984)

[17] Morpeth, F.F.: Biochim. Biophys. Acta,744,328–334 (1983)

[18] Johnson, J.L., Hainline, B.E., Rajagopalan, K.V.: J. Biol. Chem.,255,1783–1786 (1980)

[19] Briley, M.S., Eisenthal, R.: Biochem. J.,147,417–423 (1975)

[20] Waud, W.R., Brady, F.O., Wiley, R.D., Rajagopalan, K. V.: Arch. Biochem. Biophys., 169,695–701 (1975)

[21] Waud, W.R., Rajagopalan, K.V.: Arch. Biochem. Biophys.,172,354–364 (1976)

[22] Taneja, R., Taneja, V.: Biochim. Biophys. Acta,485,489–491 (1977)

[23] Woolfolk, C.A., Downard, J.S.: J. Bacteriol.,135,422–428 (1978)

[24] Nathans, G.R., Hade, E.P.K.: Biochim. Biophys. Acta,526,328–344 (1978)

[25] Machida, Y., Nakanishi, T.: Agric. Biol. Chem.,45,425–432 (1981)

[26] Bruder, G., Heid, H., Jarasch, E.-D., Keenan, T.W., Mather, I.H.: Biochim. Biophys. Acta,701,357–369 (1982)

[27] Krenitsky, T.A., Spector, T., Hall, W.W.: Arch. Biochem. Biophys.,247,108–119 (1986)

[28] Della Corte, E., Stirpe, F.: Biochem. J.,126,739–745 (1972)

[29] Clynes, M.M., Hurley, M.P., Shannon, M.F.: Biochem. Soc. Trans.,7,72–74 (1979)

1 NOMENCLATURE

EC number
1.1.3.23

Systematic name
Thiamine:oxygen 5-oxidoreductase

Recommended name
Thiamine oxidase

Synonymes
Thiamine dehydrogenase
Thiamin oxidase
Thiamin:oxygen 5-oxidoreductase

CAS Reg. No.
96779-44-1

2 REACTION AND SPECIFICITY

Catalysed reaction
Thiamine + 2 O_2 →
→ thiamine acetic acid + 2 H_2O_2

Reaction type
Redox reaction

Natural substrates
Thiamine + O_2 [1–3]

Substrate spectrum
1 Thiamine + O_2 (reaction via an aldehyde intermediate which is not released [1]) [1–3]
2 Oxythiamine + O_2 [1–3]
3 Pyrithiamine + O_2 [1–3]

Product spectrum
1 Thiamine acetate + H_2O_2 [1–4]
2 Oxidized oxythiamine + H_2O_2 [1–3]
3 Oxidized pyrithiamine + H_2O_2 [1–3]

Inhibitor(s)
Pyrithiamine [1]; Semicarbazide [1, 3]; p-Chloromeruribenzoate [3];
Quinacrine [3]; Hg^{2+} [3]; Cu^{2+} [3]

Cofactor(s)/prosthetic group(s)/activating agents
FAD (8-alpha-N(1)-histidyl-FAD [1, 2], 1 mol/mol enzyme [1]) [1–3]

Metal compounds/salts

Turnover number (min^{-1})
300 [1]; 100 [3]

Specific activity (U/mg)
30.0 [2]; 2.18 [3]

K_m-value (mM)
0.005–0.006 (thiamine) [1]; 1.72 (thiamine) [3]

pH-optimum
8.0–9.0 [3]

pH-range

Temperature optimum (°C)

Temperature range (°C)

3 ENZYME STRUCTURE

Molecular weight
41500–52000 (soil bacterium, gel filtration) [2, 3]

Subunits
Monomer (1 × 49000–50000, soil bacterium, SDS-PAGE) [2]

Glycoprotein/Lipoprotein
–

4 ISOLATION/PREPARATION

Source organism
Unidentified soil bacterium (ATCC 25589 [1]) [1–3]

Source tissue

Localisation in source

Purification
Unidentified soil bacterium [2, 3]

Crystallization

–

Cloned

–

Renaturated

–

5 STABILITY

pH

Temperature (°C)
65 (not stable above) [3]; 50 (stable for 30 min) [3]

Oxidation

Organic solvent

General stability information

Storage
–70°C, 3 months [3]; 4°C, 1 week [3]

6 CROSSREFERENCES TO STRUCTURE DATABANKS

PIR/MIPS code

Brookhaven code

7 LITERATURE REFERENCES

[1] Gomez-Moreno, C., Edmondson, D.E.: Arch. Biochem. Biophys.,239,46–52 (1985)
[2] Gomez-Moreno, C., Choy, M., Edmondson, D.E.: J. Biol. Chem.,254,7630–7635
(1979)
[3] Neal, R.A.: J. Biol. Chem.,245,2599–2604 (1970)

1 NOMENCLATURE

EC number
1.1.3.24

Systematic name
L-Galactono-1,4-lactone:oxygen 3-oxidoreductase

Recommended name
L-Galactonolactone oxidase

Synonymes
L-Galactono-1,4-lactone oxidase
More (enzyme similar to but not identical with EC 1.1.3.8, cf. EC 1.3.2.3)

CAS Reg. No.

2 REACTION AND SPECIFICITY

Catalysed reaction
L-Galactono-1,4-lactone + O_2 →
→ L-ascorbate + H_2O_2

Reaction type
Redox reaction

Natural substrates
L-Galactono-1,4-lactone + O_2 [1–4]

Substrate spectrum
1 L-Galactono-1,4-lactone + O_2 (similar to but not identical with EC 1.1.3.8)
   [1–4]
2 1,4-Lactone of D-altronate + O_2 [1]
3 1,4-Lactone of L-fuconate + O_2 [1]
4 1,4-Lactone of D-arabinate + O_2 [1]
5 1,4-Lactone of D-threonate + O_2 [1]

Product spectrum
1 L-Ascorbate + H_2O_2 [1–4]
2 ?
3 ?
4 ?
5 ?

Inhibitor(s)
L-Gulonolactone [1]; D-Galactono-1,4-lactone [1]; p-Chloromercuriphenyl sulfonate [1]; Iodoacetamide [1, 2]; N-Ethylmaleimide [1, 2]; Sulfite [1]; Sulfide [1]; $HgCl_2$ [2]; p-Chloromercuribenzoate [2, 4]; 4,4'-Dipyridyldisulfide [2]; 2,2'-Dipyridyldisulfide [2]; 5,5'-Dithiobis(2-nitrobenzoate) [2]

Cofactor(s)/prosthetic group(s)/activating agents
FAD (covalently bound [1–4], 8alpha-[N(1)histidyl]FAD [3]) [1–4]

Metal compounds/salts
Iron-sulfur cluster [1]

Turnover number (min⁻¹)

Specific activity (U/mg)
3.448 [1]; 0.76 [4]

K_m-value (mM)
0.3 (L-galactonolactone) [1]; 0.18 (oxygen) [1]; 0.36 (L-fuconolactone) [1]; 2.0 (D-altronolactone) [1]; 0.16 (D-arabinolactone) [1]; 15 (D-threonolactone) [1]

pH-optimum
8.9 [1]

pH-range

Temperature optimum (°C)

Temperature range (°C)

---

3 ENZYME STRUCTURE

Molecular weight
290000 (Saccharomyces cerevisiae, gel filtration) [4]
70000–74000 (Saccharomyces cerevisiae, gel filtration, gel electrophoresis) [1]

Subunits
Tetramer (4 × 56000, Saccharomyces cerevisiae, SDS-PAGE [4], 4 × 18000, Saccharomyces cerevisiae, SDS-PAGE [1]) [1, 4]

Glycoprotein/Lipoprotein
–

---

4 ISOLATION/PREPARATION

Source organism
Saccharomyces cerevisiae [1–4]

Source tissue

Localisation in source
Mitochondria [1, 4]

Purification
Saccharomyces cerevisiae [1–4]

Crystallization
–

Cloned
–

Renaturated
–

---

5 STABILITY

pH

Temperature (°C)

Oxidation

Organic solvent

General stability information

Storage
5°C, 6 months, 50% loss of activity [1]

---

6 CROSSREFERENCES TO STRUCTURE DATABANKS

PIR/MIPS code

Brookhaven code

---

7 LITERATURE REFERENCES

[1] Bleeg, H.S., Christensen, F.: Eur. J. Biochem.,127,391–396 (1982)
[2] Noguchi, E., Nishikimi, M., Yagi, K.: J. Biochem.,90,33–38 (1981)
[3] Kenney, W.C., Edmondson, D.E., Singer, T.P., Nishikimi, M., Noguchi, E., Yagi, K.:
    FEBS Lett.,97,40–42 (1979)
[4] Nishikimi, M., Noguchi, E., Yagi, K.: Arch. Biochem. Biophys.,191,479–486 (1978)

Enzyme Handbook © Springer-Verlag Berlin Heidelberg 1995
Duplication, reproduction and storage in data banks are only
allowed with the prior permission of the publishers

1 NOMENCLATURE

EC number
1.1.3.25

Systematic name
Cellobiose:oxygen 1-oxidoreductase

Recommended name
Cellobiose oxidase

Synonymes
Oxidase, cellobiose

CAS Reg. No.
66458-10-4

2 REACTION AND SPECIFICITY

Catalysed reaction
Cellobiose + O_2 →
→ cellobiono-1,5-lactone + H_2O_2 (mechanism [3])

Reaction type
Redox reaction

Natural substrates
Cellobiose + O_2 (role in cellulose degradation) [4, 5]

Substrate spectrum
1 Cellobiose + O_2 [1–4]
2 Cellodextrin + O_2 [1, 4]
3 Lactose + O_2 [1, 2, 4]
4 4-beta-Glucosylmannose + O_2 [1, 4]
5 Maltose + O_2 [2]
6 More (also oxidized slowly: cellulose, chitin, xylan, agarose, artficial electron acceptors: 2,6-dichlorophenolindophenol, benzyl viologen, not: FMN,
  FAD, riboflavin) [2]

Enzyme Handbook © Springer-Verlag Berlin Heidelberg 1995
Duplication, reproduction and storage in data banks are only
allowed with the prior permission of the publishers

Product spectrum
1 Cellobiono-1,5-lactone + H_2O_2 [4]
2 Corresponding aldonic acid + H_2O_2 [4]
3 4-O-(beta-D-Galactopyranosyl)-D-glucono-1,5-lactone + H_2O_2
4 4-beta-Mannosyl-glucono-1,5-lactone + H_2O_2
5 4-O-beta-Glucosyl-glucono-1,5-lactone + H_2O_2 [2]
6 More (superoxide could be the primary reduced oxygen species produced by the enzyme, H_2O_2 is possibly formed by dismutation of superoxide anion) [2]

Inhibitor(s)
NaN_3 [1, 4]; NaCN [1, 4]; 2,2-Bipyridine [1, 4]; Cellobiose (substrate inhibition at high concentration) [2]

Cofactor(s)/prosthetic group(s)/activating agents
Heme (a flavohemoprotein) [1, 4]; Cytochrome b (as prosthetic group) [2, 3]; Flavin (a flavohemoprotein [1, 4], one flavin per polypeptide [1, 4]) [1–4]; FAD [2]

Metal compounds/salts

Turnover number (min^{-1})

Specific activity (U/mg)
0.417 [1]; More [2, 4]

K_m-value (mM)
0.042 (cellobiose) [2]; 0.580 (lactose) [2]; 6.6 (maltose) [2]; 0.017 (2,6-dichlorophenol-indophenol in absence of O_2) [2]; 0.012 (2,6-dichlorophenolindophenol in presence of O_2) [2]; 3.0 (benzyl viologen) [2]

pH-optimum
5.0 [1, 4]

pH-range
4.1–5.7 (4.1 and 5.7: about 50% of maximum activity) [4]

Temperature optimum (°C)
30 (assay at) [1, 4]

Temperature range (°C)

3 ENZYME STRUCTURE

Molecular weight
74400 (Sporotrichum pulverulentum, sedimentation equilibrium centifugation) [2]

93000 (Sporotrichum pulverulentum, sedimentation equilibrium centrifugation) [1, 4]

Subunits
Monomer (1 × 102000, Sporotrichum pulverulentum, SDS-PAGE) [1, 2]

Glycoprotein/Lipoprotein
Glycoprotein (carbohydrate content: 11.9% [2]) [1, 2, 4]

---

4 ISOLATION/PREPARATION

Source organism
Sporotrichum pulverulentum (white-rot fungus) [1–5]

Source tissue
Culture filtrate [1]

Localisation in source
Extracellular [4]

Purification
Sporotrichum pulverulentum [1, 2, 4]

Crystallization
–

Cloned
–

Renaturated
–

---

5 STABILITY

pH

Temperature (°C)

Oxidation

Organic solvent

General stability information

Storage
0°C or –20°C, pH 5.0, 50 mM sodium acetate buffer, less than 10% loss of activity per week [1]

---

6 CROSSREFERENCES TO STRUCTURE DATABANKS

PIR/MIPS code

Brookhaven code

7 LITERATURE REFERENCES

[1] Ayers, A., Eriksson, K.-E.: Methods Enzymol.,89,129–135 (1982) (Review)
[2] Morpeth, F.F.: Biochem. J.,228,557–564 (1985)
[3] Jones, G.D., Wilson, M.T.: Biochem. J.,256,713–718 (1988)
[4] Ayers, A.R., Ayers, S.B., Eriksson, K.-E.: Eur. J. Biochem.,90,171–181 (1978)
[5] Eriksson, K.-E.: Biotechnol. Bioeng.,20,317–332 (1978) (Review)

1 NOMENCLATURE

EC number
1.1.3.26

Systematic name
Columbamine:oxygen oxidoreductase (cyclizing)

Recommended name
Columbamine oxidase

Synonymes
Berberine synthase
Synthase, berberine

CAS Reg. No.
95329-18-3

2 REACTION AND SPECIFICITY

Catalysed reaction
2 Columbamine + O_2 →
→ 2 berberine + 2 H_2O (oxidation of the O-methoxyphenol structure forms the methylenedioxy group of berberine)

Reaction type
Redox reaction

Natural substrates
Columbamine + O_2 (berberine biosynthesis) [1]

Substrate spectrum
1 Columbamine + O_2 [1]
2 More (not: (R)-tetrahydrocolumbamine, (S)-tetrahydrocolumbamine) [1]

Product spectrum
1 Berberine + H_2O
2 ?

Inhibitor(s)
Cyanide [1]; p-Phenanthroline [1]

Cofactor(s)/prosthetic group(s)/activating agents

Metal compounds/salts
Fe^{2+} (an iron protein, Fe^{2+} restores activity after dialysis against o-phenanthroline) [1]

Turnover number (min^{-1})

Specific activity (U/mg)

K_m-value (mM)
0.002 (columbamine) [1]

pH-optimum
8.9 [1]

pH-range

Temperature optimum (°C)
70 [1]

Temperature range (°C)

---

3 ENZYME STRUCTURE

Molecular weight
320000 (Berberis stolonifera) [1]

Subunits

Glycoprotein/Lipoprotein
–

---

4 ISOLATION/PREPARATION

Source organism
Berberis stolonifera [1]

Source tissue

Localisation in source

Purification
Berberis stolonifera [1]

Crystallization
–

Cloned
–

Renaturated
–

5 STABILITY

pH

Temperature (°C)

Oxidation

Organic solvent

General stability information

Storage

6 CROSSREFERENCES TO STRUCTURE DATABANKS

PIR/MIPS code

Brookhaven code

7 LITERATURE REFERENCES

[1] Rueffer, M., Zenk, M.H.: Tetrahedron Lett.,26,201–202 (1985)

1 NOMENCLATURE

EC number
1.1.3.27

Systematic name
L-2-Hydroxyphytanate:oxygen 2-oxidoreductase

Recommended name
Hydroxyphytanate oxidase

Synonymes
L-2-Hydroxyphytanic acid oxidase

CAS Reg. No.

2 REACTION AND SPECIFICITY

Catalysed reaction
L-2-Hydroxyphytanate + O_2 →
→ 2-oxophytanate + H_2O_2

Reaction type
Redox reaction

Natural substrates
L-2-Hydroxyphytanate + O_2 [1, 2]

Substrate spectrum
1 L-2-Hydroxyphytanate + O_2 [1, 2]

Product spectrum
1 2-Oxophytanate + H_2O_2 [1, 2]

Inhibitor(s)
L-2-Hydroxyisocaproate [2]

Cofactor(s)/prosthetic group(s)/activating agents

Metal compounds/salts

Turnover number (min^{-1})

Specific activity (U/mg)

K_m-value (mM)
0.15 (L-2-hydroxyphytanate) [1, 2]

pH-optimum
8.5 [2]

pH-range

Temperature optimum (°C)

Temperature range (°C)

---

3 ENZYME STRUCTURE

Molecular weight

Subunits

Glycoprotein/Lipoprotein
–

---

4 ISOLATION/PREPARATION

Source organism
Rat [1, 2]

Source tissue
Kidney cortex [1, 2]

Localisation in source
Peroxisomes [1, 2]

Purification

Crystallization
–

Cloned
–

Renaturated
–

---

5 STABILITY

pH

Temperature (°C)

Oxidation

Organic solvent

General stability information

Storage

6 CROSSREFERENCES TO STRUCTURE DATABANKS

PIR/MIPS code

Brookhaven code

7 LITERATURE REFERENCES

[1] Vamecq, J., Draye, J.P.: Biomed. Environ. Mass Spectrom.,15,345–351 (1988)
[2] Draye, J.P., van Hoof, F., de Hoffmann, E., Vamecq, J.: Eur. J. Biochem.,167,573–578
    (1987)

1 NOMENCLATURE

EC number
1.1.3.28

Systematic name
Nucleoside:oxygen 5'-oxidoreductase

Recommended name
Nucleoside oxidase

Synonymes
Oxidase, nucleoside

CAS Reg. No.
82599-71-1

2 REACTION AND SPECIFICITY

Catalysed reaction
2 Inosine + O_2 →
→ 2 5'-oxoinosine + 2 H_2O

Reaction type
Redox reaction

Natural substrates

Substrate spectrum
1 Inosine + O_2 [1, 3]
2 5'-Oxoinosine + O_2 (i.e. inosin-5'-aldehyde) [1–4]
3 Xanthosine + O_2 (oxidized at 125% the rate of inosine) [3]
4 Guanosine + O_2 (oxidized at 121% the rate of inosine [3]) [2, 3]
5 Adenosine + O_2 (oxidized at 97.3% the rate of inosine [3]) [2, 3]
6 Uridine + O_2 (oxidized at 92.8% the rate of inosine [3]) [2, 3]
7 Deoxyguanosine + O_2 (oxidized at 92.3% the rate of inosine) [3]
8 Deoxyadenosine + O_2 (oxidized at 81.4% the rate of inosine) [3]
9 Deoxyinosine + O_2 (oxidized at 80.7% the rate of inosine [3]) [2, 3]
10 Cytidine + O_2 (oxidized at 77.5% the rate of inosine) [3]
11 Deoxycytidine + O_2 (oxidized at 77.3% the rate of inosine) [3]
12 Thymidine + O_2 (oxidized at 55.8% the rate of inosine) [3]
13 Hydroquinone + O_2 (oxidized only in the presence of nucleoside substrates, the O_2-consumption is doubled in the presence of high concentrations of hydroquinone) [2]

14 More (several p-, o-, and m-substituted phenolic compounds are also
co-oxidized in the presence of nucleosides, the enzyme catalyzes the
oxidative coupling reaction of phenolic compounds with 4-aminoantipyri-
ne [2], no substrates are nucleotides, bases and ribose [3]) [2, 3]

Product spectrum
1 5'-Oxoinosine + H_2O (i.e. inosin-5'-aldehyde) [1]
2 Inosine-5-carboxylate + H_2O [1]
3 ?
4 ?
5 ?
6 ?
7 ?
8 ?
9 ?
10 ?
11 ?
12 ?
13 p-Quinone + H_2O [2]
14 ?

Inhibitor(s)
N-Bromosuccinimide (strong) [3]; KCN (strong) [3]; NaN_3 [3]; $HgCl_2$ (slight)
[3]; $Pb(CH_3COO)_2$ (slight) [3]; More (no inhibition by EDTA, 1,10-phenanthro-
line, 2,2'-bipyridyl, quinacrine, hydrazine sulfate, iodoacetate, PCMB, PMSF)
[3]

Cofactor(s)/prosthetic group(s)/activating agents
FAD (flavoprotein, covalently bound to subunit alpha, 1 mol/mol enzyme
protein) [3]; Dichlorophenolindophenol (activation, 5% as effective as O_2)
[1]; More (ineffective as cofactor are ferricyanide, cytochrome c, NAD^+,
$NADP^+$) [1]

Metal compounds/salts
Fe (FeS-heme-flavoprotein, a total content of 3 gatom iron/mol enzyme protein with 2 gatom non-heme iron/mol enzyme protein, 1 mol heme per mol
enzyme and 2 mol labile sulfide per mol enzyme) [3]

Turnover number (min^{-1})

Specific activity (U/mg)
15.3 [1]

K_m**-value** (mM)
0.044 (inosine) [3]

pH-optimum
More (pI: 5.3) [3]; 5.0–6.0 [3, 4]

pH-range
4.0–6.8 (about 75% of maximal activity at pH 4.0 and 6.8) [3]

Temperature optimum (°C)
25 (assay at) [1, 4]

Temperature range (°C)

---

3 ENZYME STRUCTURE

Molecular weight
130000 (Pseudomonas maltophilia, gel filtration) [3]

Subunits
Tetramer (1 × 14000 + 1 × 18000 + 1 × 33000 + 1 × 76000, Pseudomonas
maltophilia, SDS-PAGE) [3]

Glycoprotein/Lipoprotein
–

---

4 ISOLATION/PREPARATION

Source organism
Pseudomonas maltophilia (strain LB-86 [1–4], highest specific activity in
cells grown at 26°C for 22–24 h [4]) [1–4]

Source tissue
Cell [1–3]

Localisation in source

Purification
Pseudomonas maltophilia [1, 3]

Crystallization
–

Cloned
–

Renaturated
–

5 STABILITY

pH
  5.0–6.0 (most stable at 37°C [3], crude [4]) [3, 4]; 7.0 (unstable above) [3]

Temperature (°C)
  60 (stable at least 15 min below) [3]; 75 (complete inactivation after 15 min) [3]

Oxidation

Organic solvent

General stability information

Storage

6 CROSSREFERENCES TO STRUCTURE DATABANKS

PIR/MIPS code

Brookhaven code

7 LITERATURE REFERENCES

[1] Isono, Y., Sudo, T., Hoshino, M.: Agric. Biol. Chem.,53,1663–1669 (1989)
[2] Isono, Y., Hoshino, M.: Agric. Biol. Chem.,53,2197–2203 (1989)
[3] Isono, Y., Sudo, T., Hoshino, M.: Agric. Biol. Chem.,53,1671–1677 (1989)
[4] Hoshino, M., Isono, Y., Sudo, T.: Agric. Biol. Chem.,53,399–403 (1989)

1 NOMENCLATURE

EC number
1.1.3.29

Systematic name
N-Acyl-D-hexosamine:oxygen 1-oxidoreductase

Recommended name
N-Acylhexosamine oxidase

Synonymes
N-Acyl-D-hexosamine oxidase
N-Acyl-beta-D-hexosamine:oxygen 1-oxidoreductase
Oxidase, N-acyl-D-hexosamine

CAS Reg. No.
121479-58-1

2 REACTION AND SPECIFICITY

Catalysed reaction
N-Acetyl-D-glucosamine + O_2 →
→ N-acetyl-D-glucosaminate + H_2O_2

Reaction type
Redox reaction

Natural substrates

Substrate spectrum
1 N-Acetyl-beta-D-glucosamine + O_2 (alpha-anomer is not oxidized) [1]
2 N-Acetyl-D-galactosamine + O_2 [1]
3 N-Glycolyl-D-glucosamine + O_2 [1]
4 D-Glucosamine + O_2 (0.1% of activity with N-acetyl-D-glucosamine) [1]
5 N-Acetyl-D-mannosamine + O_2 (poor substrate) [1]

Product spectrum
1 N-Acetyl-D-glucosaminolacton + H_2O_2 (in presence of water:
N-acetyl-D-glucosaminate + H_2O_2) [1]
2 ?
3 ?
4 ?
5 ?

Inhibitor(s)
Zn^{2+} [1]; Cd^{2+} (weak) [1]; Mn^{2+} [1]; Ni^{2+} [1]; Cu^{2+} [1]; Hg^{2+} [1]; Oxalate [1]; PCMB [1]

Cofactor(s)/prosthetic group(s)/activating agents
More (probably contains unidentified flavin component) [1]

Metal compounds/salts

___

Turnover number (min^{-1})

Specific activity (U/mg)
70.4 (N-acetyl-D-glucosamine) [1]

K_m-value (mM)
0.1 (N-acetyl-D-galactosamine) [1]; 0.24 (N-acetyl-D-glucosamine) [1]; 2.6 (N-glycolyl-D-glucosamine) [1]; 40 (N-acetyl-D-mannosamine) [1]

pH-optimum
8.0 [1]

pH-range
7–10 (90% of maximum activity at pH 7, 75% of maximum activity at pH 10) [1]

Temperature optimum (°C)
37 (assay at) [1]

Temperature range (°C)

___

3 ENZYME STRUCTURE

Molecular weight
170000 (Pseudomonas sp., gel filtration) [1]

Subunits
Tetramer ($1 \times 63000 + 1 \times 44000 + 1 \times 36000 + 1 \times 22000$, Pseudomonas sp., SDS-PAGE) [1]

Glycoprotein/Lipoprotein
–

___

4 ISOLATION/PREPARATION

Source organism
Pseudomonas sp. [1]

Source tissue
Cell [1]

Localisation in source

Purification
  Pseudomonas sp. [1]

Crystallization
  –

Cloned
  –

Renaturated
  –

---

5 STABILITY

pH
  3–9 (stable at 45°C for 10 min) [1]

Temperature (°C)
  45 (10 min, stable) [1]; 70 (100% remaining activity after 10 min) [1]; 83
  (50% remaining activity after 10 min) [1]

Oxidation

Organic solvent

General stability information

Storage

---

6 CROSSREFERENCES TO STRUCTURE DATABANKS

PIR/MIPS code

Brookhaven code

---

7 LITERATURE REFERENCES

[1] Horiuchi, T.: Agric. Biol. Chem.,53,361–368 (1989)

---

1 NOMENCLATURE

EC number
1.1.3.30

Systematic name
Polyvinyl-alcohol:oxygen oxidoreductase

Recommended name
Polyvinyl-alcohol oxidase

Synonymes
Dehydrogenase, polyvinyl alcohol
PVA oxidase [1–3]

CAS Reg. No.
119940-13-5

2 REACTION AND SPECIFICITY

Catalysed reaction
Polyvinyl alcohol + O_2 →
→ oxidized polyvinyl alcohol + H_2O_2

Reaction type
Redox reaction

Natural substrates

Substrate spectrum
1 Polyvinyl alcohol + O_2 [1–3]
2 2-Hexanol + O_2 [3]
3 4-Heptanol + O_2 [3]

Product spectrum
1 Oxidized polyvinyl alcohol + H_2O_2 [1–3]
2 ?
3 ?

Inhibitor(s)

Cofactor(s)/prosthetic group(s)/activating agents
More (PVA oxidase has no requirement for pyrroloquinoline quinone (PQQ) as does the PQQ-dependent PVA dehydrogenase) [2, 3]

Metal compounds/salts

Turnover number (min⁻¹)

Specific activity (U/mg)
1.68 (polyvinyl alcohol) [3]

K_m-value (mM)

pH-optimum

pH-range

Temperature optimum (°C)
30 (assay at) [3]

Temperature range (°C)

---

3 ENZYME STRUCTURE

Molecular weight
31000 (Pseudomonas sp., gel filtration) [3]

Subunits

Glycoprotein/Lipoprotein
–

---

4 ISOLATION/PREPARATION

Source organism
Pseudomonas sp. (in symbiotic culture with Pseudomonas putida during PVA utilization) [1–3]

Source tissue
Cell [1–3]; Culture supernatant [1–3]

Localisation in source
Extracellular (amount depending on growth rate) [1–3]; Periplasm (membrane-bound, amount depending on growth rate) [1–3]

Purification
Pseudomonas sp. [1–3]

Crystallization
–

Cloned
–

Renaturated
–

5 STABILITY

pH

Temperature (°C)

Oxidation

Organic solvent

General stability information

Storage

6 CROSSREFERENCES TO STRUCTURE DATABANKS

PIR/MIPS code

Brookhaven code

7 LITERATURE REFERENCES

[1] Shimao, M., Nishimura, Y., Kato, N., Sakazawa, C.: Appl. Environ. Microbiol.,49,8–10
     (1985)
[2] Shimao, M., Onishi, S., Kato, N., Sakazawa, C.: Appl. Environ. Microbiol.,55,275–278
     (1989)
[3] Shimao, M., Ninomiyaaa, K., Kuno, O., Kato, N., Sakazawa, C.: Appl. Environ. Micro-
     biol.,51,268–275 (1986)

1 NOMENCLATURE

EC number
1.1.3.31

Systematic name
Methanol:oxygen oxidoreductase

Recommended name
Methanol oxidase

Synonymes
Oxidase, methanol
More (this enzyme is probably identical with alcohol oxidase EC 1.1.3.13)

CAS Reg. No.
56379-53-4

2 REACTION AND SPECIFICITY

Catalysed reaction
Methanol + $O_2 \rightarrow$
$\rightarrow$ formaldehyde + H_2O_2 (mechanism [9, 12, 16], ping-pong mechanism [10])

Reaction type
Redox reaction

Natural substrates
Methanol + O_2 (key enzyme of methanol metabolism) [1–8, 14, 15]

Substrate spectrum
1 Methanol + O_2 (best substrate) [1–17]
2 Ethanol + O_2 (oxidation at 28% [1], 82% [3], 92% [7, 14, 15] the rate of methanol oxidation) [1–4, 6–8, 14, 15]
3 n-Propanol + O_2 (oxidation at 74% [7], 53% [1], 38% [3], 34.4% [14, 15] the rate of methanol oxidation) [1, 3–8, 14, 15]
4 n-Butanol + O_2 (oxidation at 2.1% [1], 10.6% [14, 15], 27% [3], 52% [7] the rate of methanol oxidation) [1, 3, 4, 7, 8, 14, 15]
5 1-Pentanol + O_2 (i.e. n-amyl alcohol, oxidation at 1% [1], 21% [3], 30% [7] the rate of methanol oxidation) [1, 3, 7]
6 1-Hexanol + O_2 (oxidation at 4% the rate of methanol oxidation) [7]
7 2-Propen-1-ol + O_2 (i.e. allyl alcohol, oxidation at 17% [1], 82% [3] the rate of methanol oxidation) [1, 3, 8, 10]

8 Isopropanol + O_2 (oxidation at 21% [3], 1.7% [15] the rate of methanol oxidation, oxidation only at high concentrations [4]) [3, 4, 15]

9 2-Buten-1-ol + O_2 [8]

10 Chloroethanol + O_2 (oxidation at 6% [1], 70% [7] the rate of methanol oxidation) [1, 7, 8]

11 Bromoethanol + O_2 (oxidation at 4% the rate of methanol oxidation) [1]

12 3-Chloro-1-propanol + O_2 (oxidation at 22% the rate of methanol oxidation) [7]

13 4-Chloro-1-butanol + O_2 (oxidation at 11% the rate of methanol oxidation) [7]

14 Isobutanol + O_2 (poor substrate) [7]

15 Formaldehyde + O_2 (oxidation at 15% [7], 23% [3] the rate of methanol oxidation, no substrate for Phanerochaete chrysosporium [15]) [3, 7, 8, 11]

16 2-Propyn-1-ol + O_2 (i.e. propargyl alcohol, oxidized at 45% the rate of methanol oxidation [1], oxidized to mechanism-based irreversible inactivator: propynal) [1, 4, 5]

17 1,4-Butynediol + O_2 (oxidized to mechanism-based irreversible inactivator: 4-hydroxy-2-butynal) [5]

18 Ethylene glycol + O_2 (oxidation at 1.7% the rate of methanol oxidation) [15]

19 Ethylene glycol mono-methylether + O_2 (oxidation at 10% the rate of methanol oxidation) [15]

20 More (lower primary alcohols are oxidized to the corresponding aldehydes [1–10, 14, 15], unsaturated [1, 2] and halogenated alcohols [1, 2, 10] are also good substrates, isopropanol, benzyl alcohol and 2-mercaptoethanol are oxidized only at high concentrations [4], propyn-1-ol and methylenecyclopropyl alcohol are not suicide substrates [9], no substrates: longer-chain [8], branched-chain [1, 2], secondary [1, 2, 7, 8], tertiary [7, 8], cyclic [7], aromatic [7, 8], C-2-substituted alcohols [1], diols [7], aldehydes [7], DL-serine, propane diols [1], glycerol, vanillyl alcohol, formaldehyde [15]) [1–10, 14, 15]

Product spectrum

1 Formaldehyde + H_2O_2 [1–11, 14, 15]

2 Acetaldehyde + H_2O_2 [1, 7]

3 n-Propanal + H_2O_2 [1, 7]

4 n-Butanal + H_2O_2 [1, 7]

5 1-Pentanal + H_2O_2

6 1-Hexanal + H_2O_2

7 Acrolein + H_2O_2 (leads in vivo to cell intoxination) [10]

8 Isopropanal + H_2O_2

9 2-Buten-1-al + H_2O_2

10 Chloroethanal + H_2O_2

11 Bromoethanal + H_2O_2

2

12 3-Chloro-1-propanal + H_2O_2
13 4-Chloro-1-butanal + H_2O_2
14 Isobutanal + H_2O_2
15 Formate + H_2O_2 [3, 7, 8, 11]
16 Propynal + H_2O_2 (mechanism-based inhibitor) [5]
17 4-Hydroxy-2-butynal + H_2O_2 (mechanism-based inhibitor) [5]
18 ?
19 ?
20 ?

Inhibitor(s)

PCMB (complete inhibition [3, 4, 7, 8, 10], $t_{1/2}$: 11 min, beta-mercaptoacetic acid restores, substrates or products slow the inactivation rate, not tert-butanol [12]) [3, 4, 7, 8, 10, 12]; Mercuric acetate (strong) [3]; Mercuric chloride [7]; 5,5-Dithiobis(2-nitrobenzoate) [7]; Phenylhydrazine (strong) [3]; o-Phenanthroline [3]; KCN [3, 10]; Iodoacetate (time-dependent inactivation, reversible on removal of excess inhibitor [12]) [7, 12]; Hg^{2+} (strong [3], reversible by mercaptoethanol [12]) [3, 12]; Mo^{6+} (strong) [3]; Cu^{2+} (reversible by EDTA [12]) [3, 7, 8, 12]; Sodium acetate (competitive) [12]; Cd^{2+} [3]; Ag^{2+} [12]; NaN_3 (strong, competitive to methanol [4]) [4, 10]; H_2O_2 (reversible [6], irreversible, pseudo-first order saturation kinetic, methanol or catalase protects partially [9]) [6, 7, 9, 10]; Diethyldicarbonate [12]; Urea (6 M at 30°C for 1–2 h: partial inactivation, urea + KBr: complete inactivation) [10]; KBr (3.5 M at 30°C for 1–2 h: partial inactivation, KBr + urea: complete inactivation) [10]; Formaldehyde (Hansenula polymorpha [11]) [10, 11]; 2-Aminoethanol (substrate analogue) [7]; 2-Propyn-1-al (mechanism-based inactivator, GSH or anaerobiosis prevents [5]) [5, 10]; 4-Hydroxy-2-butynal (product of butynediol oxidation, mechanism-based inactivator, GSH or anaerobiosis prevents [5]) [5, 10]; Hydroxylamine (slight [7], Hansenula polymorpha [11]) [7, 11]; Thiosemicarbazide (slight) [7]; Imidazol (slight) [7]; Metal chelators (slight) [7]; Cyclopropanol (inactivation, suicide substrate) [10, 17]; Cyclopropanone (suicide substrate [10], mechanism [13]) [10, 13]; More (no inhibition by NaCN [4], NEM, N-butylmaleimide, acrylonitrile, DTNB, methylmethanesulfonate [12], 2-propyn-1-ol, methylenecyclopropyl alcohol are no suicide substrates [9]) [4, 9, 12]

Cofactor(s)/prosthetic group(s)/activating agents

FAD (requirement, flavoprotein, 1 mol FAD per mol subunit [3, 4, 8], 5.62 mol per mol enzyme [14, 15], 6 FAD per octamer, mechanism [16], 5 oxygen-stable red flavin semiquinones (ESR-spectroscopy) plus 2 oxidized flavins per octamer, the reconstituted enzyme which lacks the flavin semiquinones is still half as active as the native enzyme [17], noncovalently bound [10]) [1–11, 14–17]

Metal compounds/salts

More (no iron cofactor [4], no metal ion requirement [8]) [4, 8]

Turnover number (min^{-1})
3100 (1H-methanol) [9]; 6000 (2H-methanol) [9]

Specific activity (U/mg)
More [1, 9]; 2.8 [3]; 4.5 [9]; 6.6 [7]; 9.5 [2]; 10.94 [8]; 11.9 (Pichia pastoris)
[6]; 16.5 [15]; 17.3 [14]; 20.7 [4]; 23.2 [10]

K$_m$-value (mM)
0.019 (methanol) [3]; 0.13 (ethanol) [3]; 0.2 (methanol) [4]; 0.4–0.44 (O$_2$ [6,
10], 4-hydroxy-2-butynal [5]) [5, 6, 10]; 0.5 (methanol) [7]; 0.7 (O$_2$ (+ 0.01 M
methanol) [6]; 0.785 (methanol) [15]; 1.0 (ethanol [4], O$_2$ (+ 100 mM methanol) [6]) [4, 6]; 1.4 (methanol (+ 0.19 mM O$_2$)) [6]; 1.97 (ethanol) [15]; 2.6
(formaldehyde) [10]; 3.0 (methanol) [5]; 3.1 (methanol (+ 0.93 mM O$_2$)) [6];
3.2 (2-propyn-1-ol) [4]; 3.5 (formaldehyde) [7]; 6.1 (formaldehyde) [4]; 8.3
(n-propanol) [4]; 10.0 (2-propyn-1-ol) [5]; 13.8 (n-propanol) [15]; 21.3
(n-butanol) [4]; 25.0 (isopropanol) [4]; 27.2 (2-mercaptoethanol) [4]; 35.9
(n-butanol) [15]; 36.0 (1,4-butynediol) [5]

pH-optimum
More (pI: 4.4–4.5 [3], pI: 5.4 [14, 15], pI: 6.2 [10], pI: 6.3 [6]) [3, 6, 10, 14,
15]; 5.5–8.5 (purified) [10]; 6.0 (in vivo, owing to acidic nature of peroxisomal matrix) [10]; 6.0–10.5 (plateau) [14, 15]; 6.3–9.0 (plateau) [1]; 7.5 (Pichia pastoris [6]) [3, 6]; 8.5 (Hansenula polymorpha) [6]; 7.5–9.0 [8]; 9.0 [7]

pH-range
5.5–11.0 (about 6% of maximal activity at pH 5.5 and 12% at pH 11.0) [15];
6.0 (below, activity drops rapidly) [14]; 6.0–10.0 (about 65% of maximal activity at pH 6.0 and 85% at pH 10.0) [7]; 6.5–8.3 (about half-maximal activity
at pH 6.5 and 8.3, Pichia pastoris) [6]; 6.7–9.8 (about half-maximal activity
at pH 6.7 and 9.8, Hansenula polymorpha) [6]; 7.0 (below, activity drops rapidly) [8]

Temperature optimum (°C)
25 (assay at) [1, 4]; 30 [8]; 37.0 (Pichia pastoris) [6]; 37.5 [3]

Temperature range (°C)
18–45 (half-maximal activity at 18°C and 45°C, Pichia pastoris) [6]

---

3 ENZYME STRUCTURE

Molecular weight
300000 (Pichia sp., gel filtration) [7]
310000 (Phanerochaete chrysosporium, gel filtration) [14, 15]
520000 (Candida sp., sedimentation equilibrium centrifugation) [3]
600000 (Candida boidinii, sedimentation equilibrium [8], Hansenula polymorpha, gel filtration [10]) [8, 10]
610000 (Poria contigua, sedimentation equilibrium ultracentrifugation) [4]
675000 (Pichia pastoris, meniscus depletion method) [6]

Subunits

Tetramer (4 × 75000, Phanerochaete chrysosporium, SDS-PAGE) [14, 15]
Octamer (8 × 65000, Candida sp., SDS-PAGE [3], 8 × 74000, Candida boidinii, SDS-PAGE [8], Hansenula polymorpha, calculated from amino acid sequence derived from nucleotide sequence of a structural gene [10],
8 × 79000, Poria contigua, SDS-PAGE [4], 8 × 80000, Pichia pastoris,
SDS-PAGE [6]) [3, 4, 6, 8, 10]
More (electron microscopic analysis: the complete structure is an octad aggregate composed of 2 tetragons face-to-face [4], the subunits are noncovalently associated [8], minimal molecular weight calculated from amino
acid analysis: 81000 [6]) [4, 6, 8]

Glycoprotein/Lipoprotein

No carbohydrates [1]

---

4 ISOLATION/PREPARATION

Source organism

Basidiomycete (strain B191039 [2]) [1, 2]; Polyporus obtusus [2]; Radulum
casearium [2]; Candida sp. (strain 25-A) [3]; Poria contigua (brown rot fungus) [4]; Candida boidinii [5, 8, 9, 11–13, 16]; Pichia pastoris (strain IFP 206
[6]) [6, 11]; Pichia sp. [7]; Hansenula polymorpha (strain DL-1 [17]) [6, 10,
11, 17]; Phanerochaete chrysosporium (syn. Sporotrichum pulverulentum,
strain K-3) [14, 15]

Source tissue

Mycelium [1, 2, 4]; Cell [3, 5–13]

Localisation in source

Cytoplasm [4, 6]; Microbodies [4, 8]; Peroxisomes (synthesis of enzyme monomers in cytoplasm) [10, 11]

Purification

Basidiomycete [1, 2]; Candida sp. [3]; Poria contigua [4]; Pichia pastoris
[6]; Pichia sp. [7]; Candida boidinii [8]; Phanerochaete chrysosporium [14,
15]

Crystallization

(Basidiomycete [1], Candida sp. [3], Pichia sp. [7]) [1, 3, 7]

Cloned

–

Renaturated

–

5 STABILITY

pH
More (unstable at acidic pH-values) [6]; 5.0 (complete inactivation after 3 h
at 25°C) [1]; 6.0–8.0 (45 min stable at 30°C) [3]; 6.5 (below: rapid inactivation) [1, 2]; 6.5–8.5 (stable, below: rapid loss of activity, even at 0°C) [12];
7.0–8.0 (stable, Pichia pastoris) [6]; 7.0–9.0 (stable) [1]; 7.0–10.0 (stable)
[8]; 8.0–10.0 (stable, Hansenula polymorpha) [6]; 9.0 (and above inactivation, Pichia pastoris) [6]

Temperature (°C)
20 (24 h stable) [10]; 25 (complete inactivation at pH 5.0 after 3 h) [1]; 30
(retains full activity for 10 min [3], activation occurs at pH 7.5–8.5 [10],
about 30% loss of activity after 15 min [15]) [3, 10, 15]; 38 (50% loss of activity) [6]; 40 (about 85% loss of activity after 15 min) [15]; 45 (1 h stable)
[10]; 50 (complete inactivation after 15 min) [15]; 60 (inactivation after 10
min) [3]

Oxidation

Organic solvent
Acetonitril, 2% v/v, up to 2 h stable at 0°C [12]

General stability information
Polyethylene glycol stabilizes [2]; 2-Mercaptoethanol does not stabilize, it is
oxidized as substrate analogue [10]; Sucrose stabilizes during freezing and
thawing [10]; Freezing leads to loss of activity [2]; Freezing at –20°C and
thawing leads to 25% loss of activity [4]; Lyophilization leads to buffer-unso-
luble, powdered protein [6]

Storage
–20°C, stable [8]; –20°C, 25% loss of activity after thawing [4]; 0°C, stable
at pH 6.5–8.5 [12]; 4°C, purified, stable for 2 weeks [2, 4]; 4°C, lyophilized,
20% loss of activity within a year [2]; Crystalline stable for at least 5 months
in 0.05 M phosphate buffer, pH 7.7 [1]; Crystalline or precipitate, in phosphate buffer, pH 7.5 with 60–80% ammonium sulfate, loss of 20% activity
per month [10]

---

6 CROSSREFERENCES TO STRUCTURE DATABANKS

PIR/MIPS code

Brookhaven code

7 LITERATURE REFERENCES

[1] Janssen, F.W., Ruelius, H.W.: Biochim. Biophys. Acta,151,330–342 (1968)
[2] Janssen, F.W., Kerwin, R.M., Ruelius, H.W.: Methods Enzymol.,41B,364–369 (1975) (Review)
[3] Yamada, H., Shin, K.-C., Kato, N., Shimizu, S., Tani, Y.: Agric. Biol. Chem.,43,877–878 (1979)
[4] Bringer, S., Sprey, B., Sahm, H.: Eur. J. Biochem.,101,563–570 (1979)
[5] Nichols, C.S., Cromartie, T.H.: Biochem. Biophys. Res. Commun.,97,216–221 (1980)
[6] Couderc, R., Baratti, J.: Agric. Biol. Chem.,44,2279–2289 (1980)
[7] Patel, R.N., Hou, C.-T., Laskin, A.I., Derelanko, P.: Arch. Biochem. Biophys.,210, 481–488 (1981)
[8] Sahm, H., Schütte, H., Kula, M.-R.: Methods Enzymol.,89,424–428 (1982) (Review)
[9] Geissler, J., Ghisla, S., Kroneck, P.M.H.: Eur. J. Biochem.,160,93–100 (1986)
[10] Van der Klei, I.J., Bystryck, L.V., Harder, W.: Methods Enzymol.,188,420–427 (1990)
[11] Bystryck, L.V., Dijkhuizen, L., Harder, W.: J. Gen. Microbiol.,137,2381–2386 (1991)
[12] Cromartie, T.H.: Biochemistry,20,5416–5423 (1981)
[13] Cromartie, T.H.: Biochem. Biophys. Res. Commun.,105,785–790 (1982)
[14] Eriksson, K.-E., Nishida, A.: Methods Enzymol.,161,322–326 (1988)
[15] Nishida, A., Eriksson, K.-E. : Biotechnol. Appl. Biochem.,9,325–338 (1987)
[16] Geissler, J., Hemmerich, P.: FEBS Lett.,126,152–156 (1981)
[17] Mincey, T., Tayrtien, G., Mildvan, A.S., Abeles, R.H.: Proc. Natl. Acad. Sci. USA,77,7099–7101 (1980)

1 NOMENCLATURE

EC number
1.1.4.1

Systematic name
2-Methyl-3-phytyl-1,4-naphthoquinone:oxidized-dithiothreitol oxidoreductase

Recommended name
Vitamin-K-epoxide reductase (warfarin sensitive)

Synonymes
Reductase, phylloquinone epoxide
Phylloquinone epoxide reductase
Vitamin K epoxide reductase
Vitamin K_1 epoxide reductase
More (cf. EC 1.1.4.2)

CAS Reg. No.
55963-40-1

2 REACTION AND SPECIFICITY

Catalysed reaction
2-Methyl-3-phytyl-1,4-naphthoquinone + oxidized dithiothreitol + H_2O →
→ 2,3-epoxy-2,3-dihydro-2-methyl-3-phytyl-1,4-naphthoquinone + 1,4-dithio-
threitol (ping-pong mechanism [1, 7])

Reaction type
Redox reaction

Natural substrates
Vitamin K 2,3-epoxide + DTT (reaction in metabolic pathway of vitamin K required for its biological activity [5], thioredoxin is the possible physiological
electron acceptor [9]) [5, 9]

Substrate spectrum
1 Vitamin K 2,3-epoxide + DTT [1–9]
2 2-Hydroxymethyl-vitamin K 2,3-epoxide + DTT [8]
3 Vitamin K 2,3-epoxide analogs + DTT (such as hydroxymethyl-, chlorome-
  thyl-, fluoromethyl-, difluoromethyl-, and formyl-analogs) [8],
4 More (2-mercaptoethanol, GSH, Cys, 1,6-hexanedithiol are inactive as acceptors) [5]

Product spectrum
1 Vitamin K + oxidized DTT + H_2O [1–10]
2 2-Hydroxymethyl-vitamin K + oxidized DTT + H_2O [8]
3 ?
4 ?

Inhibitor(s)
Warfarin (i.e. 3-(alpha-acetonylbenzyl)-4-hydroxycoumarin, inhibition decrea-
sed by DTT [3]) [1, 3, 4, 6, 9]; Cholate (high concentration) [1]; Coumarin
anticoagulants [1, 2, 4, 6]; Deriphat 160 (detergent, inactivation) [1]; DTT
(high concentration) [1]; Lapachol
(2-hydroxy-3-(3-methyl-2-butenyl)-1,4-naphthoquinone, fully reversible) [2];
Imidazopyridines [4]; 2,3,5,6-Tetrachloro-4-pyridinol [4]; NEM (inhibition increased if the enzyme is prereduced by DTT, 1,4-butanedithiol or 1,2-etha-
nediol [5], protected by vitamin K 2,3-epoxide [7]) [5, 7]; IAA (not: iodoacetic acid [5]) [5]; Vitamin K 2,3-epoxide analogs (hydroxymethyl-, chlorome-
thyl-, fluoromethyl-, difluoromethyl-, formyl-analogs and analogs with modified phytyl chain, competitive inhibition) [8]; Difenacoum [6]

Cofactor(s)/prosthetic group(s)/activating agents
DTT (DL-DTT and L-DTT equally effective [5]) [1, 2, 5, 6, 9, 10]; Dithioerythritol (equally effective as DTT) [5]; 1,2-Ethanedithiol (activation) [5]; Lipoic
acid (activation, less effective than DTT [9]) [5, 6, 9]; 1,4-Butanedithiol (activation) [5]; Reduced thioredoxin (likely physiological cofactor, more potent
than DTT [9], together with thioredoxin reductase plus protein-disulfide isomerase still more effective than DTT [10]) [9, 10]; Reduced ribonuclease
(activation, together with thio-redoxin/thioredoxin reductase, EC 1.6.4.5 plus
protein-disulfide isomerase more potent than DTT) [10]; GSH (activation, 6%
as effective as DTT) [9]; Protein factor (activation, two different protein factors from rat liver cytosol) [3]; Protein disulfide isomerase (activation, together with thioredoxin/thioredoxin reductase and reduced ribonuclease)
[10]; Glycerol (activation) [11]

Metal compounds/salts
NaCl (activation at high concentration) [1]; LiCl (activation at high concentration) [1]; K_2SO_4 (activation at high concentration) [1]; KCl (activation at
high concentration) [1]

Turnover number (min⁻¹)

Specific activity (U/mg)
0.000009 (thioredoxin of bovine thymus) [10]; 0.000014 (thioredoxin of E.
coli) [10]; 0.000026 (DTT) [10]; 0.00005 (thioredoxin of bovine thymus plus
protein-disulfide isomerase) [10]; 0.000057 (thioredoxin of E. coli plus protein-disulfide isomerase) [10]; 0.00136 [1]

K_m-value (mM)
0.0025 (thioredoxin, bovine) [3, 9]; 0.008 (commercial thioredoxin, rat [9], thioredoxin plus protein-disulfide isomerase, EC 5.3.4.1 [10]) [9, 10]; 0.009 (vitamin K 2,3-epoxide) [1]; 0.016 (commercial thioredoxin, bovine) [9]; 0.24 (DTT, rat) [9]; 0.32 (lipoic acid, rat) [9]; 0.47 (DTT, bovine) [9]; 0.54–0.6 (DTT) [1]; 0.59 (lipoic acid, bovine) [9]; 9.1–10.0 (vitamin K 2,3-epoxide) [1]; More (different K_m-values at different cholate concentrations, rat liver extract) [1]

pH-optimum
9.0 [1]

pH-range
8.5–9.8 (about half-maximal activity at pH 8.5 and 9.8) [1]

Temperature optimum (°C)
25 [1]

Temperature range (°C)
18–35 (about 55% of maximal activity at 18°C and 35°C) [1]

---

3 ENZYME STRUCTURE

Molecular weight

Subunits

Glycoprotein/Lipoprotein
–

---

4 ISOLATION/PREPARATION

Source organism
Rat (Holtzman strain [1–4, 6], Wistar strain [5, 9]) [1–9]; Bovine [7–11]

Source tissue
Liver [1–10]

Localisation in source
Microsomes (integral membrane protein) [1–9]

Purification
Rat [1, 2]

Crystallization
–

Cloned
–

Renaturated

–

---

5 STABILITY

pH

Temperature (°C)
0 (considerably more stable than at 22°C) [1]; 15 (rapid inactivation) [1]; 22 (rapid inactivation) [1]

Oxidation

Organic solvent
Ethanol, up to 1% v/v, stable [1]

General stability information
KCl stabilizes [1]; Sodium cholate, low-concentrated, stabilizes [1]; Deriphat 160 stabilizes during purification [1]

Storage
4°C, acetone powder, stable [4]

---

6 CROSSREFERENCES TO STRUCTURE DATABANKS

PIR/MIPS code

Brookhaven code

---

7 LITERATURE REFERENCES

[1] Hildebrandt, E.F., Preusch, P.C., Patterson, J.C., Suttie, J.W.: Arch. Biochem. Biophys.,228,480–492 (1984)
[2] Preusch, P.C., Suttie, J.W.: Arch. Biochem. Biophys.,234,405–412 (1984)
[3] Siegfried, C.M.: Arch. Biochem. Biophys.,223,129–139 (1983)
[4] Friedman, P.A., Griep, A.E.: Biochemistry,19,3381–3386 (1980)
[5] Lee, J.J., Fasco, M.J.: Biochemistry,23,2246–2252 (1984)
[6] Whitlan, D.S., Sadowski, J.A., Suttie, J.W.: Biochemistry,17,1371–1377 (1978)
[7] Silverman, R.B., Nandi, D.L.: J. Enzyme Inhib.,3,289–294 (1990)
[8] Ryall, R.P., Nandi, D.L., Silverman, R.B.: J. Med. Chem.,33,1790–1797 (1990)
[9] Silverman, R.B., Nandi, D.R.: Biochem. Biophys. Res. Commun.,155,1248–1254 (1988)
[10] Soute, B.M., Groenen-van Dooren, M.M.C.L., Holmgren, A., Lundström, J., Vermeer, C.: Biochem. J.,281,255–259 (1992)
[11] Mukharji, I., Silverman, R.B.: Proc. Natl. Acad. Sci. USA,82,2713–2717 (1985)

1 NOMENCLATURE

EC number
1.1.4.2

Systematic name
3-Hydroxy-2-methyl-3-phytyl-2,3-dihydronaphthoquinone:oxidized-dithiothreitol oxidoreductase

Recommended name
Vitamin-K-epoxide reductase (warfarin-insensitive)

Synonymes
Reductase, vitamin K epoxide (warfarin-insensitive)
Vitamin K 2,3 epoxide reductase
Vitamin K epoxide reductase (warfarin–insensitive)
Vitamin KO reductase

CAS Reg. No.
97089-80-0

2 REACTION AND SPECIFICITY

Catalysed reaction
3-Hydroxy-2-methyl-3-phytyl-2,3-dihydronaphthoquinone + oxidized dithiothreitol + $H_2O \rightarrow$
$\rightarrow$ 2,3-epoxy-2,3-dihydro-2-methyl-3-phytyl-1,4-naphthoquinone + 1,4-dithiothreitol

Reaction type
Redox reaction

Natural substrates
2,3-Epoxy-2,3-dihydro-2-methyl-3-phytyl-1,4-naphthoquinone + DTT (i.e. vitamin K 2,3-epoxide, the enzyme is supposed to catalyze the reduction of the epoxide to quinone and of the quinone to vitamin K hydroquinone [3]) [1–5]

Substrate spectrum
1 2,3-Epoxy-2,3-dihydro-2-methyl-3-phytyl-1,4-naphthoquinone + DTT (vitamin K 2,3-epoxide) [1, 2, 4, 5]
2 More (2-mercaptoethanol, reduced GSH, 1,2-ethanediol or reduced lipoic acid is ineffective as cofactor) [1]

Product spectrum
1 3-Hydroxy-2-methyl-3-phytyl-2,3-dihydronaphthoquinone + oxidized DTT + H$_2$O (i.e. 3-hydroxy-2,3-dihydro-vitamin K) [1, 2, 4, 5]
2 ?

Inhibitor(s)
Glycerol (concentrations up to 30% v/v inhibit enzyme activity up to 88%) [1]; Cholate (high concentration) [2]; Chloromenaquinone-2 [3]; Coumarin anticoagulants [3, 4]; Lapachol (i.e. 2-hydroxy-3-(3-methyl-2-butenyl)-1,4-naphthoquinone, fully reversible) [4]; 2,3,5,6-Tetrachloro-4-pyridinol [5]; Imidazopyridines [5]; Difenacoum [6]; More (not inhibited by warfarin) [1]

Cofactor(s)/prosthetic group(s)/activating agents
DTT (non-physiological cofactor) [1]; Thioredoxin (probable physiological cofactor) [3]; 3-[(3-Cholamidopropyl)-dimethyl-ammonio]1-propane sulfonate (CHAPS, activation) [1]; Lipoic acid [6]

Metal compounds/salts
NaCl (activation at high concentration) [2]; LiCl (activation at high concentration) [2]; KCl (activation at high concentration) [2]; K$_2$SO$_4$ (activation at high concentration) [2]

Turnover number (min^{-1})

Specific activity (U/mg)
0.235 [1]

K$_m$-value (mM)
0.004 (vitamin K epoxide) [2]; 0.16 (DTT) [2]

pH-optimum
9.0 [2]

pH-range
8.5–9.8 (about half-maximal activity at pH 8.5 and 9.8) [2]

Temperature optimum (°C)
25 [1, 2]

Temperature range (°C)
7–35 (about half-maximal activity at 7°C and 35°C) [2]

3 ENZYME STRUCTURE

Molecular weight
25000 (bovine, gel filtration) [1]

Subunits
Dimer (2 × 12400, bovine, SDS-PAGE) [1]

Glycoprotein/Lipoprotein

–

---

4 ISOLATION/PREPARATION

Source organism
Bovine [1]; Rat (warfarin-resistant rat [6]) [2–6]

Source tissue
Liver [1–6]

Localisation in source
Microsomes [1–6]

Purification
Bovine [1]; Rat (partial [2]) [2, 4]

Crystallization

–

Cloned

–

Renaturated

–

---

5 STABILITY

pH

Temperature (°C)
0 (considerably stable) [2]; 15 (rapid inactivation) [2]; 22 (rapid inactivation) [2]

Oxidation

Organic solvent
Ethanol, 1% v/v, stable [2]

General stability information
High ionic strength inactivates during purification [1, 2]; Omission of salt improves storage stability [2]; Physiological salt concentration stimulates [2]; Deriphat 160, detergent, solubilizes and inactivates [2]

Storage
–70°C, microsomal preparation, stable for months [1]

6 CROSSREFERENCES TO STRUCTURE DATABANKS

PIR/MIPS code

Brookhaven code

7 LITERATURE REFERENCES

[1] Mukharji, I., Silverman, R.B.: Proc. Natl. Acad. Sci. USA,82,2713–2717 (1985)
[2] Hildebrandt, E.F., Preusch. P.C., Patterson, J.L., Suttie, J.W.: Arch. Biochem. Bio-
    phys.,228,480–492 (1984)
[3] Gardill, S.L., Suttie, J.W.: Biochem. Pharmacol.,40,1055–1061 (1990)
[4] Preusch, P.C., Suttie, J.W.: Arch. Biochem. Biophys.,234,405–412 (1984)
[5] Friedman, P.A., Griep, A.E.: Biochemistry,19,3381–3386 (1980)
[6] Whitlan, D.S., Sadowski, J.A., Suttie, J.W.: Biochemistry,17,1371–1377 (1978)

4

1 NOMENCLATURE

EC number
1.1.5.1

Systematic name
Cellobiose:quinone 1-oxidoreductase

Recommended name
Cellobiose dehydrogenase (quinone)

Synonymes
Dehydrogenase, cellobiose
Cellobiose:quinone oxidoreductase [1]
Cellobiose-quinone oxidoreductase

CAS Reg. No.
54576-85-1

2 REACTION AND SPECIFICITY

Catalysed reaction
Cellobiose + a quinone →
→ cellobiono-1,5-lactone + a phenol

Reaction type
Redox reaction

Natural substrates
Cellobiose + quinone (cellulose biodegradation [3, 4], lignin biodegradation [3, 4, 7]) [3, 4, 7]

Substrate spectrum
1 Cellobiose + quinone (e.g.: o- or p-quinones [2, 4, 6], 3-methoxy-5-tert-bu-tylbenzoquinone [4, 7]) [1–7]
2 Cellopentaose + quinone [2, 5]
3 Lactose + quinone (slowly [6]) [2, 3, 5, 6]
4 4beta-Glucosylmannose + quinone [2, 5]
5 More (low activity: glucose, maltose [3], not: crystalline cellulose [3], cellulose [5]) [3, 5]

Product spectrum
1 Cellobiono-1,5-lactone + phenol (superoxide and H_2O_2 produced [3])
  [1-7]
2 Cellopentaono-1,5-lactone + phenol
3 4-O-(beta-D-Galactopyranosyl)-D-glucono-1,5-lactone + phenol (superoxide and H_2O_2 produced [3])
4 ?
5 ?

Inhibitor(s)
2,6-Dichlorophenolindophenol (above 0.2 mM) [3]

Cofactor(s)/prosthetic group(s)/activating agents
FAD (flavoprotein with FAD as prosthetic group) [2-5]; 6-Hydroxy-FAD (contains variable amounts of a green chromophore (6-hydroxy-FAD)) [3]

Metal compounds/salts

Turnover number (min^{-1})

Specific activity (U/mg)
More [2, 5]

K_m-value (mM)
0.046 (cellobiose, enzyme form 1a) [3]; 0.024 (cellobiose, enzyme form 1b)
[3]; 0.045 (cellobiose, enzyme forms 2a, 2b) [3]; 0.700 (lactose, enzyme
forms 1a, 1b) [3]; 0.680 (lactose, enzyme form 1b) [3]; 0.740 (lactose, enzyme form 2a) [3]; 3.8 (maltose, enzyme form 1a) [3]; 3.65 (maltose, enzyme
form 1b) [3]; 400 (glucose, enzyme form 1a) [3]; 380 (glucose, enzyme form
1b) [3]; 0.013 (2,6-dichlorophenolindophenol, enzyme form 1a) [3]; 0.015
(2,6-dichlorophenolindophenol, enzyme form 1b) [3]; 0.014 (2,6-dichloro-
phenolindophenol, enzyme form 2a) [3]; 0.1111 (benzoquinone) [3]; 0.130
(benzoquinone, enzyme form 1b) [3]; 0.102 (benzoquinone, enzyme form
2a) [3]; More [3, 7]

pH-optimum
4.5-5.0 [6]

pH-range
3.5-7.0 (3.5: about 60% of activity maximum, 7.0: about 20% of activity maximum) [6]

Temperature optimum (°C)
25 (assay at) [3]

Temperature range (°C)

3 ENZYME STRUCTURE

Molecular weight
58000 (Sporotrichum pulverulentum, sedimentation equilibrium) [2, 3]

Subunits
Monomer (1 × 60000, Sporotrichum pulverulentum, SDS-PAGE) [3]

Glycoprotein/Lipoprotein
Glycoprotein (Sporotrichum pulverulentum, neutral sugar content of 4 enzyme forms: 1a (0.3%), 1b (0.8%), 2a (0.9%), 2b (1.6%)) [3]

---

4 ISOLATION/PREPARATION

Source organism
Polyporus vesicolor [2, 6, 7]; Phanerochaete chrysosporium [1]; Sporotrichum pulverulentum (anamorph of Phanerochaete chrysosporium [2]) [2–5]; Chrysosporium lignorum [6]

Source tissue
Culture medium [2]

Localisation in source
Extracellular [2, 4, 6, 7]

Purification
Phanerochaete chrysosporium [1]; Sporotrichum pulverulentum (4 forms differ in neutral-sugar content: 1a, 1b, 2a, 2b [3]) [2, 3, 5]

Crystallization
–

Cloned
–

Renaturated
–

---

5 STABILITY

pH

Temperature (°C)

Oxidation
Rapidly reduced by dithionite [3]

Organic solvent

General stability information
Freezing inactivates purified enzyme [5]

Storage

---

6 CROSSREFERENCES TO STRUCTURE DATABANKS

PIR/MIPS code

Brookhaven code

---

7 LITERATURE REFERENCES

[1] Renganathan, V., Usha, S.N., Lindenburg, F.: Appl. Microbiol. Biotechnol.,32, 609–613 (1990)
[2] Westermark, U., Eriksson, K.-E.: Methods Enzymol.,160,463–468 (1988)
[3] Morpeth, F.F., Jones, G.D.: Biochem. J.,236,221–226 (1986)
[4] Eriksson, K.-E.: Biotechnol. Bioeng.,20,317–332 (1978) (Review)
[5] Westermark, U., Eriksson, K.-E.: Acta Chem. Scand., B29,419–424 (1975)
[6] Westermark, U., Eriksson, K.-E.: Acta Chem. Scand., B28,209–214 (1974)
[7] Westermark, U., Eriksson, K.-E.: Acta Chem. Scand., B28,204–208 (1974)

1 NOMENCLATURE

EC number
1.1.99.1

Systematic name
Choline:(acceptor) 1-oxidoreductase

Recommended name
Choline dehydrogenase

Synonymes
Oxidase, choline
Choline oxidase
Choline–cytochrome c reductase
Dehydrogenase, choline
Choline:(acceptor) oxidoreductase [4]

CAS Reg. No.
9028-67-5 (identical with CAS Reg. No. of EC 1.1.3.17)

2 REACTION AND SPECIFICITY

Catalysed reaction
Choline + acceptor →
→ betaine aldehyde + reduced acceptor (mechanism [3])

Reaction type
Redox reaction

Natural substrates
Choline + coenzyme Q [4]

Substrate spectrum
1 Choline + acceptor (r, reverse reaction at 5.2% of forward reaction [4], choline is the sole substrate [1], not: 2-dimethylaminoethanol, monoethanolamine [4], acceptor: phenazine methosulfate (100% relative activity [2]) [2–5], O_2 (73% relative activity) [2], ubiquinone-2 [3], coenzyme Q (primary electron acceptor in vivo) [4], cytochrome c (63% relative activity) [2], ferricyanide (42% relative activity) [2], methylene blue (4% relative activity) [2], 2,6-dichlorophenolindophenol (27% relative activity) [2], absolute requirement for an electron acceptor other than O_2 [4], cytochrome c, ferricyanide, methylene blue and 2,6-dichlorophenolindophenol show no direct reaction with the enzyme [2]) [1–5]

Product spectrum

1 Betaine aldehyde + reduced acceptor

Inhibitor(s)

$AgNO_3$ [5]; $CuSO_4$ [5]; $MnCl_2$ [5]; $HgCl_2$ [5]; Mercuric acetate [5]; Betaine hydrochloride (slight) [5]; DL-Carnitine hydrochloride (slight) [5]; Glycine (slight) [5]; Dimethylglycine (slight) [5]; Tetra-n-propylammonium iodide (slight) [5]; Triethylbenzylammonium iodide (slight) [5]; Tetraethylammonium iodide (slight) [5]; Phenyltriethylammonium iodide (slight) [5]; Tetramethyl-ammonium iodide (slight) [5]; Tetra-n-butylammonium iodide (slight) [5]; Benzyltrimethylammonium chloride (slight) [5]; Sodium cyanide [6]; Iodoacetic acid [5]; 2-Dimethyl aminoethanol [4]; Semicarbazide [1, 4]; Mono-ethanolamine [4]; Betaine aldehyde (choline dehydrogenase activity) [4]; p-Chloromercuribenzoate [5]

Cofactor(s)/prosthetic group(s)/activating agents

Pyrroloquinoline quinone (as prosthetic group) [1]; Coenzyme Q (primary electron acceptor in vivo) [4]; More (no dissociable coenzyme) [5]

Metal compounds/salts

Turnover number (min^{-1})

Specific activity (U/mg)

6.3 [4]

K_m-value (mM)

0.14 (phenazine methosulfate) [5]; 1.1 (phenazine methosulfate (+ choline)) [4]; 1.7 (choline) [5]; 4.8 (choline) [1]; 7 (choline) [2, 4]; More [3]

pH-optimum

7.5–8.0 [1]; 7.6–8.2 (crude enzyme extract) [2]; 9 [4, 5]

pH-range

7.5–10 (about 50% of activity maximum at pH 7.5 and 10.0) [5]

Temperature optimum (°C)

45 [4]

Temperature range (°C)

---

3 ENZYME STRUCTURE

Molecular weight

Subunits

Glycoprotein/Lipoprotein

–

4 ISOLATION/PREPARATION

Source organism
Dog [1]; Rat [2–4]; Pseudomonas aeruginosa (A-16 [5, 6]) [5–7]

Source tissue
Liver [1–4]; Cell [5, 6]

Localisation in source
Mitochondria [1–4]; Particulate [5, 6]; Membrane (peripheral membrane protein) [7]

Purification
Dog [1]; Rat [4]; Pseudomonas aeruginosa (A-16, partial) [5]

Crystallization
–

Cloned
–

Renaturated
–

5 STABILITY

pH
8.0–9.5 (10 min, 37°C, stable) [6]; More [6]

Temperature (°C)
35 (10 min, pH 8.3, 10% loss of activity) [5]; 37 (pH 8.0–9.5, 10 min, stable) [6]; 40 (10 min, pH 8.3, 50% loss of activity) [5]; 50 (10 min, pH 8.3, 100% loss of activity) [5]

Oxidation

Organic solvent

General stability information
Detergents, e.g. Brij 58, Triton X-100, cholate, deoxycholate or beta-octylglucoside stabilize [1]; Glycerol, 30%, stabilizes partially purified enzyme [3]; Repeated freezing and thawing: rapid loss of activity [4]; 25 Cycles of freezing/thawing cause 40% loss of activity [2]

Storage
–15°C, 1 month, 50% loss of activity [4]; –20°C, stable, partially purified enzyme [4]; 5°C, pH 7.4, 5–6 days, complete loss of activity, partially purified enzyme [5]; –15°C, pH 7.4 as mitochondrial acetone powder, 20% loss of activity after 6 days [2]

Enzyme Handbook © Springer-Verlag Berlin Heidelberg 1995
Duplication, reproduction and storage in data banks are only
allowed with the prior permission of the publishers

6 CROSSREFERENCES TO STRUCTURE DATABANKS

PIR/MIPS code

Brookhaven code

7 LITERATURE REFERENCES

[1] Ameyama, M., Shinagawa, E., Matsushita, K., Takimoto, K., Nakashima, K., Adachi, O.: Agric. Biol. Chem.,49,3623–3626 (1985)
[2] Rendina, G., Singer, T.P.: J. Biol. Chem.,234,1605–1610 (1959)
[3] Barrett, M.C., Dawson, A.P.: Biochem. J.,151,677–683 (1975)
[4] Tsuge, H., Nakano, Y., Onishi, H., Futamura, Y., Ohashi, K.: Biochim. Biophys. Acta,614,274–284 (1980)
[5] Nagasawa, T., Mori, N., Tani, Y., Ogata, K.: Agric. Biol. Chem.,40,2077–2084 (1976)
[6] Nagasawa, T., Kawabata, Y., Tani, Y., Ogata, K.: Agric. Biol. Chem.,39,1513–1514 (1975)
[7] Bater, A.J., Venables, W.A.: Biochim. Biophys. Acta,468,209–226 (1977)

1 NOMENCLATURE

EC number
   1.1.99.2

Systematic name
   (S)-2-Hydroxyglutarate:(acceptor) 2-oxidoreductase

Recommended name
   2-Hydroxyglutarate dehydrogenase

Synonymes
   Dehydrogenase, 2-hydroxyglutarate
   alpha-Ketoglutarate reductase [8]
   alpha-Hydroxyglutarate dehydrogenase
   L-alpha-Hydroxyglutarate dehydrogenase
   Hydroxyglutaric dehydrogenase [1]
   alpha-Hydroxyglutarate oxidoreductase [7]
   L-alpha-Hydroxyglutarate:NAD$^+$ 2-oxidoreductase [3]
   alpha-Hydroxyglutarate dehydrogenase (NAD$^+$ specific) [4]
   More (enzymes described in references [2–6, 8, 9] are specific for NAD(H),
   correctly they must be listed under EC 1.1.1.–, enzymes tested in references
   [4–7] are specific for D-2-hydroxyglutarate (similar enzymes), references
   without remark to isomer (R- or S-form) [2, 8, 9], D-2-hydroxyglutarate (specific for D-isomer) + electron acceptor (flavin or nonheme iron) [7]) [2–9]

CAS Reg. No.
   9028-80-2

2 REACTION AND SPECIFICITY

Catalysed reaction
   (S)-2-Hydroxyglutarate + acceptor →
   → 2-oxoglutarate + reduced acceptor (mechanism [4])

Reaction type
   Redox reaction

Natural substrates
   2-Hydroxyglutarate + NAD$^+$ (one of the key enzymes in 2-hydroxyglutarate
   metabolism, possible involvement in metabolism of L-lysine in some bacteria) [3]

Substrate spectrum

1 (S)-2-Hydroxyglutarate + acceptor (r [1, 3], acceptors: pyocyanine and a few closely related phenazine derivatives [1], oxidizes D-2-hydroxygluta-rate with about 1/10 of the rate for L-form oxidation) [1, 3]

2 2-Oxoglutarate + NADH (r [3, 8]) [2–6, 8, 9]

3 More (enzymes described in references [2–6, 8, 9] are specific for NAD(H), correctly they must be listed under EC 1.1.1.–, enzymes in refe-rences [4–7] are specific for D-2-hydroxyglutarate (similar enzymes), refe-rences without remark to isomer (R- or S-form) [2, 8, 9], D-2-hydroxygluta-rate (specific for D-isomer) + electron acceptor (flavin or nonheme iron) [7], low activity with 2-oxobutyrate, 2-oxo-n-pentanoate, 2-oxo-n-hexanoate (about 10% the rate of 2-oxoglutarate reduction), 2-oxomalonate (about 30% the rate of 2-oxoglutarate reduction [3])) [2–9]

Product spectrum

1 2-Oxoglutarate + reduced acceptor (r [1, 3])

2 L-2-Hydroxyglutarate + NAD$^+$ (r [3, 8]) [2–6, 8]

3 ?

Inhibitor(s)

2-Oxoglutarate (product inhibition [8], substrate inhibition [8], weak substra-te inhibition [3], above 1–2 mM [5], competitive inhibition of reductase reac-tion [5]) [3, 5, 8]; NADH (strong substrate inhibition above 0.5 mM [3, 5]) [3, 5, 8]; D-2-Hydroxyglutarate (weak substrate inhibition) [3]; SDS [4]; Fe^{2+} [4, 5]; AgNO$_3$ [4]; HgCl$_2$ [4]; NAD$^+$ (competitive inhibition of reductase reaction) [5]; Sn^{2+} [5]; Zn^{2+} [5]; p-Chloromercuribenzoate [2, 5]; N-Ethylmaleimide (slight) [5]

Cofactor(s)/prosthetic group(s)/activating agents

NADH (enzymes described in references [2–6, 8, 9] are specific for NAD(H), correctly they must be listed under EC 1.1.1.–) [2–6, 8]; NADPH (8% the rate of reaction with NADH [2], 2.9% the rate of reaction with NADH [5], 3–4% of reaction with NADH [8]) [2, 5, 8]; NAD$^+$ (enzymes described in references [2–6, 8, 9] are specific for NAD(H), correctly they must be listed under E.C. 1.1.1.–) [2–6, 8, 9]; NADP$^+$ (3% the rate of reaction with NAD$^+$ [8], inactive [9]) [8]; Flavin (prosthetic group) [5]

Metal compounds/salts

Turnover number (min^{-1})

Specific activity (U/mg)

More [3, 4, 8]; 43 [4]

K_m-value (mM)
  0.064 (2-oxoglutarate) [2]; 0.032 (NADH) [2]; 0.12 (2-oxoglutarate) [3, 4];
  0.14 (NADH) [3]; 1.7 (L-2-hydroxyglutarate) [3]; 5.3 (NAD⁺) [3]; 15 (D-2-hy-
  droxyglutarate) [3]; 0.028 (NADH) [4]; 0.67 (2-hydroxyglutarate) [4]; 0.077
  (NAD⁺) [4]

pH-optimum
  11.5 (L-2-oxoglutarate + NAD⁺) [3]; 9.5 (2-hydroxyglutarate + NAD⁺) [4]; 8.8
  (2-oxoglutarate + NADH) [4]; 8.0–8.5 [1]; 7.2–7.4 [2]; 7.2 (2-oxoglutarate +
  NADH) [3]

pH-range
  9.1–9.6 [5]; 6.0–9.7 (6.0: about 30% of activity maximum, 9.7: about 15% of
  activity maximum) [1]; 5.4–10.2 (5.4: about 20% of activity maximum, 10.2:
  about 50% of activity maximum) [2]; 4.5–8.5 (at pH 4.5 and 8.5: about 25%
  of activity maximum) [3]

Temperature optimum (°C)
  46–50 [5]; 60 [4]

Temperature range (°C)
  25–68 (25°C: about 25% of activity maximum, 68°C: about 10% of activity
  maximum) [4]

---

3 ENZYME STRUCTURE

Molecular weight
  75000 (Micrococcus aerogenes, gel filtration) [8]
  67000 (Fusobacterium sp., gel filtration) [2]
  580000 (Peptococcus aerogenes, gel filtration) [5]
  53000 (Peptococcus aerogenes, sedimentation equilibrium) [4]

Subunits
  Dimer (2 × 32000, Peptococcus aerogenes, SDS-PAGE [4],
  2 × 36000–38000, Micrococcus aerogenes, SDS-PAGE [8]) [4, 8]

Glycoprotein/Lipoprotein
  –

4 ISOLATION/PREPARATION

Source organism
Acidaminococcus fermentans [6]; Rhodopseudomonas sphaeroides [9];
Pseudomonas putida P2 (inducible enzyme) [7]; Micrococcus aerogenes
[8]; Clostridium tetanomorphum [6]; Pig [1]; Rabbit [1]; Rat [1]; Fusobacterium sp. (from human dental plaque) [2]; Alcaligenes sp. [3]; Peptococcus
aerogenes [4, 5]; More (enzymes described in references [2–6, 8, 9] are
specific for NAD(H), correctly they must be listed under EC 1.1.1.–, enzymes described in references [4–7] are specific for D-2-hydroxyglutarate
(similar enzymes), references without remark to isomer (R- or S-form) [2, 8,
9], D-2-hydroxyglutarate (specific for D-isomer) + electron acceptor (flavin
or nonheme iron) [7]) [2–9]

Source tissue
Heart [1]; Liver [1]; Cell [3, 9]

Localisation in source
Membrane-bound [7]

Purification
Micrococcus aerogenes [8]; Fusobacterium sp. (from human dental plaque,
partial) [2]; Peptococcus aerogenes [4, 5]

Crystallization
[3, 4]

Cloned
–

Renaturated
–

5 STABILITY

pH
6.5–7.5 (20°C, 20 h, stable) [4]

Temperature (°C)
30 (pH 7.0, 30 min, stable below) [4]; 40 (pH 7.0, 30 min, 12% loss of activity) [4]; 45 (pH 7.4, 10 min, stable) [2]; 50 (pH 7.4, 10 min, 70% loss of activity [2], pH 7.0, 30 min, 43% loss of activity [4]) [2, 4]; 60 (pH 7.0, 30 min,
85% loss of activity [4], 10 min, complete inactivation [5]) [4, 5]

Oxidation

Organic solvent

General stability information

2-Oxoglutarate, 10 mM, stabilizes [8]; Glycerol, 20%, stabilizes [8]; Phosphate, 0.1 M, partially stabilizes [8]; $MgCl_2$, 0.1 M, partially stabilizes [8]; KCl, 0.1 M, partially stabilizes [8]; Dithiothreitol, 1 mM, partially stabilizes [8]

Storage

0°C, at high protein concentration, several weeks [8]; –20°C, at high protein concentration, several months [8]; 4°C, dilute solutions lose activity [8]

---

6 CROSSREFERENCES TO STRUCTURE DATABANKS

PIR/MIPS code

Brookhaven code

---

7 LITERATURE REFERENCES

[1] Weil-Malherbe, H.: Biochem. J.,31,2080–2094 (1937)
[2] Hayashi, K., Kuwata, K., Tanaka, H.: J. Nihon Univ. Sch. Dent.,28,12–21 (1986)
[3] Suzuki, T., Uozomi, T., Beppu, T.: Agric. Biol. Chem.,49,2939–2947 (1985)
[4] Otawara, S., Oshima, T., Esaki, N., Soda, K.: Agric. Biol. Chem.,48,1713–1719 (1984)
[5] Johnson, W.M., Westlake, D.W.S.: Can. J. Microbiol.,18,881–892 (1972)
[6] Buckel, W., Miller, S.L.: Eur. J. Biochem.,164,565–569 (1987)
[7] Reitz, M.S., Rodwell, V.W.: J. Bacteriol.,100,708–714 (1969)
[8] Lerud, R.F., Whiteley, H.R.: J. Bacteriol.,106,571–577 (1971)
[9] Okuyama, M., Tsuiki, S., Kikuchi, G.: Biochim. Biophys. Acta,110,66–80 (1965)

1 NOMENCLATURE

EC number
1.1.99.3

Systematic name
D-Gluconate:(acceptor) 2-oxidoreductase

Recommended name
Gluconate 2-dehydrogenase

Synonymes
Dehydrogenase, gluconate 2-
Gluconate oxidase
Gluconate dehydrogenase
Gluconic dehydrogenase
D-Gluconate dehydrogenase
GADH (though GADH has been tentatively classified as EC 1.1.99.3, exi-
stence and function of cytochrome gave the favorable data that the enzyme
should be classified in EC 1.1.2.) [2]
Gluconic acid dehydrogenase
2-Ketogluconate reductase
D-Gluconate dehydrogenase, 2-keto-D-gluconate yielding [5]

CAS Reg. No.
9028-81-3

2 REACTION AND SPECIFICITY

Catalysed reaction
D-Gluconate + acceptor →
→ 2-dehydro-D-gluconate + reduced acceptor

Reaction type
Redox reaction

Natural substrates
2-Dehydro-D-gluconate + NADPH (physiological role: reduction of
2-keto-D-gluconate to D-gluconate to supply a substrate for glucokinase and
a carbon source for growth and to regenerate NADP+ [11]) [11, 12]

Substrate spectrum

1 D-Gluconate + acceptor (r [1, 10, 11], high rate of reduction of 2-ketoglu-
conate to gluconate, poor rate of oxidation of gluconate to 2-ketoglucona-
te [10], acceptor: NADP+ (not [15]) [1, 9–12], 2,6-dichlorophenolindophe-
nol [2, 4–6, 15], phenazine methosulfate [5, 6], ferricyanide [4–6], coenzyme Q_1 [4, 6], NAD+ (not [4, 11, 15], enzyme from Brevibacterium helvolum: with NADH 6% of the rate of that with NADPH, enzyme from Gluconobacter oxydans: no reaction with NADH [12]) [12], pyocyanine [15],
methylene blue (very low) [15], no reaction: cytochrome c [15], menadione [4, 6], FMN [15], FAD [15], specific for D-gluconate [2, 4–6, 15]) [1–15]

2 2-Ketogalactonate + acceptor (reverse reaction [12]) [10, 11]

3 Hydroxypyruvate + acceptor [10, 11]

4 Glyoxylate + acceptor [10, 11]

5 D-Xylonate + acceptor (Gluconobacter oxydans, enzyme from Brevibacterium helvolum: very low activity) [12]

6 5-Ketogluconate + acceptor (Gluconobacter oxydans, enzyme from Brevibacterium helvolum: not) [12]

7 6-Phosphogluconate + acceptor (enzyme from Brevibacterium helvolum:
slight activity, enzyme from Gluconobacter oxydans: not) [12]

8 More (substrate for oxidation (Gluconobacter oxydans) needs a
L-threo-configuration [12], overview: substrate specificity of 2-ketogluco-
nate reductase of acetic acid bacteria [13]) [12, 13]

Product spectrum

1 2-Dehydro-D-gluconate + reduced acceptor (r [1, 10, 11]) [1–15]

2 Galactonate + reduced acceptor

3 Pyruvate + reduced acceptor

4 Glycolate + reduced acceptor

5 2-Dehydro-D-xylonate + reduced acceptor

6 2-Dehydro-5-ketogluconate + reduced acceptor

7 2-Dehydro-6-phosphogluconate + reduced acceptor

8 ?

Inhibitor(s)

Quinine hydrochloride [2, 6]; 2-Ketogluconate (product inhibition) [2, 4];
p-Chloromercuribenzoate (1 mM: not [2]) [1, 11]; Hg^{2+} [1, 10, 11]; Al^{3+} [1];
Potassium ferricyanide [1]; Pyruvate (competitive [2]) [2, 4, 6]; Oxalate
(mixed-type inhibition [6]) [4, 6, 10]; 2-Oxoglutarate [6]; Dichlorophenolindophenol (high concentration) [6]; Ferricyanide (high concentration) [6]; Zn^{2+}
[10]; Co^{2+} [10]; Cd^{2+} [10]; Oxamate (slight inhibition, 2-dehydro-D-gluconate
+ NADPH [10], noncompetitive [2]) [2, 4, 6, 10]; Tartronate (slight inhibition,
2-dehydro-D-gluconate + NADPH) [10]

Cofactor(s)/prosthetic group(s)/activating agents
Heme (heme content (nmol/mg of protein): 14.7 (Pseudomonas aeruginosa)
[4], 14.6 (Pseudomonas fluorescens) [4], 16.8 (Klebsiella pneumoniae) [4],
2 mol of heme per mol of enzyme [5]) [4, 5]; Cytochrome c_1 (dehydrogenase protein tightly bound to cytochrome c_1 [2], enzyme contains a cytochrome c_1-$c_{554(551)}$, might be a diheme cytochrome [3, 4]) [2–4]; Phospholipid
(cardiolipin, in presence of Triton X-100: stimulation) [3, 4]; Flavin (flavoprotein [2, 3, 5, 6], covalently bound flavin [3, 5, 6], the flavin is 8alpha-(N^3-histi-
dyl)riboflavin, the flavin is in the dinucleotide form [6], prosthetic group is
8alpha-[N(1)-histidyl]-FAD (Pseudomonas fluorescens) [4]) [2–6]; CN⁻
(1 mM, stimulation) [15]; Cysteine (1 mM, stimulation) [15]; Glutathione (1
mM, stimulation) [15]

Metal compounds/salts
Mn^{2+} (stimulation [1], no effect: Mn [15]) [1]; More (metal is not a cofactor)
[15]

Turnover number (min⁻¹)

Specific activity (U/mg)
237 [5]; 61.03 [9]; 187.0 [13]; 502 [2]; 561.5 [2]; 336 [10]; 470 (Pseudomonas fluorescens) [4]; 513 (Klebsiella pneumoniae) [4]

K_m-value (mM)
0.8 (gluconate, pH 6, Pseudomonas aeruginosa, Serratia marcescens,
Klebsiella pneumoniae) [4]; 0.3 (pH 6, gluconate, Pseudomonas fluores-
cens) [4]; 0.15 (phenazine methosulfate) [6]; 0.21 (dichlorophenolindophe-
nol) [6]; 2.5 (ferricyanide) [6]; 0.11 (coenzyme Q_1) [6]; 0.083 (NADP⁺ (+
D-gluconate)) [10]; 0.01 (NADPH (+ 2-keto-D-gluconate)) [10]; 0.062
(NADPH (+ 2-keto-D-galactonate)) [10]; 0.11 (NADPH (+ 2-keto-L-gulonate))
[10]; 0.86 (2-keto-D-gluconate (+ NADPH)) [10]; 16 (2-keto-D-galactonate (+
NADPH)) [10]; 91 (2-keto-L-gulonate) [10]; 0.065 (hydroxypyruvate (+
NADPH)) [10]; 0.38 (glyoxylate (+ NADPH)) [10]

pH-optimum
4.0 (gluconate + ferricyanide, Klebsiella pneumoniae) [4]; 5.0 (acceptor:
ferricyanide [6], gluconate + ferricyanide (Pseudomonas aeruginosa, Pseudomonas fluorescens, Serratia marcescens) [4]) [2, 4, 6]; 5.5 (acceptor:
coenzyme Q_1) [6]; 6.0 (reduction of 2-keto-D-gluconate [11]) [5, 11, 13]; 6.5
(reduction of 2-dehydro-D-gluconate) [1]; 7.0 (reduction of 2-ketogluconate)
[10]; 7–7.3 (reduction of 2-ketogluconate, Brevibacterium helvolum) [12];
7–7.5 (reduction of 2-ketogluconate, Gluconobacter oxidans) [12]; 10 (oxidation of gluconate, Brevibacterium helvolum) [12]; 10.5 (oxidation of gluconate [11, 13], Gluconobacter liquefaciens [11], oxidation of gluconate [1])
[1, 11, 13]; 11.0 (oxidation of gluconate, Aerobacter ascendens) [11]; 11.3
(oxidation of gluconate, Gluconobacter oxidans) [12]

pH-range
4–7 (4: about 40% of activity maximum, 7: about 25% of activity maximum) [2]; 4–8.5 (4: about 45% of activity maximum, 8.5: about 50% of activity maximum, reduction of 2-ketogluconate) [1]; 4–9 (4: about 25% of activity maximum, 9: about 40% of activity maximum, reduction of 2-ketogluconate) [13]; 5–9 (at pH 5 and 9: about 60% of activity maximum, reduction of 2-ketogluconate) [10]; 9–12 (9: about 35% of activity maximum, 12: about 50% of activity maximum, oxidation of gluconate) [1]; 10–12 (at pH 10 and 12: about 45% of activity maximum) [13]; 11–12.5 (11: about 55% of activity maximum, 12.5: about 15% of activity maximum, oxidation of gluconate) [10]

Temperature optimum (°C)
30 [6]; 40 [2, 5]; 50 (oxidation and reduction) [1, 10]; 55 (oxidation and reduction) [13]

Temperature range (°C)
20–70 (at 20°C and 70°C: about 40% of activity maximum) [1]; 20–50 (at 20°C and 50°C: about 40% of activity maximum) [2]; 30–65 (at 30°C and 65°C: about 45% of activity maximum, oxidation and reduction) [13]

---

3 ENZYME STRUCTURE

Molecular weight
110000 (Gluconobacter liquefaciens, gel filtration) [1]
120000 (Gluconobacter liquefaciens, gel filtration, SDS-PAGE [9], Acetobacter rancens, gel filtration [10]) [9, 10]
124000–131000 (Pseudomonas aeruginosa, sucrose density gradient centrifugation) [3, 4]
132000–138000 (Pseudomonas aeruginosa, PAGE in presence of Triton X-100) [4]
135000 (Serratia marcescens, gel filtration) [2]

Subunits
Monomer (1 × 120000, Gluconobacter liquefaciens, SDS-PAGE) [9]
Trimer (2 × 42000–43000 + 1 × 34000, Gluconobacter liquefaciens, SDS-PAGE [10, 11], 3 × 40000, Acetobacter ascendens, SDS-PAGE [13]) [10, 11, 13]
Octamer (8 × 15000, Acetobacter rancens, SDS-PAGE, amino acid analysis) [10, 11]
? (x × 68000 + x × 52500 + x × 15500, Serratia marcescens, SDS-PAGE) [2]
More (in absence of Triton X-100 association to a dimer [3, 4], in presence of SDS, enzyme dissociates into 3 components [3–5], with MW of 66000 (flavoprotein), 50000 (cytochrome c_1) and 22000 (unclear function) [3], MW 64000, 45000 and 21000 [5]) [3–5]

4

Glycoprotein/Lipoprotein
More (enzyme has a hydrophobic and a phospholipid-interacting domain)
[4]

4 ISOLATION/PREPARATION

Source organism
Klebsiella pneumoniae [4]; Pseudomonas fluorescens [4, 7]; Gluconobacter liquefaciens [1, 9–11, 15]; Serratia marcescens (IFO 3054) [2]; Pseudomonas aeruginosa [3, 4, 6, 7, 15]; Gluconobacter dioxyaceticus [5]; Gluconobacter melanogenus [7]; Rhizobium sp. (cowpea rhizobia) [8]; Acetobacter rancens [10, 11]; Acetobacter ascendens [11, 13]; Brevibacterium helvolum [12]; Acetobacter cloacae [12]; Bacillus cereus [12]; Debaryomyces hansenii [12]; Aspergillus nidulans [12]; Gluconobacter oxydans [12]; More (overview: distribution in acetic acid bacteria) [14]

Source tissue
Cell [1, 2, 5, 11, 13, 15]

Localisation in source
Membrane (bound [3], outer surface of cytoplasmic membrane [4]) [3, 4, 6]

Purification
Klebsiella pneumoniae [4]; Acetobacter rancens [10, 11]; Gluconobacter liquefaciens [1, 9, 11]; Serratia marcescens (IFO 3054) [2]; Pseudomonas aeruginosa [3, 6, 15]; Pseudomonas fluorescens [4]; Gluconobacter dioxyaceticus [5]; Acetobacter ascendens [11, 13]; Brevibacterium helvolum [12]; Gluconobacter oxydans [12]

Crystallization
[9, 11, 13]

Cloned
–

Renaturated
–

5 STABILITY

pH
4.5–7.0 (5°C, 16 h, stable) [2]; 5.0–11.0 (stable) [1]

Temperature (°C)
35 (10 min, pH 6.0, stable up to) [2]; 50 (10 min, pH 6.0, stable below [1], 10 min, pH 6.0, 50% loss of activity [2]) [1, 2]; 60 (pH 6.0, 10 min, 90% loss of activity) [2]; 70 (pH 6.0, 10 min, complete loss of activity) [1, 9]; 50 (5 min, complete inactivation) [6]; 55 (10 min, stable up to) [10]; 65 (10 min, complete inactivation) [10]

Oxidation

Organic solvent

General stability information
Dialysis against distilled water: gradual loss of activity after 8–10 h, against neutral phosphate or Tris buffer, 24 h in the cold: stable [15]; Sodium gluconate, 10 mM, stabilizes [6]; Triton X-100, 0.2%, stabilizes [6]

Storage
–20°C, 10 mM sodium gluconate, 5 mM $MgCl_2$, 20% sucrose [6]; 4°C, protein concentration: 1 mg/ml, 0.01 M potassium phosphate buffer, pH 6.0, 5 mM $MgCl_2$, 20% loss of activity after 5 days, 50% loss of activity after 10 days [6]; –20°C, freezing in presence of sulfhydryl compounds and ammonium sulfate [10]; 0°C, 0.01 M buffer, pH 6.0, 1 mM 2-mercaptoethanol, stable for 1 month or more [10]; –10°C, 0.1 M Tris buffer, pH 7.0, stable for several weeks [15]; 4°C, 0.1 M Tris buffer, pH 7.0, stable for 3–4 days [15]

---

6 CROSSREFERENCES TO STRUCTURE DATABANKS

PIR/MIPS code

Brookhaven code

7 LITERATURE REFERENCES

[1] Chiyonobu, T., Adachi, O., Ameyama, M.: Agric. Biol. Chem.,37,2871–2878 (1973)
[2] Shinagawa, E., Matsushita, K., Adachi, O., Ameyama, M. : Agric. Biol. Chem.,42,2355–2361 (1978)
[3] Matsushita, K., Shinagawa, E., Adachi, O., Ameyama, M. : J. Biochem.,85, 1173–1181 (1979)
[4] Matsushita, K., Shinagawa, E., Ameyama, M.: Methods Enzymol.,89,187–193 (1982) (Review)
[5] Shinagawa, E., Matsushita, K., Adachi, O., Ameyama, M.: Agric. Biol. Chem.,48, 1517–1522 (1984)
[6] Matsushita, K., Shinagawa, E., Adachi, O., Ameyama, M. : J. Biochem.,86,249–256 (1979)
[7] McIntire, W., Singer, T.P., Ameyama, M., Adachi, O., Matsushita, K., Shinagawa, E.: Biochem. J.,231,651–654 (1985)
[8] Stowers, M.D., Elkan, G.H.: Arch. Microbiol.,137,3–9 (1984)
[9] Chiyonobu, T., Shinagawa, E., Adachi, O., Ameyama, M.: Agric. Biol. Chem.,39,2263–2264 (1975)
[10] Chiyonobu, T., Shinagawa, E., Adachi, O., Ameyama, M. : Agric. Biol. Chem.,40,175–184 (1976)
[11] Ameyama, M., Adachi, O.: Methods Enzymol.,89,203–210 (1982) (Review)
[12] De Ley, J.: Methods Enzymol.,9,196–200 (1966) (Review)
[13] Adachi, O., Chiyonobu, T., Shinagawa, E., Matsushita, K., Ameyama, M.: Agric. Biol. Chem.,42,2057–2062 (1978)
[14] Shinagawa, E., Chiyonobu, T., Adachi, O., Ameyama, M. : Agric. Biol. Chem.,40,475–483 (1976)
[15] Ramakrishnan, T., Campbell, J.J.R.: Biochim. Biophys. Acta,17,122–127 (1955)

1 NOMENCLATURE

EC number
1.1.99.4

Systematic name
2-Dehydro-D-gluconate:acceptor 2-oxidoreductase

Recommended name
Dehydrogluconate dehydrogenase

Synonymes
Ketogluconate dehydrogenase
Dehydrogenase, ketogluconate
alpha-Ketogluconate dehydrogenase
2-Keto-D-gluconate dehydrogenase
2-Oxogluconate dehydrogenase

CAS Reg. No.
9028-82-4

2 REACTION AND SPECIFICITY

Catalysed reaction
2-Dehydro-D-gluconate + acceptor →
→ 2,5-didehydro-D-gluconate + reduced acceptor

Reaction type
Redox reaction

Natural substrates
2-Dehydro-D-gluconate + acceptor (enzyme is the primary dehydrogenase
of the 2-keto-D-gluconate oxidizing system) [5]

Substrate spectrum
1 2-Dehydro-D-gluconate + acceptor (acceptors: ferricyanide [4–6], phenazine methosulfate [5, 6], nitroblue tetrazolium [5], 2,6-dichlorophenolindo-
phenol [5, 6, 8]) [4–6, 8]
2 More (high specificity for 2-keto-D-gluconate [5, 6], not: NAD^+, $NADP^+$, O_2
[5, 6], 5-keto-D-gluconate [6], 2-keto-D-galactonate [6], 2-keto-D-gulonate
[6]) [5, 6]

Product spectrum
1 2,5-Didehydro-D-gluconate + reduced acceptor
2 ?

Inhibitor(s)
CN⁻ [5]; Oxamate [5]; Succinate [5]; Citrate [5]; Oxalate [5]; More (not: sulf-hydryl reagents [5, 6], metal chelators [6]) [5, 6]

Cofactor(s)/prosthetic group(s)/activating agents
Flavin (flavoprotein [3, 6], contains 8alpha-(N^3-histidyl) riboflavin in the dinucleotide form [3], covalently bound flavin [6]) [3, 6]; Cytochrome (con-tains a cytochrome component [5], tightly bound c-type cytochrome existing as a dehydrogenase-cytochrome complex [6]) [5, 6]

Metal compounds/salts

Turnover number (min^{-1})

Specific activity (U/mg)
More [5, 6]

K_m-value (mM)
50 (2-keto-D-gluconate (+ ferricyanide)) [5, 6]

pH-optimum
4.0 [5, 6]

pH-range

Temperature optimum (°C)
39 [5, 6]

Temperature range (°C)

3 ENZYME STRUCTURE

Molecular weight
133000 (Gluconobacter melanogenus, calculated from the sum of the MW of the subunits) [5, 6]

Subunits
Trimer (1 × 61000 (flavoprotein), 1 × 47000 (cytochrome component), 1 × 25000 (function unclear), SDS-PAGE, Gluconobacter melanogenus) [5, 6]

Glycoprotein/Lipoprotein
–

4 ISOLATION/PREPARATION

Source organism
Tatumella ptyseos [2]; Pseudomonas putida [1]; Erwinia cypripedii [2];
Escherichia blattae [2]; Ewingella americana [2]; Serratia grimesii [2]; Serratia liquefaciens [2]; Serratia marcescens [2]; Acetobacter melanogenum [8];
Rahnella aquatilis [2]; Gluconobacter melanogenus [3, 5–7]; Gluconobacter liquefaciens [5, 7]; Gluconobacter sphaericus [5, 7]; Gluconobacter sp. [4]

Source tissue

Localisation in source
Membrane (outer surface of cytoplasmic membrane) [5]

Purification
Gluconobacter melanogenus [5, 6]; Acetobacter melanogenum [8]

Crystallization
–

Cloned
–

Renaturated
–

5 STABILITY

pH
5.0–8.0 [5, 6]

Temperature (°C)

Oxidation

Organic solvent

General stability information
Triton X-100 stabilizes [5, 6]

Storage

6 CROSSREFERENCES TO STRUCTURE DATABANKS

PIR/MIPS code

Brookhaven code

7 LITERATURE REFERENCES

[1] Gallego-Iniesta, M., Madero-Madero, E., Medina-Puerta, M.M., Garrido-Pertierra, A.: Curr. Microbiol.,18,279–284 (1989)
[2] Bouvet, O.M.M., Lenormand, P., Grimont, P.A.D.: Int. J. Syst. Bacteriol.,39,61–67 (1989)
[3] McIntire, W., Singer, T.P., Ameyama, M., Adachi, O., Matsushita, K., Shinagawa, E.: Biochem. J.,231,651–654 (1985)
[4] Ameyama, M.: Methods Enzymol.,89,20–29 (1982) (Review)
[5] Shinagawa, E., Ameyama, M.: Methods Enzymol.,89,194–198 (1982) (Review)
[6] Shinagawa, E., Matsushita, K., Adachi, O., Ameyama, M.: Agric. Biol. Chem.,45,1079–1085 (1981)
[7] Shinagawa, E., Chiyonobu, T., Adachi, O., Ameyama, M.: Agric. Biol. Chem.,40,475–483 (1976)
[8] Datta, A.G., Katznelson, H.: Arch. Biochem. Biophys.,65,576–578 (1956)

1 NOMENCLATURE

EC number
1.1.99.5

Systematic name
sn-Glycerol-3-phosphate:(acceptor) 2-oxidoreductase

Recommended name
Glycerol-3-phosphate dehydrogenase

Synonymes
EC 1.1.2.1 (formerly)
sn-Glycerol-3-phosphate dehydrogenase [17]
L-Glycerol-3-phosphate dehydrogenase [1]
Glycerol 3-phosphate dehydrogenase
alpha-Glycerophosphate dehydrogenase
L-3-Glycerophosphate-ubiquinone oxidoreductase [9]
sn-Glycerol 3-phosphate oxidase [15]
Dehydrogenase, glycerol phosphate (acceptor)
Glycerophosphate dehydrogenase
Glycerol phosphate dehydrogenase
NAD-independent glycerol phosphate dehydrogenase
Glycerol 3-phosphate cytochrome c reductase
Flavoprotein-linked L-glycerol 3-phosphate dehydrogenase
alpha-Glycerophosphate dehydrogenase (acceptor)
FAD-dependent glycerol-3-phosphate dehydrogenase
FAD-linked glycerol 3-phosphate dehydrogenase
Glycerol phosphate dehydrogenase (FAD)
Glycerol phosphate dehydrogenase (acceptor)
sn-Glycerol-3-phosphate oxidase (EC 1.1.99.5)
FAD-dependent sn-glycerol-3-phosphate dehydrogenase
L-Glycerophosphate dehydrogenase [2, 6]
Glycerol-3-phosphate dehydrogenase (flavin-linked) [4]
L-Glycerol 3-phosphate dehydrogenase [7]
Flavin-linked glycerol-3-phosphate dehydrogenase [8]
NAD-independent glycerol-phosphate dehydrogenase [11]
Pyridine nucleotide-independent L-glycerol 3-phosphate dehydrogenase
[13]
DL-Glycerol 3-phosphate oxidase [14]
alpha-Glycerophosphate oxidase [16]
FAD-linked L-glycerol-3-phosphate dehydrogenase [28, 29]

CAS Reg. No.
9001-49-4

---

2 REACTION AND SPECIFICITY

Catalysed reaction
sn-Glycerol 3-phosphate + acceptor →
→ glycerone phosphate + reduced acceptor (ping-pong mechanism [2])

Reaction type
Redox reaction

Natural substrates
sn-Glycerol 3-phosphate + acceptor (enzyme interacts with the mitochondrial electron transport system at the level of ubiquinone [1], enzyme involved
in dihydroxyacetone phosphate:glycerol-3-phosphate cycle in Trypanosoma
brucei [23]) [1, 23, 25]

Substrate spectrum
1 sn-Glycerol 3-phosphate + acceptor (electron acceptors: long-chain ubiquinone [2], short-chain ubiquinone [2], ubiquinone analogs [6], phenazine methosulfate (amphipaths may specifically induce a phenazine methosulfate binding site [20]) [1, 2, 7, 11, 13, 18, 20, 28], ferricyanide [1, 2, 5,
  11, 18], 2,6-dichlorophenolindophenol [1, 2, 5–7, 9, 11, 18], methylene
  blue (low activity [11]) [1, 11, 18], duroquinone [7], menadione [18], ubi-
  quinone-5 (Q-5) [7], ubiquinone (–0, –1 and –2) [9], enzyme can also in-
  teract directly with molecular oxygen to form H_2O_2 [1], molecular oxygen
  is a poor electron acceptor (production of H_2O_2) [11], not reduced: cytochrome c (from yeast) [11], cytochrome c_2 (from R. rubrum) [11], specific
  for L-3-glycerophosphate as electron donor [2, 11], Longissimus dorsi:
  alpha-glycerophosphate oxidase system with 3 phosphorylation sites
  [16], distinct binding sites for hydrophobic and hydrophilic electron acceptors [28], active with analogs in which the C-1 hydroxy-group has
  been replaced by hydrogen or fluorine, not active with analogs in which
  the C-2 hydroxy-group is in the D-configuration or has been replaced by
  hydrogen or fluorine [19]) [1, 2, 5–7, 9, 11, 13, 15, 16, 18–20, 28]
2 Deoxy-sn-glycerol 3-phosphate + acceptor [19]
3 Deoxy-1-fluoro-sn-glycerol 3-phosphate + acceptor [19]

Product spectrum
1 Dihydroxyacetone phosphate + reduced acceptor
2 ?
3 ?

Inhibitor(s)

Dihydroxyacetone phosphate [1, 4, 18]; EDTA (slight [4]) [1, 4]; D-3-Gly-ceraldehyde phosphate [1, 18]; D-3-Phosphoglyceric acid [1, 18]; D-3-Gly-cerophosphate [1]; D-2-Phosphoglyceric acid [18]; Glyceraldehyde 3-phos-phate [4]; 1,2-Dihydroxybenzene 3,5-disulfonic acid (Tiron) [2]; 1-(2-Thenoyl)-3,3,3-trifluoroacetone (low inhibition with 2,6-dichlorophenolin-dophenol, inhibition with ubiquinone-1 [9]) [2, 9, 11]; p-Chloromercuribenzo-ate [1, 7, 11, 17, 18]; Phosphoenolpyruvate [4, 18]; Phosphoglycolic acid [4, 18]; Na$^+$ (slight) [4]; K$^+$ (slight) [4]; EGTA (slight) [4]; Cu^{2+} [4, 11]; Co^{2+} [4]; Ni^{2+} [4]; Zn^{2+} [4, 7]; Mg^{2+} [4]; Ca^{2+} [4]; Triton X-100 (inhibits in presence of phospholipids) [7]; Bathophenanthroline (inhibits activity for ubiquinone-5 but not for phenazine methosulfate) [7]; 1,10-Phenanthroline (inhibits activity for ubiquinone-5 but not for phenazine methosulfate [7]) [7, 18]; Hg^{2+} [7, 11]; Ag$^+$ [7]; Chloroquine (no inhibition with 2,6-dichlorophenolindophenol, inhibition with ubiquinone-1) [9]; m-Chlorohydroxamate (low inhibition with 2,6-dichlorophenolindophenol, inhibition with ubiquinone-1) [9]; Salicylhy-droxamate (low inhibition with 2,6-dichlorophenolindophenol, inhibition with ubiquinone-1) [9]; Amytal [11]; Dicumarol [11]; Dipyridyl [11]; Cd^{2+} (weak) [11]; N-Ethylmaleimide (weak [11]) [11, 18]; H$_2$O$_2$ [11]; Rotenone [16]; Pieri-cidin A [16]; Dithionitrobenzoate [17]; ATP [18]; GTP [18]; Analogs of sn-gly-cerol 3-phosphate (in which the C-2 hydroxy-group is in the D-configuration or has been replaced by hydrogen or fluorine) [19]

Cofactor(s)/prosthetic group(s)/activating agents

FAD (flavoprotein [1–3, 5–7, 13], contains 1 mol FAD per 100000 MW [8], en-zyme contains 1 mol of acid-liberable FAD per 55000 g of protein [2], 0.7–0.9 mol noncovalently bound FAD per mol of enzyme [4], 3.5 flavins per 520000 MW [5], 0.5 mol of FAD per protein monomer [18], 1 mol of noncovalently bound FAD per dimer [17], addition of exogenous FAD: stimulation (activity with methylene blue and phenazine methosulfate) [17]) [1–7, 13, 17, 18]; FMN (addition of exogenous FMN, stimulation of activity with phenazine me-thosulfate, inhibition of activity with methylene blue) [17]; Phospholipid (de-tergent-depleted enzyme requires exogenous phospholipid or non-denatu-ring detergent) [18]; Non-denaturing detergent (detergent-depleted enzyme requires exogenous phospholipid or non-denaturing detergent) [18]

Metal compounds/salts

Iron (contains non-heme iron [1, 4, 6–8], contains 1 gatom of acid liberable iron per 310000 g of protein [2], 1 mol per 100000 MW protein [8], 1.1 mol of non-heme iron per mol of enzyme [4], 1 nmol of non-heme per mg of protein [9], enzyme is unlikely to contain iron-sulfur centres, any iron present is due to contaminating proteins [9], 2 mol of non-heme iron per dimer [17]) [1, 4–7, 9, 17]; Copper (enzyme contains copper) [5]; NaCl (200 mM NaCl or KCl re-quired for maximum activation) [7]; KCl (200 mM NaCl or KCl required for maximum activation) [7]; Ca^{2+} (stimulation [21], alteration of enzyme confor-mation [19, 21]) [19, 21]

Turnover number (min^{-1})
2000 (sn-glycerol 3-phosphate, in presence of flavins) [17]

Specific activity (U/mg)
1.15 [1]; 32 (Acidiphilium sp. 63) [3]; 27 (Acidiphilium sp. 24R) [3]; 7.5 [10];
More [4, 6, 7, 12, 13]

K$_m$-value (mM)
10 (L-3-glycerophosphate (+ dichlorophenolindophenol, phenazine methosulfate, Q_0, Q_1 or Q_6)) [2]; 6.2 (L-3-glycerophosphate (+ Q_{10})) [2]; 0.125
(dichloroindophenol, Q_0) [2]; 0.52 (Q_1) [2]; 0.023 (Q_2) [2]; 0.11 (Q_6) [2]; 0.38
(Acidiphilium sp. 63, glycerol 3-phosphate) [3]; 1.3 (Acidiphilium sp. 63, glycerol 3-phosphate) [3]; 6 (DL-alpha-glycerol 3-phosphate) [4]; 0.007 (FAD)
[4]; 5.17 (DL-alpha-glycerophosphate) [5]; 4.02 (L-glycerol 3-phosphate (+
phenazine methosulfate)) [7]; 0.13 (phenazine methosulfate) [7]; 3.7 (L-glycerol 3-phosphate (+ ferricyanide)) [7]; 0.55 (ferricyanide) [7]; 1.02
(deoxy-sn-glycerol 3-phosphate, enzyme of intact mitochondria) [19]; 5.58
(deoxy-sn-glycerol 3-phosphate, enzyme of Triton-solubilized mitochondria)
[19]; 1.25 (deoxy-1-fluoro-sn-glycerol 3-phosphate, enzyme of intact mitochondria) [19]; 3.69 (deoxy-1-fluoro-sn-glycerol 3-phosphate, enzyme of Triton-solubilized mitochondria) [19]; 0.050 (Q_{10}) [2]; More (effect of adryamycin on K$_m$ [28]) [1, 6–9, 11, 13, 17, 19, 20, 28]

pH-optimum
6.8 (Acidiphilium sp. 63) [3]; 7.0–7.4 [5]; 7–9.5 (broad) [18]; 7.4 (glycerol
3-phosphate + dichlorophenolindophenol) [11]; 7.6 (Acidiphilium sp. 24R)
[3]; 8.5–9.0 [7]

pH-range

Temperature optimum (°C)
44 [18]

Temperature range (°C)

---

3 ENZYME STRUCTURE

Molecular weight
80000 (E. coli, gel filtration) [13]
93500 (E. coli, sedimentation equilibrium studies) [17]
108000 (Acidiphilium sp., FPLC gel filtration) [3]
130000 (E. coli, gel filtration) [18]
250000 (rat, gel filtration) [4]
300000 (Vibrio alginolyticus, gel filtration) [7]
330000–350000 (rabbit, gel filtration, disc gel electrophoresis) [8]
500000 (Candida utilis, gel electrophoresis) [5]

Subunits
Monomer (1 × 75000, pig, SDS-PAGE) [6]
Dimer (2 × 54000, Acidiphilium sp., SDS-PAGE [3], 2 × 35000, E. coli,
SDS-PAGE [13], 1 × 62000 + 1 × 43000, E. coli, SDS-PAGE, sedimentation
equilibrium studies, chemical cross-linking studies [17]) [3, 13, 17]
? (x × 74000, rat, SDS-PAGE, HPLC gel filtration in presence of 0.1% SDS
[4], x × 76000, rabbit, SDS-PAGE [8], x × 58000, E. coli, SDS-PAGE [18]) [4,
8, 18]

Glycoprotein/Lipoprotein
Lipoprotein (contains 0.4 mol of phospholipid per mol of protein [4], contains cardiolipin (essential) [28]) [4, 28]

4 ISOLATION/PREPARATION

Source organism
Pig [1, 2, 6, 9, 12]; Candida utilis [5, 25]; Musca domestica [1]; Streptococcus faecalis [1]; Propionibacterium arabinosum [1]; Rat [4, 15, 21, 22, 24,
26–29]; Vibrio alginolyticus [7]; Rabbit [8]; Trypanosoma brucei [23]; Acidiphilium sp. (strain 63 and 24R) [3]; Propionibacterium arabinosum [10, 11];
Bacillus megaterium KM [14]; Longissimus dorsi (alpha-glycerophosphate
oxidase system sensitive to rotenone and piericidin A) [16]; E. coli [13, 17,
18, 20]; Locust [19]

Source tissue
Vegetative cells [14]; Muscle (flight muscle [16], skeletal muscle [8]) [8, 16,
19]; Brown adipose tissue [24]; Brain [1, 2, 6, 8, 9, 12, 26]; Liver [4, 15, 21,
22, 26, 28, 29]; Bloodstream form of Trypanosoma brucei [23]

Localisation in source
Membrane (only of vegetative cells not of spores [14], bound [13]) [6, 13,
14, 18]; Mitochondria (outer face of inner membrane [1, 27], tightly associated with a specific lipid in membrane [3]) [1–4, 6–9, 12, 19, 21–23, 25–29]

Purification
Acidiphilium sp. (strain 63 and 24R) [3]; Propionibacterium arabinosum
[10]; E. coli [20]; Pig (partial [2, 12]) [2, 9, 12]; Rat [4]; Candida utilis [5]; Vibrio alginolyticus (partial) [7]; Rabbit [8]; Trypanosoma brucei [23]; More
(cell-free synthesis of a putative precursor of rat liver mitochondrial enzyme)
[22]

Crystallization
–

Cloned
–

Renaturated

–

---

5 STABILITY

pH
6 (irreversible denaturation) [4]; 6.0–8.5 (destabilization below pH 6.0 and above pH 8.5, half-life: below 8 h) [17]; 7.4 (rather unstable) [11]; 7–8 (stable) [4]

Temperature (°C)
4 (half-life: 30 h) [17]; 37 (45 min, stable) [4]; 45 (5 min, stable) [4]; 50 (biphasic thermal inactivation) [4, 8]; 60 (45 min, complete inactivation) [4]; 70 (5 min, complete inactivation) [4]

Oxidation

Organic solvent

General stability information
Repeated freezing and thawing inactivates [12]; Ethylene glycol, 20%, stabilizes [17]; Glycerol, 20%, stabilizes [17]; NaCl, 400 mM, destabilizes, half-life: 30 h [17]; Divalent cations destabilize, half-life: below 5 h [17]

Storage
0°C or –20°C, pH 5–6, fairly stable [11]; –10°C, 0.3 M KH_2PO_4-NaOH buffer, pH 7.6, 10 days [12]; –20°C, acetone powder: stable for 4 months, solubilized enzyme: stable for 1 week [2]; –20°C, 50 mM citrate buffer, 1 mM EDTA, 3 weeks, less than 10% loss of activity [11]; 3°C, 30% ethylene glycol, 4 days [13]; –70°C, 30% ethylene glycol, 1 month [13]; 4°C, 50% glycerol, 1 week, 40% loss of activity [4]; –20°C, 50% glycerol, 1 week, 12% loss of activity [4]; –70°C, 50% glycerol, 1 week, 0.5% loss of activity [4]; –70°C or 4°C, protein concentration 4–10 mg/ml, stable [17]

---

6 CROSSREFERENCES TO STRUCTURE DATABANKS

PIR/MIPS code
PIR1:DEECGD ((aerobic) Escherichia coli); PIR1:DEECNB ((anaerobic) chain B Escherichia coli); PIR1:DEECNC ((anaerobic) chain C Escherichia coli); PIR2:C45868 (Bacillus subtilis); PIR3:S18565 (Bacillus subtilis); PIR1:DEECNA (chain A anaerobic Escherichia coli); PIR2:S38190 (precursor mitochondrial yeast (Saccharomyces cerevisiae)); PIR2:PQ0148 (mitochondrial yeast (Saccharomyces cerevisiae) (fragment))

Brookhaven code

7 LITERATURE REFERENCES

[1] Hatefi, Y., Stiggall, D.L. in "The Enzymes" (Boyer, P. D., ed.) 13,3rd Ed.,175–297 (1976) (Review)
[2] Dawson, A.P., Thorne, C.J.R.: Methods Enzymol.,41B,254–259 (1978)
[3] Hatta, T., Inagaki, K., Sugio, T., Kishimoto, N., Tano, T.: Agric. Biol. Chem.,53,651–658 (1989)
[4] Garrib, A., McMurray, W.C.: J. Biol. Chem.,261,8042–8048 (1986)
[5] Halsey, Y.D.: Biochim. Biophys. Acta,682,387–394 (1982)
[6] Cottingham, I.R., Ragan, C.I.: Biochem. J.,192,9–18 (1980)
[7] Unemoto, T., Hayashi, M., Hayashi, M.: J. Biochem.,90,619–628 (1981)
[8] Cole, E.S., Lepp, C.A., Holohan, P.D., Fondy, T.P.: J. Biol. Chem.,253,7952–7959 (1978)
[9] Cottingham, I.R., Ragan, C.I.: Biochem. Soc. Trans.,6,1307–1310 (1978)
[10] Sone, N., Kitsutani, S.: J. Biochem.,72,291–297 (1972)
[11] Sone, N.: J. Biochem.,74,297–305 (1973)
[12] Dawson, A.P., Thorne, C.J.R.: Biochem. J.,111,27–34 (1969)
[13] Weiner, J.H., Heppel, L.A.: Biochem. Biophys. Res. Commun.,47,1360–1365 (1972)
[14] Wilkinson, B.J., Ellar, D.J.: Eur. J. Biochem.,55,131–139 (1975)
[15] Bowley, M., Manning, R., Brindley, D.N.: Biochem. J.,136,421–427 (1973)
[16] Cheah, K.S.: FEBS Lett.,10,109–112 (1970)
[17] Schryvers, A., Weiner, J.H.: J. Biol. Chem.,256,9959–9965 (1981)
[18] Schryvers, A., Lohmeier, E., Weiner, J.H.: J. Biol. Chem.,253,783–788 (1978)
[19] Lloyd, W.J., Harrison, R.: Arch. Biochem. Biophys.,163,185–190 (1974)
[20] Robinson, J.J., Weiner, J.H.: Can. J. Biochem.,58,1172–1178 (1980)
[21] Beleznai, Z., Szalay, L., Jancsik, V.: Eur. J. Biochem.,170,631–636 (1988)
[22] Garrib, A., McMurray, W.C.: J. Biol. Chem.,263,19821–19826 (1988)
[23] Opperdoes, F.R., Borst, P., Bakker, S., Leene, W.: Eur. J. Biochem.,76,29–39 (1977)
[24] Hemon, P., Berbey, B.: Biochim. Biophys. Acta,170,235–243 (1968)
[25] Gancedo, C., Gancedo, J.M., Sols, A.: Eur. J. Biochem.,5,165–172 (1968)
[26] Hemon, P.: Biochim. Biophys. Acta,151,681–683 (1968)
[27] Klingenberg, M.: Eur. J. Biochem.,13,247–252 (1970)
[28] Beleznai, Z., Jancsik, V.: Biochem. Biophys. Res. Commun.,159,132–139 (1989)
[29] Beleznai, Z., Amler, E., Jancsik, V., Rauchova, H., Drahota, Z.: Biochim. Biophys. Acta,1018,72–76 (1990)

1 NOMENCLATURE

EC number
1.1.99.6

Systematic name
(R)-2-Hydroxy-acid:(acceptor) 2-oxidoreductase

Recommended name
D-2-Hydroxy-acid dehydrogenase

Synonymes
Dehydrogenase, D-2-hydroxy acid
D-2-Hydroxy acid dehydrogenase

CAS Reg. No.
9028-83-5

2 REACTION AND SPECIFICITY

Catalysed reaction
(R)-Lactate + acceptor →
→ pyruvate + reduced acceptor

Reaction type
Redox reaction

Natural substrates
(R)-Lactate + acceptor (enzyme acts on a variety of (R)-2-hydroxy acids [1, 2], cytochrome c, ubiquinone and O_2 are possible physiological electron acceptors [2]) [1, 2]

Substrate spectrum
1 (R)-Lactate + acceptor (acceptors: 2,6-dichlorophenolindophenol, phenazine methosulfate, ferricyanide [1, 2], cytochrome c, ubiquinone and O_2 are possible physiological acceptors [2]) [1, 2]
2 D-Glycerate + acceptor [2]
3 D-Malate + acceptor [2]
4 meso-Tartrate + acceptor (good substrate) [1]
5 Long-chain homologues of D-lactate + acceptor (as far as D-2-hydroxy-n-octanoate) [2]
6 D-Tartrate + acceptor (poor substrate) [2]
7 More (no substrates are glycolate, 3-methylpentanoate, phenylglycolate, phenyllactate, 3-phosphoglycerate, 6-phosphoglycerate) [2]

Product spectrum

1 Pyruvate + reduced acceptor [1, 2]
2 ?
3 Oxaloacetate + reduced acceptor [2]
4 ?
5 ?
6 ?
7 ?

Inhibitor(s)

1,10-Phenanthroline (progressive inactivation, substrate decreases rate of inactivation [2], inhibition potentiated by KCN [1], inactivation during preincubation at 0°C [2]) [1, 2]; EDTA (progressive inactivation, substrate decreases rate of inactivation [2]) [1, 2]; KCN (potentiates inhibition of 1,10-phenanthroline [1], competitive, reversible [2]) [1, 2]; p-Chloromercuribenzoate (strong) [1]; Cu^{2+} (strong) [1]; Hg^{2+} (strong) [1]; Oxalate (competitive) [2]; L-Lactate (weak) [2]; Pyruvate (competitive inhibitor to acceptor) [2]; Oxaloacetate (competitive to acceptor) [2]

Cofactor(s)/prosthetic group(s)/activating agents

2,6-Dichlorophenolindophenol [1, 2]; Methylene blue [1, 2]; Phenazine methosulfate [1, 2]; Ferricyanide [1, 2]; Cytochrome c (possible physiological acceptor) [1, 2]; FAD (flavoprotein, 2 molecules per molecule of enzyme, not replaceable by FMN or riboflavin) [2]; Ubiquinone (possible physiological acceptor) [2]; O_2 (possible physiological acceptor [2], no appreciable O_2 uptake in absence of electron carrier [1]) [1, 2]; More (ageing of extracts increases dehydrogenase activity [1], NAD^+ or $NADP^+$ do not act as acceptors [2]) [1, 2]

Metal compounds/salts

KCN (raises dehydrogenase activity of freshly prepared kidney extracts) [1]

Turnover number (min^{-1})

Specific activity (U/mg)

7.4 [2]

K_m-value (mM)

1.0 (D-malate) [2]; 2.0 (D-lactate) [2]; 3.4 (D-lactate) [1]

pH-optimum

6.5 (and below for malate as substrate) [2]; 8.1 [1]; 8.6 (D-lactate as substrate) [2]

pH-range

Temperature optimum (°C)

20 (assay at) [1]; 30 (assay at) [2]

Temperature range (°C)

3 ENZYME STRUCTURE

Molecular weight
    102000 (rabbit, gel filtration) [2]

Subunits

Glycoprotein/Lipoprotein
    –

---

4 ISOLATION/PREPARATION

Source organism
    Rabbit [1, 2]

Source tissue
    Kidney [1, 2]; Liver [1]

Localisation in source
    Mitochondria [1]; Particulate enzyme [2]

Purification
    Rabbit [1, 2]

Crystallization
    –

Cloned
    –

Renaturated
    –

---

5 STABILITY

pH
    8.0 (in 25 mM Tris buffer at 4°C stable for weeks) [2]

Temperature (°C)

Oxidation

Organic solvent

General stability information
    Oxalate stabilizes [2]

Storage
    0°C, as precipitate in ammonium sulfate stable for months [2];
    4°C, in 25 mM Tris buffer, pH 8.0, stable for weeks [2]

---

6 CROSSREFERENCES TO STRUCTURE DATABANKS

PIR/MIPS code
PIR2:D40649 (Zymomonas mobilis)

Brookhaven code

7 LITERATURE REFERENCES

[1] Tubbs, P.K., Greville, G.D.: Biochim. Biophys. Acta,34,290–291 (1959)
[2] Cammack, R.: Methods Enzymol.,41,323–329 (1975)

1 NOMENCLATURE

EC number
1.1.99.7

Systematic name
(S)-Lactate:oxaloacetate oxidoreductase

Recommended name
Lactate-malate transhydrogenase

Synonymes
Transhydrogenase, lactate-malate
Malate-lactate transhydrogenase

CAS Reg. No.
9077-15-0

2 REACTION AND SPECIFICITY

Catalysed reaction
(S)-Lactate + oxaloacetate →
→ malate + pyruvate (mechanism [3, 5])

Reaction type
Redox reaction

Natural substrates
(S)-Lactate + oxaloacetate (first step in fermentation of lactate) [2]

Substrate spectrum
1 (S)-Lactate + oxaloacetate (r [1, 7], catalyzes hydrogen transfer from C_3 or C_4(S)-2-hydroxy acids to 2 oxo-acids [1], hydrogen donors: L-malate [1], DL-lactate [1], DL-alpha-hydroxybutyrate [1], low activity with: DL-alpha-hydroxyglutarate [1], DL-alpha-hydroxyvalerate [1], DL-beta-hydroxybutyrate [1], DL-alpha-hydroxycaprylate [1], hydrogen acceptors: oxalacetate [1], low activity with: alpha-ketoglutarate [1], alpha-ketocaprylate [1], keto-tautomer is the emzymatically active form of oxalacetate [7]) [1–7]

Product spectrum
1 Malate + pyruvate

Inhibitor(s)
Malate [6]; Lactate [6]; p-Hydroxymercuribenzoate [1]; KCl [6]; Oxaloacetate [6]; Pyruvate [6]; Acetate [6]; More (non-specific inhibition by high ionic strength) [6]

Cofactor(s)/prosthetic group(s)/activating agents
NAD+ (prosthetic group [3], tightly bound in the active centre [1, 5], NAD+/NADH equivalent binding weight is 35000 [4]) [1, 3–5]; NADH [1, 3–5]

Metal compounds/salts
No metal ion requirement [2]

Turnover number (min⁻¹)

Specific activity (U/mg)
117 [1, 2]; 368 [5]

K_m-value (mM)
1.9 (L-lactate) [1]; 0.05 (oxaloacetate) [1]; 2.4 (pyruvate) [1]; 1.4 (L-malate) [1]

pH-optimum
7.5–8.5 [1, 2]

pH-range
6.3–9.5 (at pH 6.3 and 9.5 about 50% of activity maximum) [1, 2]

Temperature optimum (°C)
23 (assay at) [1]; 25 (assay at) [5]

Temperature range (°C)

3 ENZYME STRUCTURE

Molecular weight
70000 (Veillonella alcalescens, sedimentation-diffusion and high-speed equilibrium methods) [4]
99000–100000 (Micrococcus lactilyticus, sucrose gradient studies) [1–3]

Subunits
Dimer (2 × 30000–43000, Veillonella alcalescens, SDS-PAGE) [4, 5]
Trimer or tetramer (3 (or 4) × 30000, Micrococcus lactilyticus, sucrose gradient centrifugation of succinylated and urea solubilized enzyme) [3]

Glycoprotein/Lipoprotein
–

4 ISOLATION/PREPARATION

Source organism
Veillonella alcalescens [4–6]; Micrococcus lactilyticus [1–3, 7]; More (no activity in E. coli, Propionibacterium shermanii) [1]

Source tissue
Cell [1, 2, 5]

Localisation in source

Purification
Veillonella alcalescens [5]; Micrococcus lactilyticus [1]

Crystallization
–

Cloned
–

Renaturated
–

5 STABILITY

pH

Temperature (°C)

Oxidation

Organic solvent

General stability information

Storage
–15°C, as ammonium sulfate suspension, about 10–20 mg of protein per ml, 40% loss of activity after 1 year [2]; –20°C, pH 7.4 [5]

6 CROSSREFERENCES TO STRUCTURE DATABANKS

PIR/MIPS code

Brookhaven code

7 LITERATURE REFERENCES

[1] Allen, S.H.G.: J. Biol. Chem.,241,5266–5275 (1966)
[2] Allen, S.H.G.: Methods Enzymol.,13,262–269 (1969) (Review)
[3] Allen, S.H.G., Patil, J.R.: J. Biol. Chem.,247,909–916 (1972)
[4] Allen, S.H.G.: Eur. J. Biochem.,35,338–345 (1973)
[5] Allen, S.H.G.: Methods Enzymol.,89,367–376 (1982)
[6] Dolin, M.I.: J. Biol. Chem.,244,5273–5285 (1969)
[7] Dolin, M.I.: J. Biol. Chem.,243,3916–3923 (1968)

1 NOMENCLATURE

EC number
1.1.99.8

Systematic name
Alcohol:(acceptor) oxidoreductase

Recommended name
Alcohol dehydrogenase (acceptor)

Synonyms
Primary alcohol dehydrogenase [3]
MDH [8]
Dehydrogenase, alcohol (acceptor)
Quinohemoprotein alcohol dehydrogenase
Quinoprotein alcohol dehydrogenase
Quinoprotein ethanol dehydrogenase

CAS Reg. No.
37205-43-9

2 REACTION AND SPECIFICITY

Catalysed reaction
Primary alcohol + acceptor →
→ aldehyde + reduced acceptor (mechanism [10], ping pong kinetic mechanism [14])

Reaction type
Redox reaction

Natural substrates
Primary alcohol + acceptor (acetic acid bacteria: wide specificity for primary alcohol except methanol, enzyme has a role as vinegar producer by
coupling with aldehyde dehydrogenase [4], primary dehydrogenase in respiratory chain of acetic acid bacteria [15]) [4, 15]

Substrate spectrum
1 Primary alcohol + acceptor (electron acceptor: benzyl viologen [19], phenosaffarin [19], potassium ferricyanide (not [6]) [4, 19], phenazine methosulfate [2, 4, 5, 16–19, 22, 24, 26], nitro blue tetrazolium [4, 17], 2,6-dichlorophenolindophenol (not [6, 22]) [4, 17], tetramethyl-p-phenylenediamine [5, 6, 10, 16, 22, 25, 26], cytochrome c (horse heart cytochrome c, not: cytochrome c [19], cytochrome c_L is the physiological electron acceptor [26]) [5, 22, 26], 2,2'-azino-di-(3-ethylbenzylthiazoline-6-sulfonic acid) [6], phenazine ethosulfate [16, 19, 26], methylene blue [17], thionine [17], indophenol [17], nile blue [17, 19], ethylphenazine ethosulfate [19], 5-methylphenazinium methosulfate [19], neutral red [19], rhodamine B [19], janus green B [19], alizarin yellow [19], not: NAD^+ [4, 17, 19, 20, 22], $NADP^+$ [4, 17, 19, 20, 22], O_2 [4], ferricyanide [6], FAD [17, 19, 22], FMN [22], vitamin K_3 [22], alcohol: aliphatic primary alcohol with a chain length of 6 or less [4, 8], primary alcohols [5–7, 9, 18–20, 22, 25, 26], secondary alcohols (not [4, 16, 18–20]) [5, 7, 25], aromatic alcohols (not [18, 19], e.g. benzyl alcohol, vanillyl alcohol, veratryl alcohol [9]) [9], not: methanol [4, 15], higher primary alcohols (e.g. n-octanol, n-nonanol, n-decanol, n-hexadecanol) [20], active with aldehydes [5, 7, 25], formaldehyde [6–8, 16, 18, 19, 22, 26], acetaldehyde (not [8], slowly [16]) [6, 7, 16], overview [7, 17], broad specificity [3, 4, 7, 9, 17], different specificities of Rhodopseudomonas acidophila enzymes from strains 10050 grown anaerobically on methanol and M402 grown aerobically on vanillyl alcohol [24]) [1–26]

Product spectrum
1 Aldehyde + reduced acceptor

Inhibitor(s)
Cyclopropanol [5, 6, 23]; Cyanide [6]; KCN [22]; Hydroxylamine [6]; Cyclopropanone hydrate [6]; Cyclopropanone ethylhemiketal (suicide substrate) [7]; EDTA (not [17], partial [22]) [9, 19, 22]; p-Nitrophenylhydrazine [9]; Zn^{2+} [17]; N_3^- (slight) [17]; Cyclohexanol (partial) [19]; 2-Hexanol (partial) [19]; Mn^{2+} [22]; Co^{2+} [22]; 2,2'-Bipyridyl (partial) [22]; 1,10-Phenanthroline (partial) [22]; ADP [22]; ATP [22]; Cyclopropane-derived compounds (mechanism of inhibition) [11]

Cofactor(s)/prosthetic group(s)/activating agents

4,5-Dihydro-4,5-dioxo-1H-pyrrolo[2,3-f]quinoline-2,7,9-tricarboxylic acid (i.e. 2,7,9-tricarboxy-1H-pyrrolo[2,3-f]quinoline-4,5-dione [1], pyrroloquinoline quinone [1, 4–6, 8, 13, 22], methoxatin [3], as prosthetic group, 2 molecules per enzyme molecule [5–8, 22], 1 molecule per enzyme molecule [1, 25], o-quinone structure is essential for activity [1], 9-carboxylic acid group and pyrrolo ring are not essential for activity, they can be replaced by a pyridinol ring and a 9-hydroxy group [1], [13]) [1, 3–6, 13, 22, 25]; Pyrroloquinoline quinol (enzyme contains two prosthetic group molecules: pyrroloquinoline quinol and pyrroloquinoline quinone) [13]; Heme (enzyme contains 1 heme c group [25], heme c group participates in enzymatic mechanism [25], Gluconobacter suboxydans: enzyme contains 3 mol of heme c per mol of enzyme [4], Comamonas testosteroni ATCC 15667: 1 mol of heme c per apoenzyme molecule [5]) [4, 5, 25]; Cytochrome c (molar ratio of methanol dehydrogenase dimer to cytochrome c at pH 9.5 is 1 at pH 7.0 it is 2.5–3.5) [21]; Methylamine (presence of ammonium chloride or methylamine is essential for activity [22], poor activator [6]) [6, 22]; Benzyl ester of glycine (activates) [6]; Ethyl ester of glycine (activates) [6]; Phenylpropylamine (activates) [6]; Phenylbutyramine (activates) [6]; Substituted benzylamines (activate) [6]; 2-Bromoethylamine (activates) [6]

Metal compounds/salts

Ammonium salt (or primary amine salt required as activator (Pseudomonas aeruginosa) [5], no activation of Comamonas testosteroni enzyme [5], ammonium chloride activates [6], required as activator [10, 16, 17, 19, 20, 24], ammonium chloride or methylamine chloride required for activity with phenazine methosulfate, phenazine ethosulfate or tetramethyl-p-phenylenediamine, not with cytochrome c_L [26]) [5–7, 10, 14, 16, 17, 19, 20, 24, 26]; Calcium (enzyme contains one atom of tightly bound calcium per enzyme molecule [8], Ca^{2+} is essential for activity [25]) [8, 25]; Iron (enzyme contains 0.7–1.0 gatom of non-heme acid-labile iron per mol of enzyme) [12]; More (no significant metal content) [2]

Turnover number (min^{-1})

Specific activity (U/mg)

More [2, 4, 5, 7, 8, 18, 19, 22, 24, 25]

K$_m$-value (mM)

0.0035 (octanol) [23]; 0.0020 (decanol) [23]; 0.010 (propan-1-ol) [23]; 0.013 (ethanol) [23]; 0.014 (ethanol, Pseudomonas aeruginosa) [7]; 0.018 (ethanol, Pseudomonas putida) [7]; 0.020 (methanol) [16], 0.028 (ethanol) [16]; 0.038 (octanal) [23]; 0.050 (methanol) [9]; 0.07 (formaldehyde) [20]; 0.12 (acetaldehyde) [22]; 0.280 (propan-1-ol) [16]; 0.285 (butan-1-ol) [16]; 0.3 (methanol (+ tetramethyl-p-phenylenediamine + 50 mM NH$_4$Cl as activator)) [16]; 0.395 (phenazine ethosulfate) [16]; 0.44 (phenazine methosulfate) [16]; 0.5 (ethanol) [17]; 0.64 (propan-2-ol) [23]; 0.74 (ethanol) [22]; 1 (benzyl alcohol) [9]; 1.01 (formaldehyde) [16]; 1.1 (n-hexanol) [17]; 1.5 (veratryl alcohol) [9]; 1.6 (ethanol) [4]; 2 (methanol, Pseudomonas putida) [7]; 2.3 (formaldehyde) [22]; 4 (formaldehyde) [23]; More [15–17, 20, 22–24]

pH-optimum

4.0 (Acetobacter aceti) [4, 17]; 6.0 (Gluconobacter suboxydans) [4]; 7.0 (cytochrome c$_L$) [26]; 7.5 (tetramethyl-p-phenylenediamine, Comamonas testosteroni) [5]; 7.7 [25]; 9.0 (Pseudomonas aeruginosa [5], phenazine methosulfate, phenazine ethosulfate or tetramethyl-p-phenylenediamine [26]) [5, 7, 19, 22, 26]; 9.5 [18]; 9.6 [20]; 9.8 [16]

pH-range

1.5–5 (1.5: about 70% of activity maximum, 5: about 85% of activity maximum) [17]; 8–10 (8: about 40% of activity maximum, 10: about 90% of activity maximum) [19]

Temperature optimum (°C)

30 (assay at) [26]

Temperature range (°C)

---

3 ENZYME STRUCTURE

Molecular weight

60000 (Methylomonas methanica, gel filtration) [18]

61000 (Methylosinus sporium, gel filtration, sedimentation equilibrium analysis) [19]

67000 (Pseudomonas testosteroni, gel filtration of native and denatured enzyme) [25]

70000 (Comamonas testosteroni, gel filtration) [5]

100000 (Pseudomonas aeruginosa 80.53, gel filtration) [5]

101000 (Pseudomonas aeruginosa LMD, HPLC gel filtration) [23]

115000 (Methylophilus methylotrophus, gel filtration) [16]

120000 (Hyphomicrobium sp. X, gel filtration [6], Methylosinus trichosporium OB3b, gel filtration [21]) [6, 21]

128000 (Pseudomonas sp. No. 2941, gel filtration) [20]

135000 (Methylomonas sp. J, gel filtration) [22]

140000 (Methylobacillus glycogenes, gel filtration) [8]
146000 (Pseudomonas sp. M27, analytical ultracentrifugation) [2]
150000 (Gluconobacter suboxydans, gel filtration) [4]

Subunits
Monomer (1 × 100000, Pseudomonas aeruginosa [5], 1 × 67000, Pseudomonas testosteroni [25], 1 × 65000, Comamonas testosteroni [5], 1 × 60000, Methylomonas methanica [18], Methylosinus sporium [19], SDS-PAGE [5, 18, 19, 25]) [5, 18, 19, 25]
Dimer (2 × 62000, Methylophilus methylotrophus [16], Pseudomonas sp. No. 2941 [20], 2 × 60000, Pseudomonas aeruginosa [7], Pseudomonas putida [7], Methylosinus trichosporium [21], SDS-PAGE [7, 16, 20, 21]) [7, 16, 20, 21]
Trimer (1 × 85000 (dehydrogenase component) + 1 × 49000 (cytochrome component) + 1 × 14400 (component of unclear function), Gluconobacter suboxydans, SDS-PAGE) [4]
Tetramer (2 × 60000 (alpha) + 2 × 90000 (beta), Methylobacillus glycogenes [8], 1 × 63000 (dehydrogenase component) + 1 × 44000 (cytochrome c component) + 1 × 29000 + 1 × 13500, Acetobacter aceti, SDS-PAGE [4], 2 × 60000 + 2 × 8000, Hyphomicrobium sp. X, SDS-PAGE [6], 2 × 60000 (alpha) + 2 × 8460 (beta), Methylobacterium extorquens AM 1 [26]) [4, 6, 8, 26]
? (x × 60000 + x × 10000, Methylomonas sp. J, SDS-PAGE) [22]
More [17]

Glycoprotein/Lipoprotein
Glycoprotein (enzyme contains 4.1% glucosamine) [2]

---

4 ISOLATION/PREPARATION

Source organism
Comamonas testosteroni (quinohemoprotein) [5]; Hyphomicrobium sp. X [1, 6, 10, 11, 13, 14]; Methylophilus methylotrophus [1, 16]; Pseudomonas sp. (M27 [2], No. 2941 [20]) [2, 20]; Methylotrophic bacteria [3]; Acetic acid bacteria [4]; Gluconobacter suboxydans (IFO 12528 [4]) [4, 15]; Acetobacter aceti (IFO 3284 [4]) [4, 15, 17]; Pseudomonas testosteroni (quinoprotein [5]) [5, 25]; Pseudomonas aeruginosa (ATCC 17933 [7], LMD 80.53 [5]) [5, 7, 23]; Pseudomonas putida (ATCC 17421 [7]) [7]; Methylobacillus glycogenes [8]; Methylosinus trichosporium (OB3b [21]) [9, 21]; Methylomonas methanica [18]; Methylosinus sporium [19]; Methylomonas sp. J [22]; Rhodopseudomonas acidophila (strain 10050, M402 [24]) [12, 24]; Methylobacterium extorquens AM 1 [26]

Source tissue
Cell [4, 5]

Localisation in source
Cytoplasmic membrane (surface) [4]; Membrane (loosely bound to intracellular membrane [16]) [15, 16]; Periplasm [26]; Soluble [9, 18, 19]

Purification
Hyphomicrobium sp. X [6]; Methylobacillus glycogenes [8]; Methylosinus trichosporium (strain OB3b, 4 isoenzymes [21]) [9, 21]; Methylophilus methylotrophus [16]; Methylomonas methanica [18]; Methylosinus sporium [19]; Methylomonas sp. J [22]; Rhodopseudomonas acidophila (strain 10050, grown anaerobically on methanol, strain M402, grown aerobically on vanillyl alcohol) [24]; Methylobacterium extorquens AM 1 [26]; Pseudomonas aeruginosa (strain LMD 80.53) [5]; Pseudomonas sp. (M27 [2], No. 2941 [20]) [2, 20]; Gluconobacter suboxydans (IFO 12528 [4]) [4, 15]; Acetobacter aceti (IFO 3284 [4]) [4, 15, 17]; Pseudomonas testosteroni [25]

Crystallization
[4, 7, 18–22]

Cloned
–

Renaturated
–

---

5 STABILITY

pH
2.2–3.0 (complete loss of activity after 18 h at room temperature) [2]; 3 (24 h, 4°C, 40% loss of activity) [17]; 3.3 (unstable below) [18]; 4–8 (stable, Gluconobacter suboxydans) [4]; 4.0–9.0 (less than 10% loss of activity after 18 h at room temperature) [2]; 5–8 (stable, Acetobacter aceti) [4]; 6 (45 h, 30°C, 40% loss of activity) [22]; 6.5–11.6 (10 min, 55°C, stable) [20]; 7 (24 h, 4°C, no loss of activity) [17]; 8 (45 h, 30°C, 90% loss of activity) [22]; 9 (24 h, 4°C, 40% loss of activity) [17]; 10.0 (17% loss of activity after 18 h at room temperature [2], 45 h, 30°C, 70% loss of activity [22]) [2, 22]; 11.0 (29% loss of activity after 18 h at room temperature) [2]; 12.0 (81% loss of activity after 18 h at room temperature) [2]

Temperature (°C)
4 (half-life: 1–2 days) [16]; 30 (10 min, stable up to) [17]; 50 (10 min, 10 mM methanol, 0.1 M Tris-HCl buffer, 5% loss of activity [16], 10 min, pH 4, 40% loss of activity [17]) [16, 17]; 55 (10 min, stable up to [7], 30 min, stable [19], 10 min, pH 6.5–11.5, stable [20]) [7, 19, 20]; 60 (2 min, 10 mM methanol, 0.1 Tris-HCl buffer, 40% loss of activity [16], 30 min, 90% loss of activity [19]) [16, 19]; 65 (5 min, stable) [20]; 70 (10 min, 70% loss of activity of Pseudomonas putida enzyme, 50% loss of activity of Pseudomonas aeruginosa enzyme [7], 2 min, 10 mM methanol, 0.1 M Tris-HCl buffer, 90% loss of activity [16]) [7, 16]; 80 (5 min, complete loss of activity) [20]; 22 (18 h, pH 4.0–9.0, stable) [1]

Oxidation

Organic solvent

General stability information
Glycerol stabilizes [4]; Sucrose stabilizes [4]; Polyethylene glycol stabilizes [4]; Triton X-100 stabilizes [4]; Methanol stabilizes [16, 20]

Storage
2°C, 20 mM Tris-HCl buffer, pH 8.0, stable for 1 week [2]; –22°C, pH 7–8, 10 mM methanol, 6 months, 80% of activity retained [16]; –22°C, 20 mM Tris-HCl buffer, pH 8.0, about 15% loss of activity after 2 months [2]; 4°C, 0.1% Triton X-100, pH 6.0, several weeks [4]; 4°C, pH 7–8, 10 mM methanol, several days, 80% of activity retained [16]; –70°C, stable for a long time [25]

---

6 CROSSREFERENCES TO STRUCTURE DATABANKS

PIR/MIPS code
PIR2:JQ0706 (66K chain precursor Methylobacterium extorquens)

Brookhaven code
8ADH (Horse (Equus caballus) liver); 3AAH (Methylophilis w3A1)

Enzyme Handbook © Springer-Verlag Berlin Heidelberg 1995
Duplication, reproduction and storage in data banks are only
allowed with the prior permission of the publishers

7 LITERATURE REFERENCES

[1] Duine, J.A., Frank, J., Verwiel, P.E.J.: Eur. J. Biochem.,108,187–192 (1980)
[2] Anthony, C., Zatman, L.J.: Biochem. J.,104,953–959 (1967)
[3] Salisbury, S.A., Forrets, H.S., Cruse, W.B.T., Kennard, O.: Nature,280,843–844 (1979)
[4] Ameyama, M., Adachi, O.: Methods Enzymol.,89,450–457 (1982) (Review)
[5] Groen, B.W., Duine, J.A.: Methods Enzymol.,188,33–39 (1990) (Review)
[6] Frank, J., Duine, J.A.: Methods Enzymol.,188,202–209 (1990)
[7] Görisch, H., Rupp, M.: Antonie Leeuwenhoek,56,35–45 (1989)
[8] Adachi, O., Matsushita, K., Shinagawa, E., Ameyama, M. : Agric. Biol. Chem.,54,3123–3129 (1990)
[9] Mountfort, D.O.: J. Bacteriol.,172,3690–3694 (1990)
[10] Frank, J., Dijkstra, M., Duine, J.A., Balny, C.: Eur. J. Biochem.,174,331–338 (1988)
[11] Frank, J., van Krimpen, S.H., Verwiel, P.E.J., Jongejan, J.A., Mulder, A.C., Duine, J.A.: Eur. J. Biochem.,184,187–195 (1989)
[12] Bamforth, C.W., Quayle, J.R.: Biochem. J.,181,517–524 (1979)
[13] Duine, J.A., Frank, J., Verwiel, P.E.J.: Eur. J. Biochem.,118,395–399 (1981)
[14] Duine, J.A., Frank, J.: Biochem. J.,187,213–219 (1980)
[15] Matsushita, K., Takaki, Y., Shinagawa, E., Ameyama, M., Adachi, O.: Biosci. Biotechnol. Biochem.,56,304–310 (1992)
[16] Ghosh, R., Quayle, J.R.: Biochem. J.,199,245–250 (1981)
[17] Muraoka, H., Watabe, Y., Ogasawara, N., Takahashi, H.: J. Ferment. Technol.,60, 41–50 (1982)
[18] Patel, R.N., Hou, C.T., Felix, A.: J. Bacteriol.,133,641–649 (1978)
[19] Patel, R.N., Felix, A.: J. Bacteriol.,128,413–424 (1976)
[20] Yamanaka, K., Matsumoto, K.: Agric. Biol. Chem.,41,467–475 (1977)
[21] Parker, M.W., Cornish, A., Gossain, V., Best, D.J.: Eur. J. Biochem.,164,223–227 (1987)
[22] Ohta, S., Fujita, T., Tobari, J.: J. Biochem.,90,205–213 (1981)
[23] Groen, B., Frank, J., Duine, J.A.: Biochem. J.,223,921–924 (1984)
[24] Yamanaka, K., Minoshima, R.: Agric. Biol. Chem.,48,171–179 (1984)
[25] Groen, B.W., van Kleef, M.A.G., Duine, J.A.: Biochem. J.,234,611–615 (1986)
[26] Day, D.J., Anthony, C.: Methods Enzymol.,188,210–222 (1990) (Review)

1 NOMENCLATURE

EC number
1.1.99.9

Systematic name
Pyridoxine:(acceptor) 5-oxidoreductase

Recommended name
Pyridoxine 5-dehydrogenase

Synonymes
Pyridoxal-5-dehydrogenase
Dehydrogenase, pyridoxol 5-
Pyridoxol 5-dehydrogenase
Pyridoxin 5-dehydrogenase
Pyridoxine dehydrogenase
Pyridoxine 5'-dehydrogenase

CAS Reg. No.
9023-39-6

2 REACTION AND SPECIFICITY

Catalysed reaction
Pyridoxine + acceptor →
→ isopyridoxal + reduced acceptor

Reaction type
Redox reaction

Natural substrates
Pyridoxine + acceptor (utilization of pyridoxine as sole source of carbon and nitrogen) [2]

Substrate spectrum
1 Pyridoxine + acceptor (acceptor: 2,6-dichlorophenolindophenol [1, 2], O_2 or quinones [2], not: oxygen [1], NAD$^+$, NADP$^+$ [2], highly specific for pyridoxine [1]) [1, 2]
2 Pyridoxamine + acceptor (low activity) [1, 2]
3 4-Deoxypyridoxine + acceptor [2]
4 Pyridoxal + acceptor [2]

Product spectrum
1 Isopyridoxal + reduced acceptor
2 4-Aminomethyl-5-formyl-2-methylpyridin-3-ol + reduced acceptor
3 3-Hydroxy-2-methylpyridine-5-carboxaldehyde + reduced acceptor
4 3-Hydroxy-2-methylpyridine-4,5-dicarbaldehyde + reduced acceptor

Inhibitor(s)
Ni^{2+} [2]; Cu^{2+} [2]; Zn^{2+} [2]; Hg^{2+} [2]; p-Chloromercuribenzoate [1, 2]; N-Ethyl-maleimide [2]; Dithionitrobenzoic acid [2]; Substrate analogs (e.g.: pyrido-xal, pyridoxamine, isopyridoxal, 4-pyridoxic acid) [2]; Sulfhydryl reagents [2]; More (not: alpha,alpha'-dipyridyl, o-, m- or p-phenanthroline) [1]

Cofactor(s)/prosthetic group(s)/activating agents
FAD (flavoprotein, FAD-dependent) [1, 2]

Metal compounds/salts

Turnover number (min^{-1})
1320 (pyridoxine + O_2) [2]

Specific activity (U/mg)
11.9 [2]

K_m-value (mM)
0.48 (pyridoxine) [1]; 0.18 (pyridoxine) [2]; 0.28 (O_2) [2]

pH-optimum
5.5 [1]; 7.0–8.3 (pyridoxine + O_2, Tris buffer, phosphate buffer) [2]

pH-range
5–7 (5: about 30% of activity maximum, 7: 17% of activity maximum) [1]; More [2]

Temperature optimum (°C)

Temperature range (°C)

---

3 ENZYME STRUCTURE

Molecular weight
112000 (Arthrobacter sp. Cr-7, gel filtration) [2]

Subunits
Dimer (2 × 58600, Arthrobacter sp. Cr-7, SDS-PAGE) [2]

Glycoprotein/Lipoprotein
–

4 ISOLATION/PREPARATION

Source organism
  Pseudomonas sp. IA [1]; Arthrobacter sp. Cr-7 (inducible enzyme) [2]

Source tissue
  Cell [1]

Localisation in source
  Soluble [2]

Purification
  Pseudomonas sp. IA (partial) [1]; Arthrobacter sp. Cr-7 [2]

Crystallization
  –

Cloned
  –

Renaturated
  –

5 STABILITY

pH

Temperature (°C)

Oxidation

Organic solvent

General stability information

Storage

6 CROSSREFERENCES TO STRUCTURE DATABANKS

PIR/MIPS code

Brookhaven code

7 LITERATURE REFERENCES

[1] Sundaram, T.K., Snell, E.E.: J. Biol. Chem.,244,2577–2584 (1969)
[2] Jong, Y.-J., Nelson, M.J.K., Snell, E.E.: J. Biol. Chem.,261,15102–15105 (1986)

1 NOMENCLATURE

EC number
1.1.99.10

Systematic name
D-Glucose:(acceptor) 1-oxidoreductase

Recommended name
Glucose dehydrogenase (acceptor)

Synonymes
Glucose dehydrogenase (Aspergillus)
Dehydrogenase, glucose (acceptor)
Dehydrogenase, glucose (Aspergillus)
Glucose dehydrogenase (decarboxylating)

CAS Reg. No.
37250-84-3

2 REACTION AND SPECIFICITY

Catalysed reaction
D-Glucose + acceptor →
→ D-glucono-1,5-lactone + reduced acceptor

Reaction type
Redox reaction

Natural substrates

Substrate spectrum
1 D-Glucose + acceptor (acceptor: 2,6-dichlorophenolindophenol [1–3], phenazine methosulfate [3], tetramethyl-p-phenylenediamine [3], coenzyme Q_1 [3], ferricyanide [3]) [1–3]
2 Fructose + 2,6-dichlorophenolindophenol (8% of the activity with D-glucose [2], not [3]) [2]
3 D-Mannose + acceptor (8.6% of the activity with D-glucose) [3]
4 D-Galactose + acceptor (6.5% of the activity with D-glucose) [3]
5 D-Xylose + acceptor (13% of the activity with D-glucose) [3]
6 L-Rhamnose + acceptor (7.5% of the activity with D-glucose) [3]
7 L-Arabinose + acceptor (2.8% of the activity with D-glucose) [3]
8 Maltose + acceptor (3.2% of the activity with D-glucose) [3]

Enzyme Handbook © Springer-Verlag Berlin Heidelberg 1995
Duplication, reproduction and storage in data banks are only
allowed with the prior permission of the publishers

Product spectrum
 1 D-Glucono-1,5-lactone + reduced acceptor (product: gluconic acid [1])
 2 ?
 3 ?
 4 ?
 5 ?
 6 ?
 7 ?
 8 ?

Inhibitor(s)
 p-Benzoquinone [3]

Cofactor(s)/prosthetic group(s)/activating agents
 FAD (contains 1 mol per mol of enzyme) [1]; More (properties of the prosthetic group) [4]

Metal compounds/salts

Turnover number (min^{-1})
 320000 (glucose) [4]

Specific activity (U/mg)
 More [1, 3]

K_m-value (mM)
 0.04 (2,6-dichlorophenolindophenol) [2]; 0.06 (coenzyme Q_1) [3]; 0.13 (phenazine methosulfate) [3]; 0.56 (tetramethyl-p-phenylenediamine) [3]; 0.69 (ferricyanide) [3]; 1.6 (2,6-dichlorophenolindophenol) [3]

pH-optimum
 4.5 (ferricyanide) [3]; 6.0 (assay at, impossible to assay below pH 5.5, 2,6-dichlorophenolindophenol) [3]; 6.5 (coenzyme Q_1 [3]) [2, 3]; 8.75 (phenazine methosulfate, tetramethyl-p-phenylenediamine) [3]

pH-range

Temperature optimum (°C)

Temperature range (°C)
 20–25 (assay at) [1]; 25 (assay at) [2]

3 ENZYME STRUCTURE

Molecular weight
86000 (Bacillus anitratum, titration of the prosthetic group with glucose) [4]
90000 (Pseudomonas sp., urea-SDS-PAGE, sucrose gradient centrifugation
+ Triton X-100) [3]
118000 (Aspergillus oryzae, gel filtration, approach to sedimentation equilibrium) [1]

Subunits
Monomer (1 × 90000, Pseudomonas sp., urea-SDS-PAGE, monomeric in presence of 1% Triton X-100, aggregation after removing the detergent) [3]
More (aggregation in medium containing no Triton X-100 to dimers, trimers
and tetramers) [3]

Glycoprotein/Lipoprotein
Glycoprotein (contains 24% carbohydrate consisting of glucose, mannose
and hexosamines) [1]

---

4 ISOLATION/PREPARATION

Source organism
Aspergillus oryzae [1]; Aspergillus niger [2]; Pseudomonas sp. [3]; Bacillus
anitratum [4]

Source tissue
Mycelium [1, 2]; Culture medium [1]; Cell [3]

Localisation in source
Membrane (bound) [3]

Purification
Aspergillus oryzae [1]; Pseudomonas sp. [3]

Crystallization
--

Cloned
-

Renaturated
-

5 STABILITY

pH
5.0–8.0 (2 h, 30°C, stable, rapid inactivation outside this range) [1]; 5.2–7.0 (frozen, stable) [1]

Temperature (°C)
30 (15 min, stable) [1]; 45 (15 min, about 40% loss of activity) [1]; 50 (15 min, about 40% loss of activity) [1]; 60 (15 min, complete loss of activity) [1]

Oxidation

Organic solvent

General stability information
Glucose, 0.2 M, protects during heat treatment [1]

Storage
–18°C, 0.05 M potassium phosphate buffer, pH 6.5, 10% loss of activity after 3 months, 50% loss of activity after 6 months [1]

6 CROSSREFERENCES TO STRUCTURE DATABANKS

PIR/MIPS code
PIR2:S06628 (fruit fly (Drosophila melanogaster) (fragment))

Brookhaven code

7 LITERATURE REFERENCES

[1] Bak, T.-G.: Biochim. Biophys. Acta,139,277–293 (1967)
[2] Müller, H.-M. : Zentralbl. Bakteriol. Parasitenkd. Infektionskr. Hyg.,132,14–24 (1977)
[3] Matsushita, K., Ohno, Y., Shinagawa, E., Adachi, O., Ameyama, M.: Agric. Biol. Chem.,44,1505–1512 (1980)
[4] Hauge, J.G.: J. Biol. Chem.,239,3630–3639 (1964)

1 NOMENCLATURE

EC number
1.1.99.11

Systematic name
D-Fructose:(acceptor) 5-oxidoreductase

Recommended name
Fructose 5-dehydrogenase

Synonymes
Dehydrogenase, fructose 5- (acceptor)
D-Fructose dehydrogenase

CAS Reg. No.
37250-85-4

2 REACTION AND SPECIFICITY

Catalysed reaction
D-Fructose + acceptor →
→ 5-dehydro-D-fructose + reduced acceptor

Reaction type
Redox reaction

Natural substrates
D-Fructose + acceptor (utilization of D-fructose by acetic acid bacteria) [3]

Substrate spectrum
1 D-Fructose + acceptor (ir [3], high specificity for D-fructose [1, 2], acceptor: e.g. ferricyanide [1, 3], nitro blue tetrazolium [1], phenazine methosulfate [1, 3], 2,6-dichlorophenolindophenol [1–3], not: NAD+ [1–3], NADP+ [1–3], O$_2$ [1, 3]) [1–3]

Product spectrum
1 5-Keto-D-fructose + reduced acceptor [1–3]

Inhibitor(s)
Phenylmercuric nitrate [2]; p-Chloromercuribenzoate [2]; Atebrine (slight) [2]; o-Phenanthroline (slight) [2]; EDTA (slight) [2]; Ag+ [2]; Hg^{2+} [2]; Cu^{2+} [2]; Zn^{2+} [2]; Cd^{2+} [2]

Cofactor(s)/prosthetic group(s)/activating agents
Cytochrome c (contains a cytochrome c like electron carrier, tightly bound)
[1, 3]

Metal compounds/salts

Turnover number (min^{-1})

Specific activity (U/mg)
172.0 [1, 3]; More [2]

K$_m$-value (mM)
10 (D-fructose [1–3], in presence of 2,6-dichlorophenolindophenol (67 mM),
pH 5.5, 15°C [2]) [1–3]

pH-optimum
4.0 [1, 3]; 4.8–5.0 [2]

pH-range
4.5–6.4 (4.5: about 85% of activity maximum, 6.4: 50% of activity maximum)
[2]

Temperature optimum (°C)
25 [1, 3]

Temperature range (°C)

3 ENZYME STRUCTURE

Molecular weight
140000 (Gluconobacter industrius, gel filtration) [1]

Subunits
Trimer (1 × 67000 (flavin dehydrogenase) + 1 × 50800 (cytochrome c) +
1 × 19700 (unknown function), Gluconobacter industrius, SDS-PAGE) [1]

Glycoprotein/Lipoprotein
–

4 ISOLATION/PREPARATION

Source organism
Gluconobacter industrius IFO 3260 [1, 3]; Gluconobacter cerinus [2]

Source tissue
Cell [1–3]

Localisation in source
Cytoplasmic membrane (outer surface) [1]; More (particulate fraction) [2]

Purification
Gluconobacter industrius IFO 3260 [1]; Gluconobacter cerinus [2]

Crystallization
–

Cloned
–

Renaturated
–

---

5 STABILITY

pH
4.5–6.0 (stable) [3]; 5.0–7.0 (1 h, 30°C, stable) [2]; 2.5–8.5 (immediately in-activated below pH 2.5 and above 8.5) [2]

Temperature (°C)
35 (rapid inactivation beyond) [3]; 40 (10 min, no loss of activity) [2]; 45 (10 min, 15% loss of activity) [2]; 50 (10 min, complete loss of activity) [2]

Oxidation

Organic solvent

General stability information
Triton X-100 is essential for preservation of purified enzyme [1]; Sucrose, 5–10%, stabilizes [1]; Glycerol, 5–10%, stabilizes [1]; Fructose, 5–10%, stabilizes [1]; Detergents stabilize [3]

Storage
0–4°C, pH 4.0–5.0, 0.1% Triton X-100, 1 mM 2-mercaptoethanol, for at least 2 weeks [1]; –20°C or below, activity completely preserved [1, 3]

---

6 CROSSREFERENCES TO STRUCTURE DATABANKS

PIR/MIPS code

Brookhaven code

---

7 LITERATURE REFERENCES

[1] Ameyama, M., Adachi, O.: Methods Enzymol.,89,154–159 (1982)
[2] Yamada, Y., Aida, K., Uemura, T.: J. Biochem.,61,636–646 (1967)
[3] Ameyama, M., Shinagawa, E., Matsushita, K., Adachi, O.: J. Bacteriol.,145,814–823 (1981)

---

1 NOMENCLATURE

EC number
1.1.99.12

Systematic name
L-Sorbose:(acceptor) 5-oxidoreductase

Recommended name
Sorbose dehydrogenase

Synonymes
Dehydrogenase, sorbose

CAS Reg. No.
37250-86-5

---

2 REACTION AND SPECIFICITY

Catalysed reaction
L-Sorbose + acceptor →
→ 5-dehydro-D-fructose + reduced acceptor

Reaction type
Redox reaction

Natural substrates

Substrate spectrum
1 L-Sorbose + 2,6-dichlorophenolindophenol [1]
2 More (not: D-fructose, sucrose, dulcitol) [1]

Product spectrum
1 5-Keto-D-fructose + reduced 2,6-dichlorophenolindophenol [1]
2 ?

Inhibitor(s)
Cu^{2+} [1]; Phenylmercuric nitrate [1]; Ag^+ [1]; Hg^{2+} [1]; p-Chloromercuriben-zoate [1]

Cofactor(s)/prosthetic group(s)/activating agents

Metal compounds/salts

Turnover number (min^{-1})

Specific activity (U/mg)

K$_m$-value (mM)
4.3 (L-sorbose) [1]

pH-optimum
6.4 [1]

pH-range
5.0–8.7 (5.0: about 40% of activity maximum, 8.7: about 30% of activity maximum) [1]

Temperature optimum (°C)
20 (assay at) [1]

Temperature range (°C)

---

3 ENZYME STRUCTURE

Molecular weight

Subunits

Glycoprotein/Lipoprotein
–

---

4 ISOLATION/PREPARATION

Source organism
Gluconobacter suboxydans [1]

Source tissue

Localisation in source
More (particulate fraction) [1]

Purification

Crystallization
–

Cloned
–

Renaturated
–

5 STABILITY

ph

6.0–6.5 (30 min, 20°C, highest stability) [1]; 4.0 (30 min, 20°C, complete inactivation) [1]; 7.5 (30 min, 20°C, complete inactivation) [1]

Temperature (°C)

40 (5 min, complete inactivation) [1]

Oxidation

Organic solvent

General stability information

Storage

6 CROSSREFERENCES TO STRUCTURE DATABANKS

PIR/MIPS code

Brookhaven code

7 LITERATURE REFERENCES

[1] Sato, K., Yamada, Y., Aida, K., Uemura, T.: J. Biochem.,66,521–527 (1969)

1 NOMENCLATURE

EC number
1.1.99.13

Systematic name
D-Aldohexoside:(acceptor) 3-oxidoreductase

Recommended name
Glucoside 3-dehydrogenase

Synonymes
Dehydrogenase, D-glucoside 3-
D-Aldohexopyranoside dehydrogenase
D-Aldohexoside:cytochrome c oxidoreductase
D-Glucoside 3-dehydrogenase
Hexopyranoside-cytochrome c oxidoreductase

CAS Reg. No.
9031-74-7

2 REACTION AND SPECIFICITY

Catalysed reaction
Sucrose + acceptor →
→ 3-dehydro-alpha-D-glucosyl-beta-D-fructofuranoside + reduced acceptor

Reaction type
Redox reaction

Natural substrates
Aldohexosides + acceptor (reaction in sugar metabolism) [1]

Substrate spectrum
1 Sucrose + acceptor (acts on carbohydrates of hexopyranose C-1 chair
form with a hemiacetal oxygen or sulfur at C-1, an equatorial hydroxy
group at C-3 and a CH_2OH-group at C-5, equatorial configuration at C-2
and C-4 is preferred to an axial configuration [4], electron acceptors:
phenazine methosulfate [2–4], 2,6-dichlorophenolindophenol [2–4], ferricyanide [2, 4], cytochrome c (from heart muscle) [2], cytochrome c_{551}
(physiological H-acceptor, membrane-bound) [5]) [1, 2, 4, 5]

2 D-Glucose + acceptor [1, 2, 4, 5]
3 Validoxylamine A + acceptor [4, 5]
4 p-Nitrophenylvalidamine + acceptor [4, 5]
5 p-Nitrophenylvalienamine + acceptor [4, 5]
6 D–Galactose + acceptor [1, 2]
7 D-Glucosides + acceptor (react more rapidly than galactosides, aldo-
   pentoses and methylpentoses are no substrates) [1]
8 D-Galactosides + acceptor [1]
9 Cellobiose + acceptor [1, 2, 4, 5]
10 Lactose + acceptor [1, 2, 4, 5]
11 Maltose + acceptor [1, 2]
12 Trehalose + acceptor [1, 2, 4, 5]
13 D-Glucose 1-phosphate + acceptor (not D-Glc-6-phosphate [1]) [1, 2]
14 Lactobionate + acceptor [2, 4]
15 Methyl-alpha-D-glucose + acceptor [2, 4]
16 Methyl-beta-D-glucose + acceptor [2, 4]
17 Maltobionate + acceptor [2]
18 beta-Melibiose + acceptor [1, 2]
19 Raffinose + acceptor [1, 2]
20 2-Deoxy-D-glucose + acceptor [1, 2]
21 D-Glucosamine + acceptor [1, 2]
22 D-Sorbitol + acceptor [1]
23 alpha-Methyl-D-glucoside + acceptor [1, 4, 5]
24 beta-Methyl-D–glucoside + acceptor [5]
25 Lactulose + acceptor [2]
26 Salicin + acceptor [1]
27 UDPglucose + acceptor [1, 5]
28 p-Arbutine + acceptor [2]
29 Cellobionate + acceptor [2]
30 Leucrose + acceptor [2]
31 Methyl-beta-thiogalactose + acceptor [2]
32 Melezitose + acceptor [2]
33 D-Mannose + acceptor [2]
34 Anhydro-1,6-D-glucose + acceptor [2]
35 2-Deoxy-D-galactose + acceptor [2]
36 TDPglucose + acceptor [5]
37 p-Nitrophenyl alpha-D-glucoside + acceptor [5]
38 p-Nitrophenyl 1-thio-beta-D-glucoside + acceptor [5]
39 1,5-Anhydro-D-glucitol + acceptor [4, 5]
40 Methyl-alpha-D-galactoside + acceptor [4]

Product spectrum

1 3-Dehydro-alpha-D-glucosyl-beta-D-furanoside + reduced acceptor
   (3-ketosucrose, alpha-D-ribo-hexopyranosyl-3-ulose-beta-fructofuranosi-
   de) [1]
2 3-Ketoglucose + reduced acceptor (D-ribohexos-3-ulose) [1]
3 3-Ketovalidoxylamine A + reduced acceptor [4, 5]
4 p-Nitrophenyl-3-ketovalidamine + reduced acceptor [4, 5]
5 p-Nitrophenyl-3-ketovalienamine + reduced acceptor [4, 5]
6 ?
7 3-Keto-glucosides + reduced acceptor [1]
8 3-Ketogalactosides + reduced acceptor [1]
9 3-Ketocellobiose + reduced acceptor [1, 2, 4, 5]
10 3-Ketolactose + reduced acceptor [1, 2, 4, 5]
11 3-Ketomaltose + reduced acceptor [1, 2]
12 3-Ketotrehalose + reduced acceptor [1, 2, 4, 5]
13 3-Keto-D-glucose 1-phosphate + reduced acceptor [1, 2]
14 ?
15 3-Keto-methyl-alpha-D-glucose + reduced acceptor [2, 4]
16 3-Keto-methyl-beta-D-glucose + reduced acceptor [2, 4]
17 ?
18 ?
19 ?
20 ?
21 ?
22 ?
23 3-Keto-alpha-methyl-glucoside + reduced acceptor [1, 4, 5]
24 3-Keto-beta-methyl-glucoside + reduced acceptor [5]
25 ?
26 ?
27 ?
28 ?
29 ?
30 ?
31 ?
32 ?
33 ?
34 ?
35 ?
36 ?
37 ?
38 ?
39 ?
40 ?

Inhibitor(s)
Ca^{2+} (complete [1], not [2, 4]) [1]; Cu^{2+} (slight [4]) [1, 4]; L-Arabinose (slight) [1]; Cyanide (70% inhibition at high concentration) [2]; Hg^{2+} [4, 5]; Mn^{2+} (slight inhibition [4]) [4, 5]; PCMB (no protection by substrate [5]) [4, 5]; More (not N_3^-, CO, EDTA [2], iodoacetic acid, NEM or DTNB [5]) [2, 5]

Cofactor(s)/prosthetic group(s)/activating agents
FAD (flavoprotein) [1–5]; Atabrine (activation at acidic pH-values) [1]

Metal compounds/salts
Fe^{2+} (non-heme iron sulfur protein) [5]

Turnover number (min^{-1})

Specific activity (U/mg)
2.94 (validoxylamine A) [4]

K_m-value (mM)
0.016 (D-glucose) [1]; 0.038 (sucrose) [1]; 0.12 (D-glucose) [4]; 0.2 (cellobiose) [2]; 0.21 (lactobionate) [2]; 0.36 (maltobionate) [2]; 1.1 (validoxylamine A) [4]; 1.7 (lactose) [2]; 2.8 (maltose) [2]; 2.9 (glucose) [2]; 4.1 (sucrose) [2]; 25 (galactose) [2]

pH-optimum
6.0 (with 2,6-dichlorphenolindophenol as acceptor [4, 5], lactose with 2,6-dichlorphenolindophenol as acceptor [2]) [2, 4, 5]; 6.0–6.7 [1]; 7.0 (lactose with cytochrome c as acceptor [2], ferricytochrome c_{551} [5]) [2, 5]; 8.0 (lactose with phenazine methosulfate as acceptor) [2]; 8–9 (phenazine methosulfate and 2,6-dichlorphenolindophenol as acceptor) [5]; 9.0 (phenazine methosulfate or 3-(4,5-dimethyl-2-thiazolyl)-2,5-di-phenyl-2H-tetrazolium as acceptor) [4]

pH-range
5.2–7.2 (85% of maximal activity at pH 5.2 and 7.2) [1]

Temperature optimum (°C)
20 (assay at) [1]; 30 (assay at) [2, 4]

Temperature range (°C)

3 ENZYME STRUCTURE

Molecular weight
65000 (Agrobacterium tumefaciens, gel filtration [3], Flavobacterium saccharophilum, gel filtration [5]) [3, 5]
73000 (Agrobacterium tumefaciens, gel filtration) [2]
85000 (Agrobacterium tumefaciens, low speed sedimentation) [2]
270000 (Flavobacterium saccharophilum, gel filtration) [4]

Subunits
Monomer (1 × 66000, Flavobacterium saccharophilum, SDS-PAGE [5],
1 × 68000, Agrobacterium tumefaciens, SDS-PAGE [3]) [3, 5]
Tetramer (4 × 66000, Flavobacterium saccharophilum, SDS-PAGE) [4]

Glycoprotein/Lipoprotein
–

4 ISOLATION/PREPARATION

Source organism
Agrobacterium tumefaciens [1–3]; Flavobacterium saccharophilum [4, 5]

Source tissue
Cell [1–5]

Localisation in source
Cytoplasm [1, 2, 5]; Periplasm [3]; Membrane-bound [4]

Purification
Agrobacterium tumefaciens (partially [1]) [1–3]; Flavobacterium saccharophilum [4, 5]

Crystallization
–

Cloned
–

Renaturated
–

5 STABILITY

pH
6.0 (and below, labile, sucrose stabilizes) [1]; 7.0–9.0 (stable) [1]

Temperature (°C)

Oxidation

Organic solvent

General stability information
Sucrose protects against inactivation in acidic media [1]

Storage
–18°C, stable for 2 weeks under nitrogen [2]; Frozen, half-life 1 week [2]

6 CROSSREFERENCES TO STRUCTURE DATABANKS

PIR/MIPS code

Brookhaven code

7 LITERATURE REFERENCES

[1] Hayano, K., Fukui, S.: J. Biol. Chem.,242,3665–3672 (1967)
[2] Van Beeumen, J., DeLey, J.: Methods Enzymol.,41,153–158 (1975) (Review)
[3] Chern, C.K., Fukui, S.: Agric. Biol. Chem.,38,2039–2040 (1974)
[4] Takeuchi, M., Ninomiya, K., Kawabata, K., Asano, N., Kameda, Y., Matsui, K.: J. Biochem.,100,1049–1055 (1986)
[5] Takeuchi, M., Asano, N., Kameda, Y, Matsui, K.: Agric. Biol. Chem.,52,1905–1912 (1988)

1 NOMENCLATURE

EC number
1.1.99.14

Systematic name
Glycolate:(acceptor) 2-oxidoreductase

Recommended name
Glycolate dehydrogenase

Synonymes
Dehydrogenase, glycolate
Glycolate oxidoreductase
Glycolic acid dehydrogenase

CAS Reg. No.
37368-32-4

2 REACTION AND SPECIFICITY

Catalysed reaction
Glycolate + acceptor →
→ glyoxylate + reduced acceptor

Reaction type
Redox reaction

Natural substrates
Glycolate + acceptor (the bacterial enzyme catalyzes phenazine methosulfate dependent reduction of 2,6-dichlorophenolindophenol or cytochrome c [1], in green algae the enzyme is part of glycolate metabolism [4]) [1, 4]

Substrate spectrum
1 Glycolate + acceptor (artificial acceptor: 2,6-dichlorophenolindophenol is reduced via phenazine methosulfate, cytochrome c can replace 2,6-dichlorophenolindophenol [1]) [1, 4]
2 D-Lactate + acceptor (best substrate [1], artificial acceptor: 2,6-dichlorophenolindophenol is reduced via phenazine methosulfate, cytochrome c can replace 2,6-dichlorophenolindophenol [1]) [1, 2]
3 L(+)-Lactate + acceptor [2]
4 Glyoxylate + acceptor (2,6-dichlorophenolindophenol as acceptor, no reaction with O_2 as acceptor) [2]

Enzyme Handbook © Springer-Verlag Berlin Heidelberg 1995
Duplication, reproduction and storage in data banks are only
allowed with the prior permission of the publishers

Product spectrum
1 Glyoxylate + reduced acceptor [1, 4]
2 2-Oxopropionate + reduced acceptor
3 2-Oxopropionate + reduced acceptor
4 Oxalate + reduced acceptor

Inhibitor(s)
$CuSO_4$ [1]; PCMB [1, 2]; KCN [1, 3]; $HgCl_2$ (complete) [1]; $ZnCl_2$ [1]; Cytochrome c (inhibits rate of phenazine methosulfate stimulated dichlorophenolindophenol reduction) [1]; $K_3[Fe(CN)]_6$ (complete) [1]; High ionic strength [1]; NEM [2]; NaN_3 [3]; Aminooxyacetate (competitive to glycolate and glyoxylate) [4]; Glyoxylate (strong, competitive to glycolate) [4]

Cofactor(s)/prosthetic group(s)/activating agents
Phenazine methosulfate (in vitro requirement [2], transfers electrons from the substrate preferentially to 2,6-dichlorophenolindophenol [1]) [1, 2]; 2,6-Dichlorophenolindophenol (in vitro requirement, terminal acceptor) [1, 2]; Cytochrome c (can replace dichlorophenolindophenol as acceptor) [1]; More (FAD, FMN, NAD^+, $NADP^+$, oxidized glutathione, KNO_3, ferricyanide, $FeCl_3$, cytochrome c do not stimulate electron transfer from glycolate to dichlorophenolindophenol) [1]

Metal compounds/salts

Turnover number (min^{-1})

Specific activity (U/mg)
0.009 [2]; 0.042 [3]; 3.1 [1]

K_m-value (mM)
0.04 (glycolate) [1]; 0.16 (glycolate) [4]; 0.7 (D-lactate) [1]

pH-optimum
8.0–8.8 [1]; 8.7 [2]

pH-range
7.3–9.0 (about half-maximal activity at pH 7.3 and 9.0) [1]

Temperature optimum (°C)
30 (assay at) [1]

Temperature range (°C)

3 ENZYME STRUCTURE

Molecular weight

Subunits

Glycoprotein/Lipoprotein

–

4 ISOLATION/PREPARATION

Source organism
E. coli K12 [1]; Thallassiosira pseudonana (Cyclotella nana, Bacillareophy-
ceae, marine diatom) [2]; Dunaliella salina (green algae) [3]; Euglena gracilis Z [4]

Source tissue
Cell [1–4]

Localisation in source
Cytoplasm [1]; Particulate [2]; Thylakoid membrane [3]

Purification
E. coli (partially, from cells grown on glycolate) [1]

Crystallization
–

Cloned
–

Renaturated
–

5 STABILITY

pH

Temperature (°C)

Oxidation

Organic solvent

General stability information

Storage
2°C, half-life of 10 days, FMN or FAD do not restore activity [1]

6 CROSSREFERENCES TO STRUCTURE DATABANKS

PIR/MIPS code

Brookhaven code

7 LITERATURE REFERENCES

[1] Lord, J.M.: Biochim. Biophys. Acta,267,227–237 (1971)
[2] Paul, J.S., Volcani, B.E.: Arch. Microbiol.,101,115–120 (1974)
[3] Sallal, A.K.-J., Al-Hasan, R.H., Nimer, N.A.: Planta,171,429–432 (1987)
[4] Yokota, A. Kitaoka, S.: Agric. Biol. Chem.,51,665–670 (1987)

4

1 NOMENCLATURE

EC number
1.1.99.16

Systematic name
(S)-Malate:(acceptor) oxidoreductase

Recommended name
Malate dehydrogenase (acceptor)

Synonymes
Dehydrogenase, malate (acceptor)
FAD-dependent malate-vitamin K reductase [1]
Malate-vitamin K reductase [2]

CAS Reg. No.
71822-24-7

---

2 REACTION AND SPECIFICITY

Catalysed reaction
(S)-Malate + acceptor →
→ oxaloacetate + reduced acceptor

Reaction type
Redox reaction

Natural substrates

Substrate spectrum
1 (S)-Malate + acceptor [1–4]

Product spectrum
1 Oxaloacetate + reduced acceptor [1–4]

Inhibitor(s)

Cofactor(s)/prosthetic group(s)/activating agents
FAD (flavoprotein, FAD required as cofactor) [1, 2, 4]; Phospholipid (required) [1, 2, 4]; Vitamin K (required) [2]; More (di(triethyleneglycol-n-tetra-decylether)phosphate and natural bovine heart cardiolipin activate lipid-depleted inactive enzyme) [4]

Metal compounds/salts

Turnover number (min⁻¹)

Specific activity (U/mg)
52.0 [2]

K_m-value (mM)

pH-optimum

pH-range

Temperature optimum (°C)

Temperature range (°C)

3 ENZYME STRUCTURE

Molecular weight
51000–53000 (Mycobacterium phlei, gel filtration, SDS-PAGE, monomeric
form, at high salt concentrations the enzyme exists as monomer, which is
more active than aggregated form) [2]
164000 (Mycobacterium phlei, gel filtration, aggregated form, at high salt
concentrations the enzyme exists as monomer, which is more active than
aggregated form) [2]

Subunits
Monomer (1 × 51000, Mycobacterium phlei, SDS-PAGE, at high salt concentrations the enzyme exists as monomer, which is more active than aggregated form) [2]
Trimer (3 × 51000, Mycobacterium phlei, aggregated form, at high salt concentrations the enzyme exists as monomer, which is more active than aggregated form) [2]

Glycoprotein/Lipoprotein
–

4 ISOLATION/PREPARATION

Source organism
Mycobacterium smegmatis [1, 4]; Mycobacterium phlei [1, 2]; Rhodococcus
sp. (An 117) [3]

Source tissue

Localisation in source

Purification
Mycobacterium phlei (partial) [2]

Crystallization

–

Cloned

–

Renaturated

–

5 STABILITY

pH

Temperature (°C)

Oxidation

Organic solvent

General stability information
Glycerol, 20%, stabilizes [2]; KCl, 0.35 M, stabilizes [2]

Storage

6 CROSSREFERENCES TO STRUCTURE DATABANKS

PIR/MIPS code

Brookhaven code

7 LITERATURE REFERENCES

[1] Reddy, T.L.P., Murthy, P.S., Venkitasubramanian, T.A.: Biochim. Biophys. Acta,376,210–218 (1975)
[2] Imai, K., Brodie, A.F.: J. Biol. Chem.,248,7487–7494 (1973)
[3] Janke, D., Maltseva, O.V., Golovleva, L.A., Fritsche, W.: Z. Allg. Mikrobiol.,24, 305–316 (1984)
[4] Imai, T.: Biochem. Int.,19,1277–1286 (1989)

1 NOMENCLATURE

EC number
1.1.99.17

Systematic name
D-Glucose:(pyrroloquinoline-quinone) 1-oxidoreductase

Recommended name
Glucose dehydrogenase (pyrroloquinoline-quinone)

Synonymes
Dehydrogenase, glucose (pyrroloquinoline-quinone)
Glucose dehydrogenase (PQQ dependent)
Quinoprotein D-glucose dehydrogenase
Quinoprotein glucose dehydrogenase

CAS Reg. No.
81669-60-5

2 REACTION AND SPECIFICITY

Catalysed reaction
D-Glucose + acceptor →
→ D-glucono-1,5-lactone + reduced acceptor (hexa uni ping-pong mechanism [3])

Reaction type
Redox reaction

Natural substrates
More (enzyme is linked to the respiratory chain of a wide variety of bacteria [9], membrane-bound enzyme functions by linking to the respiratory chain via ubiquinone, the function of the soluble enzyme remains unclear [12]) [9, 12]

Substrate spectrum
1 D-Glucose + acceptor
(4,5-dihydro-4,5-dioxo-1H-pyrrolo[2,3-f]-quinoline-2,7,9-tricarboxylic acid i.e. pyrroloquinoline-quinone, methoxatin (primary electron aceptor), acceptor: tetramethyl-p-phenylenediamine [1, 3, 4, 14], 2,6-dichlorophenolindophenol [4, 14, 16], coenzyme Q (not coenzyme Q_1 [16]) [14], ubiquinone (short-chain and long-chain) [15], phenazine methosulfate [16], potassium ferricyanide [16], cationic acceptors are active, anionic acceptors not [4], not: NAD^+, $NADP^+$, O_2 [16], broad specificity [4, 14] for aldose sugars [4], high specificity for glucose [16]) [1–16]

2 D-Xylose + acceptor (activity relative to glucose oxidation: 20% [4], 13%
   [14], 15% [12]) [4, 8, 12, 14]
3 L-Arabinose + acceptor (activity relative to glucose oxidation: 35% [4],
   2.8% [14]) [4, 8, 14]
4 Lactose + acceptor (activity relative to glucose oxidation: 65% [4], 72%
   [12]) [4, 8, 12]
5 Galactose + acceptor (activity relative to glucose oxidation: 30% [4, 12],
   6.5% [14]) [4, 8, 12, 14]
6 D-Melibiose + acceptor (activity relative to glucose oxidation: 10%) [4]
7 D-Allose + acceptor [8]
8 2-Deoxy-D-glucose + acceptor [8]
9 D-Mannose + acceptor (activity relative to glucose oxidation: 8.6% [14])
   [8, 14]
10 D-Ribose + acceptor (activity relative to glucose oxidation: 8%) [12]
11 D-Maltose + acceptor (activity relative to glucose oxidation: 93% [12],
    3.2% [14]) [12, 14]
12 D-Cellobiose + acceptor [8]
13 D-Fucose + acceptor [12]
14 Rhamnose + acceptor (activity relative to glucose oxidation: 7.5%) [14]

Product spectrum
   1 D-Glucono-1,5-lactone + reduced acceptor
   2 Xylono-1,5-lactone + reduced acceptor
   3 Arabinono-1,5-lactone + reduced acceptor
   4 ?
   5 Galactono-1,5-lactone + reduced acceptor
   6 ?
   7 ?
   8 ?
   9 ?
  10 ?
  11 ?
  12 ?
  13 ?
  14 ?

Inhibitor(s)
   Glucose (substrate inhibition at high concentration) [3];
   Tetramethyl-p-phenylenediamine (substrate inhibition at high concentration)
   [3]; o-Benzoquinone [14]; EDTA [14]

Cofactor(s)/prosthetic group(s)/activating agents

Pyrroloquinoline quinone (i.e. 4,5-dihydro-4,5-dioxo-1H-pyrrolo[2,3-f]-quinoline-2,7,9-tricarboxylic acid, PQQ, prosthetic group, 1 PQQ per subunit [3], 2 PQQ per molecule [4], 2 types of quinoprotein glucose dehydrogenase, type I: acidic proteins from which PQQ can be removed by dialysis against EDTA-containing buffers (E. coli, Klebsiella aerogenes, Pseudomonas sp.), type II: basic enzymes from which PQQ is not removed by dialysis against EDTA-containing buffers (Acinetobacter calcoaceticus, Gluconobacter oxydans [4])) [1–16]

Metal compounds/salts

Ca^{2+} (pyrroloquinoline quinone bound at the active site via a Ca^{2+} bridge) [11]

Turnover number (min^{-1})

More [10]

Specific activity (U/mg)

More [3, 4, 8, 10, 12, 14, 16]

K_m-value (mM)

0.12 (tetramethyl-p-phenylenediamine) [3]; 0.78 (tetramethyl-p-phenylene-diamine) [4]; 1.5 (D-allose) [8]; 1.7 (D-glucose) [8]; 2.7 (cellobiose) [8]; 3.2 (maltose) [8]; 3.3 (glucose) [3]; 3.5 (D-galactose) [8]; 4.2 (D-glucose, membrane-bound enzyme [10, 12], lactose [8]) [8, 10, 12]; 4.8 (L-arabinose disaccharides) [8]; 5.5 (D-xylose) [8]; 13.6 (2-deoxy-D-glucose) [8]; 19.0 (D-mannose) [8]; 22 (glucose) [4]; 24.5 (glucose, soluble enzyme) [10, 12]; 26.7 (lactose, soluble enzyme) [10, 12]; 40.0 (D-ribose) [8]

pH-optimum

3.0 (potassium ferricyanide) [16]; 6.0 (2,6-dichlorophenolindophenol [4, 16], phenazine methosulfate, tetramethyl-p-phenylenediamine [16]) [4, 16]; 6.5 (phenazine methosulfate, 2,6-dichlorophenolindophenol, soluble enzyme) [10]; 8.5 (phenazine methosulfate, 2,6-dichlorophenolindophenol, membrane-bound enzyme) [10]; 8.8 (tetramethyl-p-phenylenediamine) [14]; 9.0 (tetramethyl-p-phenylenediamine) [4]; More (acidic optimum with 2,6-dichlorophenolindophenol or ferricyanide) [14]

pH-range

Temperature optimum (°C)

25 (assay at) [3]

Temperature range (°C)

3 ENZYME STRUCTURE

Molecular weight
87000 (Gluconobacter suboxydans, urea SDS-PAGE, sedimentation data)
[16]
87000–93000 (Pseudomonas fluorescens, urea SDS-PAGE, sucrose density
gradient centrifugation) [14]
94000 (Acinetobacter calcoaceticus, gel filtration) [4]
110000 (Acinetobacter calcoaceticus, sedimentation-equilibrium centrifugation) [3]
More (MW of enzyme from E. coli and Klebsiella pneumoniae (88000) is higher than that of the enzymes in Gluconobacter suboxydans, Acetobacter
aceti, Pseudomonas aeruginosa and Acinetobacter calcoaceticus (83000).
Both enzymes of MW 88000 are apoenzymes, MW 83000 enzymes are ho-
loenzymes, apo-type GDH may have a molecular weight greater than that of
holo-type GDH.) [13]

Subunits
Monomer (1 × 83000, Acinetobacter calcoaceticus, soluble enzyme,
SDS-PAGE [10, 12], 1 × 87000, Pseudomonas fluorescens, urea SDS-PAGE
[14], Gluconobacter suboxydans, urea SDS-PAGE [16]) [10, 12, 14, 16]
Dimer (2 × 54000, Acinetobacter calcoaceticus, SDS-PAGE [3], 2 × 48000,
Acinetobacter calcoaceticus, SDS-PAGE [4]) [3, 4]

Glycoprotein/Lipoprotein
–

4 ISOLATION/PREPARATION

Source organism
Acinetobacter lwoffii [7]; Azotobacter vinelandii [7]; Agrobacterium sp. [7];
Rhizobium sp. [7]; Pseudomonas fluorescens [14, 15]; Acinetobacter calcoaceticus (LMD 79.41 [4, 12]) [1, 3, 4, 8, 10–13]; Pseudomonas aeruginosa
[2, 13]; E. coli [5, 13]; Klebsiella pneumoniae [6, 13]; Gluconobacter suboxydans [9, 13, 16]; Acetobacter aceti [13]

Source tissue

Localisation in source
Soluble [3, 10, 12]; Membrane-bound (outer surface of cytoplasmic membrane of E. coli [13], surface of cytoplasmic membrane [14]) [9, 10, 12–16];
More (Acinetobacter calcoaceticus contains a soluble and a
membrane-bound enzyme, Pseudomonas and Gluconobacter contain only
membrane-bound enzyme) [12]

Purification
Pseudomonas fluorescens [14, 15]; Acinetobacter calcoaceticus (membrane-bound and soluble form [10, 12], partial [8]) [3, 4, 8, 10, 12]; Gluconobacter suboxydans [9, 16]

Crystallization
[3]

Cloned
[5]

Renaturated
−

5 STABILITY

pH
More [16]

Temperature (°C)
35 (reversible inactivation above 35°C, Ca^{2+} necessary for reactivation) [11]

Oxidation

Organic solvent

General stability information

Storage
−80°C [3]

6 CROSSREFERENCES TO STRUCTURE DATABANKS

PIR/MIPS code
PIR2:S00943 (Acinetobacter calcoaceticus); PIR2:JV0107 (Escherichia coli); PIR2:A45997 (Escherichia coli (fragment)); PIR1:QPKEX (Gluconobacter oxydans); PIR2:S04784 (precursor Acinetobacter calcoaceticus)

Brookhaven code

7 LITERATURE REFERENCES

[1] Duine, J.A., Frank, J., van Zeeland, J.K.: FEBS Lett.,108,443–446 (1979)
[2] Duine, J.A., Frank, J., Jongejan, J.A.: Anal. Biochem.,133,239–243 (1983)
[3] Geiger, O., Görisch, H.: Biochemistry,25,6043–6048 (1986)
[4] Dokter, P., Frank, J., Duine, J.A.: Biochem. J.,239,163–167 (1986)
[5] Cleton-Jansen, A.-M., Goosen, N., Fayet, O., van de Putte, P.: J. Bacteriol.,172, 6308–6315 (1990)
[6] Hommes, R.W.J., Herman, P.T.D., Postma, P.W., Tempest, D.W., Neijssel, O.M.: Arch. Microbiol.,151,257–260 (1989)
[7] van Schie, B.J., de Mooy, O.H., Linton, J.D., van Dijken, J.P., Kuenen, J.G.: J. Gen. Microbiol.,133,867–875 (1987)
[8] Dokter, P., Pronk, J.T., van Schie, B.J., van Dijken, J.P., Duine, J.A.: FEMS Microbiol. Lett.,43,195–200 (1987)
[9] Matsushita, K., Shinagawa, E., Adachi, O., Amiyama, M. : J. Biochem.,105,633–637 (1989)
[10] Matsushita, K., Shinagawa, E., Adachi, O., Ameyama, M.: Biochemistry,28, 6276–6280 (1989)
[11] Geiger, O., Görisch, H.: Biochem. J.,261,415–421 (198)
[12] Matsushita, K., Shinagawa, E., Adachi, O., Ameyama, M.: Antonie Leeuwenhoek, 56,63–72 (1989)
[13] Matsushita, K., Shinagawa, E., Inoue, T., Adachi, O., Ameyama, M.: FEMS Microbiol. Lett.,37,141–144 (1986)
[14] Matsushita, K., Ameyama, M.: Methods Enzymol.,89,149–154 (1982) (Review)
[15] Matsushita, K., Ohno, Y., Shinagawa, E., Adachi, O., Ameyama, M.: Agric. Biol. Chem.,46,1007–1011 (1982)
[16] Ameyama, M., Shinagawa, E., Matsushita, K., Adachi, M.: Agric. Biol. Chem.,45,851–861 (1981)

1 NOMENCLATURE

EC number
1.1.99.18

Systematic name
Cellobiose:(acceptor) 1-oxidoreductase

Recommended name
Cellobiose dehydrogenase (acceptor)

Synonymes
Dehydrogenase, cellobiose
Cellobiose dehydrogenase

CAS Reg. No.
54576-85-1

2 REACTION AND SPECIFICITY

Catalysed reaction
Cellobiose + acceptor →
→ cellobiono-1,5-lactone + reduced acceptor

Reaction type
Redox reaction

Natural substrates
Cellobiose + acceptor (may play an important role in the degradation of cellulose and lignin by wood-rotting fungi) [1]

Substrate spectrum
1 Cellobiose + acceptor (electron acceptor: only those compounds having a redox potential of 0.22 V (2,6-dichlorophenolindophenol, phenol blue, cytochrome c) [1–5] Fe^{3+} [5], methylene blue [5], potassium ferricyanide (at a low rate [3]) [1–3, 5], 3,5-di-tert-butyl-1,2-benzoquinone [5], p-benzoquinone [5]) [1–7]
2 Lactose + acceptor (60% the rate of cellobiose oxidation [1, 2]) [1–3, 5]
3 4-beta-Glucosylmannose + acceptor (47% the rate of cellobiose oxidation) [1, 2, 5]
4 4-Methylumbelliferyl-beta-D-cellobioside + acceptor (70% the rate of cellobiose oxidation) [1, 2]

Enzyme Handbook © Springer-Verlag Berlin Heidelberg 1995
Duplication, reproduction and storage in data banks are only
allowed with the prior permission of the publishers

5 Cello-oligosaccharides + acceptor (67% the rate of cellobiose oxidation)
   [1, 2]
6 Cellodextrin + acceptor (cellotriose + cellohexaose [3]) [3, 5]
7 Xylobiose + acceptor (38% the rate of cellobiose oxidation) [5]

Product spectrum
1 Cellobiono-1,5-lactone + reduced acceptor
2 4-O-(beta-D-Galactopyranosyl)-D-glucono-1,5-lactone + reduced acceptor
3 ?
4 4-Methylumbelliferyl-beta-D-cellobiono-1,5-lactone + reduced acceptor
5 ?
6 ?
7 ?

Inhibitor(s)

Cofactor(s)/prosthetic group(s)/activating agents

Metal compounds/salts
   Mg^{2+} (stimulates activity up to 2-fold) [5]

Turnover number (min^{-1})

Specific activity (U/mg)
   More [3, 4]

K_m-value (mM)
   0.012 (cellobiose) [1, 2]; 0.10 (cellohexaose) [3]; 0.08 (2,6-dichlorophenolin-
   dophenol) [1, 2]; 0.0666 (cellotetraose) [3]; 0.083 (cellopentaose) [3];
   0.0416 (cellobiose) [3]; 0.0625 (cellotriose) [3]; 0.220 (2,6-dichlorophenolin-
   dophenol) [3]; 0.384 (cytochrome c) [3]; 0.0069 (cellobiose) [5]

pH-optimum
   3.5–4.5 (acceptor: methylene blue) [5]; 4.0–4.5 [1, 2]; 5.0–6.0 (acceptor:
   2,6-dichlorophenolindophenol) [5]; 7 (3,5-di-tert-butyl-1,2-benzoquinone) [5];
   8–9 (phenazine methosulfate) [5]

pH-range

Temperature optimum (°C)
   37 [3, 4]

Temperature range (°C)

3 ENZYME STRUCTURE

Molecular weight
48000 (Monilia sitophila, gel filtration) [1, 2]
63100 (Sclerotium rolfsii, gel permeation) [3]

Subunits
Monomer (1 x 64500, Sclerotium rolfsii, SDS-PAGE) [3]

Glycoprotein/Lipoprotein
Glycoprotein (8.9% carbohydrate) [3, 4]

---

4 ISOLATION/PREPARATION

Source organism
Sclerotium rolfsii [3, 4]; Sporotrichum thermophile [5–7]; Monilia sitophila (inducible enzyme) [1, 2]; Trichoderma reesei [2]

Source tissue
Culture medium (bound to mycelium from which it is released into the extracellular medium) [1–3, 5, 6]; Culture filtrate [3, 5]

Localisation in source
Extracellular (bound to mycelium from which it is released into the extracellular medium) [1–3, 5, 6]

Purification
Sclerotium rolfsii [3, 4]; Sporotrichum thermophile (enrichment) [7]

Crystallization
–

Cloned
–

Renaturated
–

---

5 STABILITY

pH
4.5–5.0 (most stable at) [3, 4]

Temperature (°C)
70 (10 min, 86% loss of activity) [3]; 100 (2 min, complete loss of activity) [3]

Oxidation

Organic solvent

General stability information
Stable to repeated freezing and thawing [3, 4]

Storage
-15°C, pH 4.5 [3, 4]

---

6 CROSSREFERENCES TO STRUCTURE DATABANKS

PIR/MIPS code

Brookhaven code

---

7 LITERATURE REFERENCES

[1] Dekker, R.F.H.: J. Gen. Microbiol.,120,309–316 (1980)
[2] Dekker, R.F.H.: Methods Enzymol.,160,454–463 (1988) (Review)
[3] Sadana, J.C., Patil, R.V.: J. Gen. Microbiol.,131,1917–1923 (1985)
[4] Sadana, J.C., Patil, R.V.: Methods Enzymol.,160,448–454 (1988) (Review)
[5] Coudray, M.-R., Canevascini, G., Meier, H.: Biochem. J.,203,277–284 (1982) (Review)
[6] Cossar, D., Canevascini, G.: Appl. Microbiol., Biotechnol.,24,306–310 (1986)
[7] Canevascini, G.: Methods Enzymol.,160,443–448 (1988) (Review)

1 NOMENCLATURE

EC number
1.1.99.19

Systematic name
Uracil:(acceptor) oxidoreductase

Recommended name
Uracil dehydrogenase

Synonymes
Uracil oxidase
Dehydrogenase, uracil
EC 1.2.99.1 (formerly)

CAS Reg. No.
9029-00-9

---

2 REACTION AND SPECIFICITY

Catalysed reaction
Uracil + acceptor →
→ barbiturate + reduced acceptor

Reaction type
Redox reaction

Natural substrates
Uracil + acceptor (reaction in two-step oxidative catabolism of uracil in pro-
karyotes [1]) [1–3]

Substrate spectrum
1 Uracil + acceptor (acceptor: 2,6-dichlorophenolindophenol or methylene
   blue act as artificial electron acceptors, O_2 only in the presence of methylene blue [2]) [1–3]
2 Thymine + acceptor [2, 3]
3 5-Aminouracil + acceptor [2]
4 More (barbiturate, isobarbiturate, 5-methylbarbiturate, 6-methyluracil,
   dihydrothymine, dihydrouracil, 2-thiouracil, 2-thiothymine, 2-thio-5-methyl-
   uracil and cytosine are no substrates) [1–3]

Product spectrum
1 Barbiturate + reduced acceptor [1–3]
2 5-Methylbarbiturate + reduced acceptor
3 ?
4 ?

Inhibitor(s)

Cofactor(s)/prosthetic group(s)/activating agents
2,6-Dichlorophenolindophenol (required as electron acceptor, promotes barbiturate formation better than methylene blue) [1]; Methylene blue (required
as electron acceptor) [1–3]; More (no oxidation in the presence of O_2 without methylene blue as electron acceptor) [2, 3]

Metal compounds/salts

Turnover number (min^{-1})

Specific activity (U/mg)
0.0115 [1]

K_m-value (mM)
0.035 (thymine) [3]; 0.131 (uracil) [3]

pH-optimum
8.5 [3]

pH-range
8.0–9.7 (about half-maximal activity at pH 8.0 and 9.7 with uracil as substrate) [3]; 8.0–10.0 (about half-maximal activity at pH 8.0 and 10 with thymine
as substrate) [3]

Temperature optimum (°C)
22–25 (assay at) [3]; 30 (assay at) [1, 2]

Temperature range (°C)

3 ENZYME STRUCTURE

Molecular weight

Subunits

Glycoprotein/Lipoprotein
–

4 ISOLATION/PREPARATION

Source organism
  Enterobacter aerogenes [1]; Aerobic soil bacterium (strain U-1) [2]; Corynebacterium sp. (strain 161) [3]; Mycobacterium sp. [3]

Source tissue
  Cell [1–3]

Localisation in source
  Cytoplasm [1–3]

Purification

Crystallization
  –

Cloned
  –

Renaturated
  –

5 STABILITY

pH
  6–7 (90% loss of activity) [3]; 9.0 (most stable at –10°C, no appreciable loss
  of activity for several months) [3]

Temperature (°C)

Oxidation

Organic solvent

General stability information

Storage
  –20°C, 80% loss of activity in crude extracts within two months [2]; –10°C,
  most stable at pH 9.0, no appreciable loss of activity for several months [3]

6 CROSSREFERENCES TO STRUCTURE DATABANKS

PIR/MIPS code

Brookhaven code

7 LITERATURE REFERENCES

[1] Bharat, N.P., West, T.P.: FEMS Microbiol. Lett.,40,33–36 (1987)
[2] Wang, T.P., Lampen, J.O.: J. Biol. Chem.,194,785–791 (1952)
[3] Hayaishi, O., Kornberg, A.: J. Biol. Chem.,197,717–732 (1952)

1 NOMENCLATURE

EC number
1.1.99.20

Systematic name
Alkan-1-ol:(acceptor) oxidoreductase

Recommended name
Alkan-1-ol dehydrogenase (acceptor)

Synonymes
Polyethylene glycol dehydrogenase
Dehydrogenase, polyethylene glycol

CAS Reg. No.
75496-55-8

2 REACTION AND SPECIFICITY

Catalysed reaction
Primary alcohol + acceptor →
→ aldehyde + reduced acceptor

Reaction type
Redox reaction

Natural substrates
Polyethylene glycol + acceptor (first and main step in aerobic PEG-metabolism [3, 4], constitutive quinoprotein enzyme, enzyme of Flavobacterium sp. No.203 inductive [4], ubiquinone possible physiological acceptor [1]) [1–6]

Substrate spectrum
1 Primary alcohol + acceptor (overview: C-2 to C-16 [1], C-3 to C-12 [5], methanol, glycerol, secondary alcohols and sugar alcohols are no substrates, ethanol is only a poor one [3], 2,6-dichlorophenolindophenol is the acceptor in vitro [1–6], coenzyme Q_1 and Q_2 can act as acceptors [1, 5]) [1–6]
2 1-Pentenol + acceptor [3, 5]
3 Aldehyde + acceptor (products of enzymatic reaction from corresponding alcohols, overview: C-4 to C-7 [1], C-3 to C-7 [5]) [1, 5]
4 Aromatic alcohols + acceptor (such as: benzyl alcohol, beta-phenethyl alcohol, cinnamyl alcohol, vannilyl alcohol) [5]

5 Diols + acceptor (overview: with primary alcoholic groups [5], except
  ethylene glycol, 1,2- and 1,3-propenediol [3, 5], diethylene glycol and
  triethylene glycol are only poor substrates [3, 5]) [3, 5]
6 Tetraethylene glycol + acceptor [2, 3, 5]
7 Monoalkylethers + acceptor (such as ethylene glycol monoethylether,
  ethyleneglycolmono-n-butylether, ethyleneglycol monophenylether, di-
  ethyleneglycol monoethylether) [5]
8 Polyethylene glycols + acceptor (various MW: overview: PEG-200 [5],
  PEG-300 [5], PEG-400, PEG-1000, PEG-4000, PEG-6000: most active
  substrate of all PEGs [1], PEG-20000) [1–3, 5]
9 Aromatic aldehydes + acceptor (such as benzaldehyde, trans-cin-
  namylaldehyde, vanillin, 2-carboxybenzaldehyde) [5]
10 Detergents + acceptor (overview: Triton X-100, Tween 40, Tween 60, Brij
   35, Epan 410 to 785) [1]

Product spectrum
  1 Aldehyde + reduced acceptor [1, 5]
  2 1-Pentanal + reduced acceptor [3, 5]
  3 ?
  4 ?
  5 ?
  6 Tetraethylene glycol aldehyde + reduced acceptor [2]
  7 ?
  8 ?
  9 ?
 10 ?

Inhibitor(s)
  Semicarbazide (weak [5]) [1, 5]; Hydrazine (weak [5]) [1, 5]; Hydroxylamine
  (weak [5]) [1, 5]; 1,4-Benzoquinone (competitive to 2,6-dichlorophenolindo-
  phenol [1, 4, 5], strong [4, 5]) [1, 4, 5]; PCMB (no inhibition with 2-mer-
  captoethanol [3]) [5]; Monoiodoacetate (weak [5]) [3, 5]; Arsenite [5]; Hg^{2+}
  (competitive [5]) [3, 5]; Cd^{2+} [3, 5]; Cu^{2+} (competitive [5]) [3, 5]; Co^{2+} [3, 5];
  Ni^{2+} [3, 5]; Mn^{2+} [3, 5]; Fe^{3+} [3, 5]; Pb^{2+} [3]; Zn^{2+} [3]

Cofactor(s)/prosthetic group(s)/activating agents
  Pyrroloquinoline quinone (not covalently bound [1]) [1, 5]; 2,6-Dichlorophe-
  nolindophenol (electron acceptor) [1–6]; Ubiquinone (in vivo electron acceptor) [1]; Coenzyme Q_1 (electron acceptor) [1, 5]; Coenzyme Q_2 (electron
  acceptor) [1, 5]; Phenazine methosulfate (activation, accelerates dichlorophenolindophenol reduction [4], mediator to dichlorophenolindophenol reduction [6], has no effect on reduction of dichlorophenolindophenol [1]) [4,
  6]; Methylamine (activation, Pseudomonas sp. strain 43) [4]; Ethylamine (activation, Pseudomonas sp. strain 43) [4]; More (phenazine methosulfate, ferricyanide [1, 5], nitroblue tetrazolium [1, 5],

3-(4,5-dimethyl-2-thiazolyl)-2,5-diphenyl-2H-tetrazolium bromide, NADP+ [5], NAD+ [5], coenzyme Q_6 and Q_{10} [5] or cytochrome c [1, 5] do not serve as electron acceptors) [1]

Metal compounds/salts
Ca^{2+} (activation, Pseudomonas sp. strain 101) [4]; More (no metal ion requirement) [3, 4]

Turnover number (min⁻¹)

Specific activity (U/mg)
0.001 (PEG-6000, Flavobacterium) [6]; 0.003 (PEG-400, Flavobacterium) [6]; 0.004 (valeraldehyde, mixed culture) [6]; 0.023 (PEG-6000, mixed culture) [6]; 0.028 (Pseudomonas sp. No.43) [4]; 0.029 (valeraldehyde, Flavobacterium) [6]; 0.035 (Flavobacterium sp. No.203, crude extract) [4]; 0.156 (valeraldehyde, mixed culture) [6]; 0.164 (Pseudomonas sp. No.103) [4]; 1.2 (mixed culture [4]) [3, 4]; 2.59 [5]; 2.68 (Flavobacterium sp. No.203, purified enzyme) [4]

K_m-value (mM)
0.14 (PEG-600 (+ 2,6-dichlorophenolindophenol) [1], n-butyraldehyde (+ 2,6-dichlorophenolindophenol) [1]) [1]; 1.0 (PEG-400) [5]; 1.7 (PEG-1000) [5]; 2.8 (PEG-4000) [5]; 3.0 (tetraethylene glycol) [3]; 5.9 (PEG-6000) [5]; 10.0 (PEG-6000) [3]

pH-optimum
7.5–8.0 [5]; 8.0 [3]

pH-range
6.5–8.5 (about 75% of maximal activity at pH 6.5 and 8.5) [5]; 7.0–9.0 (about half-maximal activity at pH 7.0 and 9.0) [3]

Temperature optimum (°C)
40 [5]; 60 [3]

Temperature range (°C)
18–50 (about half-maximal activity at 18°C and 50°C) [5]; 35–70 (66% of maximal activity at 35°C and 70°C) [3]

3 ENZYME STRUCTURE

Molecular weight
220000 (Flavobacterium sp. No.203, gel filtration) [5]
240000 (mixed culture of Flavobacterium sp. and Pseudomonas sp., gel filtration) [3]

Subunits
Tetramer (4 × 57000, Flavobacterium sp. No.203, SDS-PAGE [5], 4 × 60000, mixed culture of Flavobacterium sp. and Pseudomonas sp., SDS-PAGE [3]) [3, 5]

Glycoprotein/Lipoprotein
–

---

4 ISOLATION/PREPARATION

Source organism
Flavobacterium sp. (synergistic mixed culture with Pseudomonas sp. [1–4, 6], strains No.202 to 205 [4, 5]) [1–6]; Pseudomonas sp. (synergistic mixed culture with Flavobacterium sp. [1–4, 6], strains No.42, 43, 101, 103 [4]) [1–4, 6]

Source tissue
Cell [1–6]

Localisation in source
Periplasmic space [1]; Membranes [4, 5]

Purification
Flavobacterium sp. (mixed culture with Pseudomonas sp. [1, 3]) [1, 3, 5]; Pseudomonas sp. (mixed culture with Flavobacterium sp.) [1, 3]

Crystallization
–

Cloned
–

Renaturated
–

---

5 STABILITY

pH
7.0 (40% loss of activity at 5°C after 12 h) [3]; 7.0–8.0 (stable) [5]; 7.5–9.0 (stable at 5°C for 12 h) [3]; 10.0 (40% loss of activity at 5°C after 12 h) [3]

Temperature (°C)
35 (stable below) [3]; 40 (stable below) [5]; 60 (94% loss of activity after 15 min at pH 8.0) [3]

Oxidation

Organic solvent

General stability information

Storage

6 CROSSREFERENCES TO STRUCTURE DATABANKS

PIR/MIPS code

Brookhaven code

7 LITERATURE REFERENCES

[1] Kawai, F., Yamanaka, H., Ameyama, M., Shinagawa, E., Matsushita, K., Adachi, O.: Agric. Biol. Chem.,49,1071–1076 (1985)
[2] Kawai, F., Kimura, T., Tani, Y., Yamada, H., Ueno, T., Fukami, H.: Agric. Biol. Chem.,47,1669–1671 (1983)
[3] Kawai, F., Kimura, T., Yoshiki, T., Yamada, H., Mamoru, K.: Appl. Environ. Microbiol., 40,701–705 (1980)
[4] Yamanaka, H., Kawai, F.: J. Ferment. Bioeng.,67,300–302 (1989)
[5] Yamanaka, H., Kawai, F.: J. Ferment. Bioeng.,67,324–330 (1989)
[6] Kawai, F., Yamanaka, H.: Arch. Microbiol.,146,125–129 (1986)

1 NOMENCLATURE

EC number
1.1.99.21

Systematic name
D-Sorbitol:(acceptor) 1-oxidoreductase

Recommended name
D-Sorbitol dehydrogenase (acceptor)

Synonymes

CAS Reg. No.
9028-22-2 (EC 1.1.99.21 is undistinguishable from EC 1.1.1.15 in Chemical Abstracts)

2 REACTION AND SPECIFICITY

Catalysed reaction
D-Sorbitol + acceptor →
→ L-sorbose + reduced acceptor

Reaction type
Redox reaction

Natural substrates
D-Sorbitol + acceptor (high specificity) [1, 2]

Substrate spectrum
1 D-Sorbitol + acceptor (high specificity, the following dyes act in vitro as acceptors: 2,6-dichlorophenolindophenol/phenazine methosulfate, potassium ferricyanide, nitro blue tetrazolium or tetramethyl-p-phenylenediamine) [1, 2]
2 D-Mannitol + acceptor (oxidation at 5% the rate of D-sorbitol oxidation) [1, 2]
3 More (no oxidation of D-arabitol, L-iditol, meso-erythritol, galactitol (dulcitol), ribitol, xylitol) [1, 2]

Product spectrum
1 L-Sorbose + reduced acceptor [1, 2]
2 ?
3 ?

Inhibitor(s)

Cofactor(s)/prosthetic group(s)/activating agents
FAD (flavoprotein, covalently bound, 0.4 mol/mol enzyme) [1, 2]; Cytochrome c (tightly bound [1]: dehydrogenase-cytochrome c complex, separable by SDS-PAGE, rapid reduction in the presence of coenzyme Q_1) [1, 2]; 2,6-Dichlorophenolindophenol/phenazine methosulfate (in vitro acceptor) [1, 2]; Potassium ferricyanide (in vitro acceptor) [1, 2]; Nitro blue tetrazolium (in vitro acceptor) [1, 2]; Tetramethyl-p-phenylenediamine (in vitro acceptor) [1, 2]; More (NAD$^+$, NADP$^+$ or molecular oxygen are completely inactive) [1, 2]

Metal compounds/salts
KCl (activation) [1]

Turnover number (min^{-1})

Specific activity (U/mg)
433 [1, 2]

K_m-value (mM)
30.0 (D-sorbitol) [1, 2]

pH-optimum
4.5 [1, 2]

pH-range

Temperature optimum (°C)
25 [1, 2]

Temperature range (°C)

3 ENZYME STRUCTURE

Molecular weight
131000 (Gluconobacter suboxydans, calculated sum of each MW of the 3 subunits) [1, 2]

Subunits
More (Gluconobacter suboxydans, SDS-PAGE: dissociation into 3 components of MW 63000 (flavoprotein), 51000 (cytochrome c), 17000 (unknown polypeptide)) [1, 2]

Glycoprotein/Lipoprotein
−

4 ISOLATION/PREPARATION

Source organism
Gluconobacter suboxydans var. alpha [1, 2]

Source tissue
Cell [1, 2]

Localisation in source
Cytoplasmic membrane (outer surface) [1, 2]

Purification
Gluconobacter suboxydans [1, 2]

Crystallization
--

Cloned
--

Renaturated
--

5 STABILITY

pH

Temperature (°C)

Oxidation

Organic solvent

General stability information
D-Sorbitol stabilizes during solubilization and purification [1]; Ionic detergents, such as cetylpyridinium chloride, cetyltrimethyl ammonium bromide and sodium stearate, inactivate [1]; Triton X-100, best medium for solubilization [1]

Storage
5°C, 50% loss of activity overnight [1]; 5°C, storage for several months leads to cytochrome c decomposition [1]

6 CROSSREFERENCES TO STRUCTURE DATABANKS

PIR/MIPS code

Brookhaven code

7 LITERATURE REFERENCES

[1] Shinagawa, E., Matsushita, K., Adachi, O., Ameyama, M.: Agric. Biol.
Chem.,46,135–141 (1982)
[2] Shinagawa, E., Ameyama, M.: Methods Enzymol.,89,141–145 (1982) (Review)

1 NOMENCLATURE

EC number
1.1.99.22

Systematic name
Glycerol:(acceptor) 1-oxidoreductase

Recommended name
Glycerol dehydrogenase (acceptor)

Synonymes
More (could be identical with non-purified EC 1.1.2.2 or D-arabitol dehydrogenase [1])

CAS Reg. No.
9028-14-2 (EC 1.1.99.22 is undistinguishable from EC 1.1.1.6 in Chemical Abstracts)

2 REACTION AND SPECIFICITY

Catalysed reaction
Glycerol + acceptor →
→ glycerone + reduced acceptor

Reaction type
Redox reaction

Natural substrates

Substrate spectrum
1 Glycerol + acceptor (2,6-dichlorophenolindophenol/phenazine methosulfate act as in vitro acceptor) [1]
2 D-Sorbitol + acceptor (oxidation at 45% the rate of glycerol oxidation) [1]
3 D-Arabitol + acceptor (oxidation at 136% the rate of glycerol oxidation) [1]
4 meso-Erythritol + acceptor (oxidation at 111% the rate of glycerol oxidation) [1]
5 Adonitol + acceptor (oxidation at 42% the rate of glycerol oxidation) [1]
6 Galactitol + acceptor (i.e. dulcitol, oxidation at 29% the rate of glycerol oxidation) [1]
7 D-Mannitol + acceptor (oxidation at 26% the rate of glycerol oxidation) [1]
8 Propylene glycol + acceptor (oxidation at 54% the rate of glycerol oxidation) [1]
9 More (broad specificity, no substrates: xylitol, D-fructose, D-glucose, D-gluconate, 2-keto-D-gluconate, ethanol, acetaldehyde) [1]

Product spectrum
1 Glycerone + reduced acceptor (i.e. dihydroxyacetone) [1]
2 L-Sorbose + reduced acceptor [1]
3 ?
4 ?
5 ?
6 ?
7 D-Fructose + reduced acceptor [1]
8 ?
9 ?

Inhibitor(s)
EDTA [1]

Cofactor(s)/prosthetic group(s)/activating agents
Pyrroloquinoline quinone (quinoprotein) [1]; 2,6-Dichlorophenolindophenol
(acts in vitro as acceptor together with phenazine methosulfate) [1]; Phenazine methosulfate (acts in vitro as acceptor together with 2,6-dichlorophe-
nolindophenol, linked to electron-transport chain of the organism) [1]; More
(NAD$^+$, NADP$^+$ or molecular oxygen do not act as acceptors) [1]

Metal compounds/salts

Turnover number (min^{-1})

Specific activity (U/mg)
52.0 [1]

K_m-value (mM)
2.3 (glycerol, membrane fraction) [1]; 19.0 (meso-erythritol) [1]; 22.0 (D-arabitol) [1]; 34.0 (glycerol, solubilized enzyme) [1]

pH-optimum
7.5–8.0 [1]

pH-range

Temperature optimum (°C)
25 [1]

Temperature range (°C)

3 ENZYME STRUCTURE

Molecular weight

Subunits

Glycoprotein/Lipoprotein

–

4 ISOLATION/PREPARATION

Source organism
Gluconobacter industrius [1]; More (maximum activity at the end of expo-
nential growth phase, activity occurs in various Gluconobacter species,
such as G. cerinus, G. gluconicus, G. suboxydans, G. albidus) [1]

Source tissue
Cell [1]

Localisation in source
Membrane [1]

Purification
Gluconobacter industrius [1]

Crystallization
–

Cloned
–

Renaturated
–

5 STABILITY

pH
8.5–10.0 (stable in the presence of glycerol and detergent) [1]

Temperature (°C)

Oxidation

Organic solvent

General stability information
Glycerol stabilizes [1]; Amphoteric detergents, such as dimethyldodecylamineoxide, lauroylsarcosine, dodecylsulfobetaine, lauroylbetaine are effective solubilizers [1]; Dimethyldodecylamineoxide, 0.5%, stabilizes [1]

Storage
−20°C, solubilized, stable for 6 months [1]; 2–4°C, diluted enzyme solution, stable for 2 weeks [1]

6 CROSSREFERENCES TO STRUCTURE DATABANKS

PIR/MIPS code

Brookhaven code

7 LITERATURE REFERENCES

[1] Ameyama, M., Shinagawa, E., Matsushita, K., Adachi, O.: Agric. Biol. Chem.,49,1001–1010 (1985)

1 NOMENCLATURE

EC number
1.1.99.23

Systematic name
Polyvinyl-alcohol:(acceptor) oxidoreductase

Recommended name
Polyvinyl-alcohol dehydrogenase (acceptor)

Synonymes
PVA dehydrogenase
Dehydrogenase, polyvinyl alcohol

CAS Reg. No.
119940-13-5

2 REACTION AND SPECIFICITY

Catalysed reaction
Polyvinyl alcohol + acceptor →
→ oxidized polyvinyl alcohol + reduced acceptor

Reaction type
Redox reaction

Natural substrates
Polyvinyl alcohol + acceptor (reaction in bacterial polyvinyl alcohol degradation, couples polyvinyl alcohol dehydrogenation to cytochrome reduction)
[1]

Substrate spectrum
1 Polyvinyl alcohol + acceptor (cytochrome c is probably the in vivo acceptor, phenazine ethosulfate, phenazine methosulfate and 2,6-dichlorophe-
nolindophenol can act as in vitro acceptors) [1]

Product spectrum
1 Oxidized polyvinyl alcohol + reduced acceptor [1]

Inhibitor(s)

Cofactor(s)/prosthetic group(s)/activating agents
Pyrroquinoline quinone (quinoprotein enzyme) [1]; Phenazine ethosulfate
(requirement, electron acceptor) [1]; Cytochrome c (the enzyme couples to
electron transport chain, directly or via other electron carriers, ubiquinones
are no acceptors) [1]

Metal compounds/salts

Turnover number (min^{-1})

Specific activity (U/mg)
22.2 [1]

K_m-value (mM)

pH-optimum

pH-range

Temperature optimum (°C)
30 (assay at) [1]

Temperature range (°C)

3 ENZYME STRUCTURE

Molecular weight

Subunits

Glycoprotein/Lipoprotein
–

4 ISOLATION/PREPARATION

Source organism
Pseudomonas sp. ND1 (polyvinyl alcohol oxidase deficient mutant derived
from strain VM15C) [1]

Source tissue
Cell [1]

Localisation in source
Membrane-bound [1]

Purification
Pseudomonas sp. ND1 (partial, solubilized) [1]

Crystallization
–

Cloned
–

Renaturated
–

---

5 STABILITY

pH

Temperature (°C)

Oxidation

Organic solvent

General stability information

Storage

---

6 CROSSREFERENCES TO STRUCTURE DATABANKS

PIR/MIPS code

Brookhaven code

---

7 LITERATURE REFERENCES

[1] Shimao, M., Onishi, S., Kato, N., Sakazawa, C.: Appl. Environ. Microbiol.,55,275–278 (1989)

1 NOMENCLATURE

EC number
1.1.99.24

Systematic name
(S)-3-Hydroxybutanoate:2-oxoglutarate oxidoreductase

Recommended name
Hydroxyacid-oxoacid transhydrogenase

Synonymes
Transhydrogenase, hydroxy acid-oxo acid

CAS Reg. No.
117698-31-4

2 REACTION AND SPECIFICITY

Catalysed reaction
(S)-3-Hydroxybutanoate + 2-oxoglutarate →
→ acetoacetate + (R)-2-hydroxyglutarate

Reaction type
Redox reaction

Natural substrates

Substrate spectrum
1   4-Hydroxybutyrate + 2-oxoglutarate (hydrogen attached to hydroxyl-bea-
    ring C of 4-hydroxybutyrate is transferred to ketone-bearing C of 2-oxo-
    glutarate, the oxoacid acts as hydrogen acceptor, not NAD^+ or $NADP^+$)
    [1]
2   D-2-Hydroxyglutarate + succinic semialdehyde [1]
3   L-3-Hydroxybutyrate + 2-oxoglutarate (oxidized at 33% the rate of 4-hy-
    droxybutyrate) [1]
4   L-3-Hydroxybutyrate + succinic semialdehyde (oxidized at 62.5% the rate
    of D-2-hydroxyglutarate) [1]
5   More (oxo-acids that can act as hydrogen-acceptors are 2-oxoadipate
    (24% as effective as 2-oxoglutarate) and oxaloacetate (19% as effective
    as 2-oxoglutarate). No substrates are D-3-hydroxybutyrate, DL-2-hydroxy-
    butyrate, D/L-malate, D/L-lactate. The following acids do not act as hydro-
    gen acceptors: acetoacetate, 3-oxobutyrate, 3-oxoadipate, poor accep-
    tors are pyruvate and 2-oxobutyrate) [1]

Product spectrum
 1 Succinic semialdehyde + 2-hydroxyglutarate [1]
 2 2-Oxoglutarate + 4-hydroxybutyrate [1]
 3 Acetoacetate + 2-oxoglutarate [1]
 4 Acetoacetate + 4-hydroxybutyrate [1]
 5 ?

Inhibitor(s)
 More (no inhibition by anaerobiosis) [1]

Cofactor(s)/prosthetic group(s)/activating agents
 More (the reaction does not depend on NAD^+- or $NADP^+$-addition) [1]

Metal compounds/salts

Turnover number (min^{-1})

Specific activity (U/mg)
 0.0513 [1]

K_m-value (mM)
 0.0046 (succinic semialdehyde) [1]; 0.018 (2-oxoglutarate) [1]; 0.30 (4-hy-droxybutyrate) [1]; 0.42 (D-2-hydroxyglutarate) [1]; 3.0 (L-3-hydroxybutyrate) [1]

pH-optimum
 7.0 [1]

pH-range

Temperature optimum (°C)
 37 (assay at) [1]

Temperature range (°C)

3 ENZYME STRUCTURE

Molecular weight

Subunits

Glycoprotein/Lipoprotein
 −

4 ISOLATION/PREPARATION

Source organism
 Rat (Sprague-Dawley strain) [1]

Source tissue
  Kidney [1]; Liver [1]; Brain [1]

Localisation in source
  Mitochondria [1]

Purification
  Rat (partial) [1]

Crystallization
  –

Cloned
  –

Renaturated
  –

5 STABILITY

pH

Temperature (°C)

Oxidation

Organic solvent

General stability information
  NAD+ stabilizes during ammonium sulfate precipitation [1]

Storage

6 CROSSREFERENCES TO STRUCTURE DATABANKS

PIR/MIPS code

Brookhaven code

7 LITERATURE REFERENCES

[1] Kaufman, E.E., Nelson, T., Fales, H.M., Levin, D.M.: J. Biol. Chem.,263,16872–16879 (1988)

1 NOMENCLATURE

EC number
1.1.99.25

Systematic name
Quinate:pyrroloquinoline-quinone 5-oxidoreductase

Recommended name
Quinate dehydrogenase (pyrroloquinoline-quinone)

Synonymes
NAD(P)-independent quinate dehydrogenase
Dehydrogenase, quinate (pyrroloquinoline-quinone)

CAS Reg. No.
115299-99-5

2 REACTION AND SPECIFICITY

Catalysed reaction
Quinate + pyrroloquinoline-quinone →
→ 5-dehydroquinate + reduced pyrroloquinoline-quinone

Reaction type
Redox reaction

Natural substrates

Substrate spectrum
1 D-Quinate + tetramethyl-p-phenylenediamine (in vitro 2,6-dichlorophenol-
indophenol can also act as electron acceptor) [1]

Product spectrum
1 ?

Inhibitor(s)
EDTA (inhibits apoenzyme quinate dehydrogenase) [1]; More (quinate dehydrogenase holoenzyme and quinate dehydrogenase apoenzyme after recombination with pyrroloquinoline quinone is not affected by EDTA) [1]

Cofactor(s)/prosthetic group(s)/activating agents
Pyrroloquinoline quinone (quinoproteine) [1]

Metal compounds/salts
More (divalent metal ion seems necessary) [1]

Turnover number (min^{-1})

Specific activity (U/mg)

K$_m$-value (mM)

pH-optimum
   7.5 [1]

pH-range

Temperature optimum (°C)
   20 (assay at) [1]

Temperature range (°C)

---

3 ENZYME STRUCTURE

Molecular weight

Subunits

Glycoprotein/Lipoprotein
   −

---

4 ISOLATION/PREPARATION

Source organism
   Acinetobacter calcoaceticus [1]

Source tissue
   Cell [1]

Localisation in source
   Particle-bound (probably) [1]

Purification
   Acinetobacter calcoaceticus [1]

Crystallization
   −

Cloned
   −

Renaturated
   −

5 STABILITY

pH

Temperature (°C)

Oxidation

Organic solvent

General stability information

Storage

6 CROSSREFERENCES TO STRUCTURE DATABANKS

PIR/MIPS code

Brookhaven code

7 LITERATURE REFERENCES

[1] van Kleef, M.A.G., Duine, J.A.: Arch. Microbiol.,150,32–36 (1988)

1 NOMENCLATURE

EC number
1.1.99.26

Systematic name
3-Hydroxycyclohexanone:acceptor 1-oxidoreductase

Recommended name
3-Hydroxycyclohexanone dehydrogenase

Synonymes
Dehydrogenase, 3-hydroxycyclohexanone

CAS Reg. No.
123516-44-9

2 REACTION AND SPECIFICITY

Catalysed reaction
3-Hydroxycyclohexanone + acceptor →
→ cyclohexane-1,3-dione + reduced acceptor

Reaction type
Redox reaction

Natural substrates

Substrate spectrum
1 3-Hydroxycyclohexanone + acceptor (methylene blue or 2,6-dichlorophe-
nolindophenol can function as electron acceptor in vitro) [1]

Product spectrum
1 1,3-Cyclohexanedione + reduced acceptor [1]

Inhibitor(s)

Cofactor(s)/prosthetic group(s)/activating agents

Metal compounds/salts

Turnover number (min^{-1})

Specific activity (U/mg)
0.27 (2-cyclohexenone via 3-hydroxycyclohexanone) [1]

K_m-value (mM)

pH-optimum
   7.8 (assay at) [1]

pH-range

Temperature optimum (°C)
   30 (assay at) [1]

Temperature range (°C)

---

3 ENZYME STRUCTURE

Molecular weight

Subunits

Glycoprotein/Lipoprotein
   –

---

4 ISOLATION/PREPARATION

Source organism
   Pseudomonas sp. (enzyme induced by growth on cyclohexanone under
   anaerobic conditions) [1]

Source tissue
   Cell [1]

Localisation in source

Purification
   Pseudomonas sp. [1]

Crystallization
   –

Cloned
   –

Renaturated
   –

5 STABILITY

pH

Temperature (°C)

Oxidation

Organic solvent

General stability information

Storage
    −20°C, cell extract stable for several days [1]

6 CROSSREFERENCES TO STRUCTURE DATABANKS

PIR/MIPS code

Brookhaven code

7 LITERATURE REFERENCES

[1] Dangel, W., Tschech, A., Fuchs, G.: Arch. Microbiol.,152,273–279 (1989)